Markov Decision Processes
with Applications to Finance

Universitext

Universitext is a series of textbooks that presents material from a wide variety of mathematical disciplines at master's level and beyond. The books, often well class-tested by their author, may have an informal, personal even experimental approach to their subject matter. Some of the most successful and established books in the series have evolved through several editions, always following the evolution of teaching curricula, to very polished texts.

Thus as research topics trickle down into graduate-level teaching, first textbooks written for new, cutting-edge courses may make their way into *Universitext*.

For further volumes:
www.springer.com/series/223

Nicole Bäuerle · Ulrich Rieder

Markov Decision Processes with Applications to Finance

Nicole Bäuerle
Institute for Stochastics
Karlsruhe Institute of Technology
76128 Karlsruhe
Germany
nicole.baeuerle@kit.edu

Ulrich Rieder
Institute of Optimization
and Operations Research
University of Ulm
89069 Ulm
Germany
ulrich.rieder@uni-ulm.de

ISBN 978-3-642-18323-2 e-ISBN 978-3-642-18324-9
DOI 10.1007/978-3-642-18324-9
Springer Heidelberg Dordrecht London New York

Library of Congress Control Number: 2011929506

Mathematics Subject Classification (2010): 90C40, 93E20, 60J05, 91G10, 93E35, 60G40

Cover design: deblik

Printed on acid-free paper

Springer is part of Springer Science+Business Media (www.springer.com)

Für Rolf, Katja und Hannah.
Für Annika, Alexander und Katharina.

Preface

Models in mathematical finance, for example stock price processes, are often defined in continuous-time. Hence optimization problems like consumption-investment problems lead to stochastic control problems in continuous-time. However, only a few of these problems can be solved explicitly. When numerical methods have to be applied, it is sometimes wise to start with a process in discrete-time, as done for example in the *approximating Markov chain approach.* The resulting optimization problem is then a *Markov Decision Problem* and there is a rich toolbox available for solving these kind of problems theoretically and numerically.

The purpose of this book is to present selected parts of the theory of *Markov Decision Processes* and show how they can be applied in particular to problems in finance and insurance. We start by explaining the theory for problems with finite time horizon. Since we have financial applications in mind and since we do not want to restrict to binomial models we have to work with Borel state and action spaces. This framework is also needed for studying *Partially Observable Markov Decision Processes* and *Piecewise Determinsitic Markov Decision Processes.* In contrast to the case of a discrete (finite or countable) state space the theory is more demanding since non-trivial measurability problems have to be solved. However, we have decided to circumvent these kind of problems by introducing a so-called *structure assumption* for the model. The advantage is that in applications this structure assumption is often easily verified and avoids some of the technicalities. This makes the book accessible to readers who are not familiar with general probability and measure theory. Moreover, we present numerous different applications and show how this structure assumption can be verified. Applications range from consumption-investment problems, mean-variance problems, dividend problems in risk theory to indifference pricing and pricing of American options, just to name a few. The book is unique in the presentation and collection of these financial applications. Some of them appear for the first time in a book.

We also consider the theory of infinite horizon *Markov Decision Processes* where we treat so-called *contracting* and *negative* Markov Decision Problems in a unified framework. *Positive* Markov Decision Problems are also presented as well as *stopping problems.* A particular focus is on problems with *partial observation.* These kind of problems cover situations where the decision maker is not able to observe all driving factors of the model. Special cases are Hidden Markov Models and Bayesian Decision Problems. They include statistical aspects, in particular *filtering theory* and can be solved by so-called *filtered Markov Decision Processes.* Moreover *Piecewise Deterministic Markov Decision Processes* are discussed and we give recent applications to finance.

It is our aim to present the material in a mathematically rigorous framework. This is not always easy. For example, the last-mentioned problems with partial observation need a lot of definitions and notation. However each chapter on theory is followed by a chapter with applications and we give examples throughout the text which we hope shed some light on the theory. Also at the end of each chapter on theory we provide a list of exercises where the reader can test her knowledge.
Having said all this, not much general probability and optimization theory is necessary to enjoy this book. In particular we do not need the theory of stochastic calculus which is a necessary tool for continuous-time problems. The reader should however be familiar with concepts like *conditional expectation* and *transition kernels.* The only exception is Section 2.4 which is a little bit more demanding. Special knowledge in finance and insurance is not necessary. Some fundamentals are covered in the appendix. As outlined above we provide an example-driven approach. The book is intended for graduate students, researchers and practitioners in mathematics, finance, economics and operations research. Some of the chapters have been tried out in courses for masters students and in seminars.
Last but not least we would like to thank our friends and colleagues Alfred Müller, Jörn Sass, Manfred Schäl and Luitgard Veraart who have carefully read parts of an earlier version and provided helpful comments and suggestions. We are also grateful to our students Stefan Ehrenfried, Dominik Joos and André Mundt who gave significant input and corrected errors, as well as to the students at Ulm University and KIT who struggled with the text in their seminars. Special thanks go to Rolf Bäuerle and Sebastian Urban for providing some of the figures.

Bretten and Ulm,
September 2010

Nicole Bäuerle
Ulrich Rieder

Contents

List of Symbols

Markov Decision Model (non-stationary)

N	finite time horizon
E, $\mathfrak{E}$	state space with σ-algebra
A, $\mathfrak{A}$	action space with σ-algebra
D_n	admissible state-action pairs at time n
$Q_n(\cdot\|x,a)$	stochastic transition kernel from D_n to E
$r_n(x,a)$	one-stage reward at time n
$g_N(x)$	terminal reward
$\mathcal{Z}, \mathfrak{Z}$	disturbance space with σ-algebra
$T_n(x,a,z)$	transition function of the state process
$Q_n^Z(\cdot\|x,a)$	stochastic transition kernel from D_n to $\mathcal{Z}$
(X_n)	state process
(Z_n)	process of disturbances
F_n	set of decision rules at time n
π	$= (f_n)$ policy
h_n	$= (x_0, a_0, x_1, \ldots, x_n)$ history up to time n
Π_N	set of history-dependent N-stage policies
$\mathbb{P}_{nx}^{\pi}$	probability measure under policy π given $X_n = x$
$\mathbb{E}_{nx}^{\pi}$	expectation operator
$V_{n\pi}(x)$	expected total reward from n to N under policy π
$V_n(x)$	maximal expected total reward from n to N
$\delta_n^N(x)$	upper bound for $V_n(x)$
$(L_n v)(x,a)$	$= r_n(x,a) + \int v(x')Q_n(dx'\|x,a)$ reward operator
$(\mathcal{T}_{nf} v)(x)$	$= (L_n v)(x, f(x))$ reward operator of f
$(\mathcal{T}_n v)(x)$	$= \sup_{a \in D_n(x)} (L_n v)(x,a)$ maximal reward operator

Markov Decision Model (stationary)

E, $\mathfrak{E}$	state space with σ-algebra

A, $\mathfrak{A}$	action space with σ-algebra
D	admissible state-action pairs
$Q(\cdot\vert x,a)$	stochastic transition kernel from D to E
$r(x,a)$	one-stage reward
$g(x)$	terminal reward
β	discount factor
$\mathcal{Z}, \mathfrak{Z}$	disturbance space with σ-algebra
$T(x,a,z)$	transition function of the state process
$Q^Z(\cdot\vert x,a)$	stochastic transition kernel from D to $\mathcal{Z}$
F	set of decision rules
$J_{n\pi}(x)$	expected discounted reward over n stages under policy π
$J_n(x)$	maximal expected discounted reward over n stages
$\delta_N(x)$	upper bound for $J_N(x)$
$b(x)$	(upper) bounding function
$(Lv)(x,a)$	$= r(x,a) + \int v(x')Q(dx'\vert x,a)$ reward operator
$(\mathcal{T}_f v)(x)$	$= (Lv)(x,f(x))$ reward operator of f
$(\mathcal{T}v)(x)$	$= \sup_{a\in D(x)}(Lv)(x,a)$ maximal reward operator

Financial Markets

S_n^0	bond price at time n
i_{n+1}	interest rate in $[n, n+1)$
S_n	$= (S_n^1, \dots, S_n^d)$ stock prices at time n
$\tilde{R}_{n+1}^k$	$= \frac{S_{n+1}^k}{S_n^k}$ relative price change of asset k
R_n^k	$= \frac{\tilde{R}_n^k}{1+i_n} - 1$ relative risk process of asset k
$(\mathcal{F}_n)$	market filtration
ϕ	$= (\phi_n)$ self-financing portfolio strategy
$VaR_\gamma(X)$	Value-at-Risk at level γ
$AVaR_\gamma(X)$	Average-Value-at-Risk at level γ
dom U	domain of utility function U
CARA	constant absolute risk aversion
HARA	hyperbolic absolute risk aversion
MV	mean variance
MR	mean risk

Partially Observable Markov Decision Model

E_X	observable part of the state space
E_Y	unobservable part of the state space
A	action space
D	admissible state-action pairs

$Q(\cdot\|x,y,a)$	stochastic transition kernel from $E_Y \times D$ to $E_X \times E_Y$
Q_0	initial distribution of Y_0
$r(x,y,a)$	one-stage reward
$g(x)$	terminal reward
β	discount factor
$\mathcal{Z}$	disturbance space
$Q^{Z,Y}(\cdot\|x,y,a)$	stochastic transition kernel from $E_Y \times D$ to $\mathcal{Z} \times E_Y$
T_X	transition function of the observable state process
(X_n)	observable state process
(Y_n)	unobservable state process
(Z_n)	process of disturbances
Φ	Bayes operator
μ_n	conditional distribution of Y_n
h_n	$= (x_0, a_0, x_1, \dots, x_n)$ history up to time n
$\tilde{h}_n$	$= (x_0, a_0, z_1, x_1, \dots, z_n, x_n)$
Π_N	set of history-dependent N-stage policies
(t_n)	sequential sufficient statistic
$\hat{\Phi}$	information update operator

Markov Decision Model with infinite horizon

F^∞	set of infinite-stage policies
f^∞	$= (f, f, \dots)$ stationary policy
$J_{\infty\pi}(x)$	expected discounted reward under policy π
$J_\infty(x)$	maximal expected discounted reward
$J(x)$	$= \lim_{n\to\infty} J_n(x)$ limit value function
$\delta(x)$	upper bound for $J_\infty(x)$
$\varepsilon(x)$	upper bound for the negative part of the rewards
$b(x)$	(upper) bounding function
α_b	contraction module
$(\mathcal{T}_\circ v)(x)$	$= \sup_{a\in D(x)} \beta \int v(x')Q(dx'\|x,a)$ shift operator
LsA_n	upper limit of the set sequence (A_n)

Stopping Problems

τ	stopping time
τ_π	stopping time induced by policy π
R_τ	reward under stopping time τ
$V_N^*(x)$	maximal reward of N-period stopping problem
$V_\infty^*(x)$	maximal reward of unbounded stopping problem
$G_\pi(x)$	$= \liminf_{n\to\infty} J_{n\pi}(x)$
$G(x)$	$= \sup_\pi G_\pi$
S_n^*	optimal stopping set at time n

Special Symbols

δ_x	one-point measure in x
$x^\pm$	$= \max\{0, \pm x\}$
$\mathbb{M}(E)$	$= \{v : E \to [-\infty, \infty), \text{ measurable }\}$
$\mathbb{B}_b$	$= \{v \in \mathbb{M}(E) \mid \|v\|_b < \infty\}$
$\mathbb{B}_b^+$	$= \{v \in \mathbb{M}(E) \mid \|v^+\|_b < \infty\}$
$\mathbb{B}$	$= \{v \in \mathbb{M}(E) \mid v(x) \le \delta(x) \text{ for all } x \in E\}$
$\|v\|_b$	weighted supremum norm
$\mathcal{B}(E)$	Borel σ-algebra in E
MTP_2	multivariate total positivity of order 2
$\le_{st}$	stochastic order
$\le_{lr}$	likelihood ratio order
$\le_{cx}$	convex order
$\le_{icx}$	increasing convex order
$\mathcal{N}(\mu, \sigma^2)$	normal distribution
$Exp(\lambda)$	exponential distribution
$Be(\alpha, \beta)$	beta distribution
$B(n, p)$	binomial distribution
$Poi(\lambda)$	Poisson distribution
$f(x) \propto g(x)$	f equals g up to a constant
$x \wedge y$	$:= (\min\{x_1, y_1\}, \ldots, \min\{x_d, y_d\})$
$x \vee y$	$:= (\max\{x_1, y_1\}, \ldots, \max\{x_d, y_d\})$
∇f	gradient of f
$\square$	end of proof
$\lozenge$	end of remark
$\blacklozenge$	end of example

Chapter 1
Introduction and First Examples

Suppose a system is given which can be controlled by sequential decisions. The state transitions are random and we assume that the *system state process* is *Markovian* which means that previous states have no influence on future states. Given the current state of the system (which could be for example the wealth of an investor) the controller or decision maker has to choose an *admissible action* (for example a possible investment). Once an action is chosen there is a random system transition according to a stochastic law (for example a change in the asset value) which leads to a new state. The task is to control the process in an optimal way. In order to formulate a reasonable optimization criterion we assume that each time an action is taken, the controller obtains a certain *reward*. The aim is then to control the system in such a way that the expected total discounted rewards are maximized. All these quantities together which have been described in an informal way, define a so-called *Markov Decision Process*. The Markov Decision Process is the sequence of random variables (X_n) which describes the stochastic evolution of the system states. Of course the distribution of (X_n) depends on the chosen actions. Figure 1.1 shows the schematic evolution of a Markov Decision Process.

We summarize the main model data in the following list:

- E denotes the *state space* of the system. A state $x \in E$ is the information which is available for the controller at time n. Given this information an action has to be selected.
- A denotes the *action space*. Given a specific state $x \in E$ at time n, a certain subclass $D_n(x) \subset A$ of actions may only be admissible.
- $Q_n(B|x, a)$ is a *stochastic transition kernel* which gives the probability that the next state at time $n + 1$ is in the set B if the current state is x and action a is taken at time n.
- $r_n(x, a)$ gives the (discounted) *one-stage reward* of the system at time n if the current state is x and action a is taken.

N. Bäuerle and U. Rieder, *Markov Decision Processes with Applications to Finance*, Universitext, DOI 10.1007/978-3-642-18324-9_1,

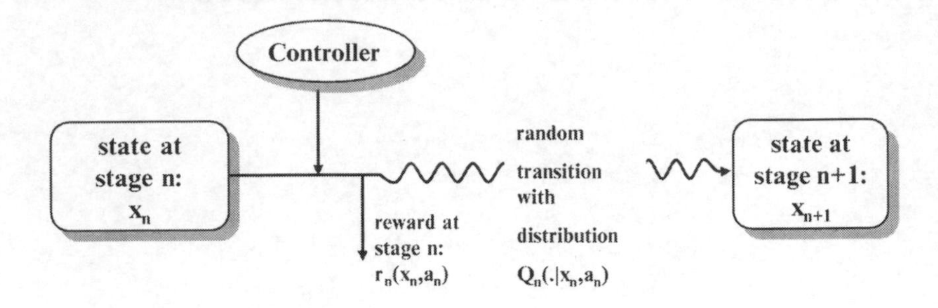

Fig. 1.1 General evolution of a Markov Decision Model.

- $g_N(x)$ gives the (discounted) *terminal reward* of the system at the end of the planning horizon.

An important assumption about these processes is that the evolution is Markovian. Since the system obeys a stochastic transition law, the sequence of visited states is not known at the beginning. Thus, a control π is a sequence of decision rules (f_n) with $f_n : E \to A$ where $f_n(x) \in D_n(x)$ determines for each possible state $x \in E$ the next action $f_n(x)$ at time n. Such a sequence $\pi = (f_n)$ is called *policy* or *strategy.* Formally the *Markov Decision Problem* is given by

$$V_0(x) := \sup_\pi \mathbb{E}_x^\pi \left[\sum_{k=0}^{N-1} r_k\big(X_k, f_k(X_k)\big) + g_N(X_N) \right], \quad x \in E,$$

where the supremum is taken over all admissible policies. Obviously the optimization problem is non-trivial since a decision today does not only determine the current reward but also has complicated influences on future states. The policies which have been defined above are Markovian by definition since the decision depends only on the current state. Indeed it suffices to search for the optimal control among all Markovian policies, though sometimes it is convenient to consider history-dependent policies.
The theory of Markov Decision Processes deals with stochastic optimization problems in *discrete time.* The time steps do not have to be equal but this is often assumed. Sometimes problems which are formulated in continuous-time can be reduced to a discrete-time model by considering an embedded state process. The theory of stochastic control problems in continuous time is quite different and not a subject of this book. However, when continuous-time problems have to be solved numerically, one way is to consider an approximation

of the state process in discrete time. This procedure is called the *approximating Markov chain approach.* The resulting problem can then be solved with the techniques presented here. Still the theory on Markov Decision Processes encompasses a lot of different models and formulations, and we will not deal with all of them. Since we have mainly applications in finance in mind we consider Markov Decision Processes with arbitrary state and action spaces (more precisely Borel spaces). Note that in these applications the spaces are often not discrete. Markov Decision Processes are also called *Markov Control Processes* or *Stochastic Dynamic Programs.* We distinguish problems with

- finite horizon $N < \infty$ – infinite horizon $N = \infty$,
- complete state observation – partial state observation,
- problems with constraints – without constraints,
- total (discounted) cost criterion – average cost criterion.

We will consider Markov Decision Problems with *finite horizon* in Chapter 2 and models with *infinite horizon* in Chapter 7. Sometimes models with infinite horizon appear in a natural way (for example when the original problem has a random horizon or when the original problem has a fixed time horizon but random time steps are used to solve it) and often these models can be seen as approximations of Markov Decision Problems with finite but large horizon. We will encounter different approaches to problems with infinite horizon in Chapter 7 and in Section 10.2 where unbounded stopping problems are treated. In any case, some convergence assumptions are needed to ensure that the infinite horizon Markov Decision Problem is well-defined. The solution of such optimization problems is then often easier because the value function can be characterized as the unique fixed point or as the smallest superharmonic function of an operator, and moreover the optimal policy is stationary. We will treat so-called *negative* Markov Decision Problems where the reward functions are negative (or zero) and *contracting* Markov Decision Problems where the maximal reward operator is contracting, in a unified framework. Besides these models we also consider so-called *positive* Markov Decision Problems where the reward functions are positive (or zero).
Another way to distinguish Markov Decision Problems is according to what can be observed by the controller. This is in contrast to deterministic control problems where the information for the controller is determined by the deterministic transition law. In stochastic control models also statistical aspects come into play. For example when a part of the state cannot be observed (e.g. some fundamental economic indicators which influence the asset price change), however due to observations of the state some information about the unobservable part is obtained. Such a model is called a *Partially Observable Markov Decision Problem.* In this case the statistical *filtering* theory has to be combined with the optimization problem. It will turn out that the Partially Observable Markov Decision Problem can be reformulated as a Markov Decision Problem with complete observation by enlarging the state space. Indeed an estimate of relevant information has to be added to the state and

updated during each new observation. Given this information a decision is taken. This is called the *separation principle of estimation and control.*
Markov Decision Problems already take some constraints about admissible actions into account. However, sometimes optimization problems arise where there are additional constraints. We do not treat these kind of problems in a systematic way but we consider for example *mean-variance problems* or *mean-risk problems* where the portfolio strategy of the investor has to satisfy some risk constraints. In Chapter 8 we consider *Piecewise Deterministic Markov Decision Processes.* These continuous-time optimization problems can be solved by discrete-time Markov Decision Processes with an action space consisting of functions. More precisely we have to introduce relaxed control functions. In Chapter 10 we deal with discrete-time *stopping problems.* Given that the underlying process is Markovian we show that stopping problems can be solved by Markov Decision Processes.

The theory of Markov Decision Processes which is outlined in this book addresses questions like: Does an optimal policy exist? Has it a particular form? Can an optimal policy be computed efficiently? Is it possible to derive properties of the optimal value function analytically? Besides developing the theory of Markov Decision Problems a main aim of this book is to show Markov Decision Problems in action. The applications are mostly taken from finance and insurance but are not limited to these areas. This book focuses on Markov Decision Processes with the total reward criterion. Problems with average-reward and risk-sensitive criteria are not treated in this book.

1.1 Applications

We will mainly focus on applications in finance, however the areas where Markov Decision Processes are used to solve problems are quite diverse. They appear in production planning, inventory control, operations management, engineering, biology and statistics, just to name a few. Let us consider some applications.

Example 1.1.1 (Consumption Problem). Suppose there is an investor with given initial capital. At the beginning of each of N periods she can decide how much of the capital she consumes and how much she invests into a risky asset. The amount she consumes is evaluated by a utility function U as well as the terminal wealth. The remaining capital is invested into a risky asset where we assume that the investor is *small* and thus not able to influence the asset price and the asset is *liquid.* How should she consume/invest in order to maximize the sum of her expected discounted utility?
The state x of the system is here the available capital. The action $a = f(x)$ is the amount of money which is consumed, where it is reasonable to assume

that $0 \le a \le x$. The reward is given by $U(a)$ and the terminal reward by $U(x)$. Hence the aim is to maximize

$$\mathbb{E}_x^\pi \left[\sum_{k=0}^{N-1} U\big(f_k(X_k)\big) + U(X_N) \right]$$

where the maximization is over all policies $\pi = (f_0, \dots, f_{N-1})$. This problem is solved in Section 4.3. ◆

Example 1.1.2 (Cash Balance or Inventory Problem). Imagine a company which tries to find the optimal level of cash over a finite number of N periods. We assume that there is a random stochastic change in the cash reserve each period (due to withdrawal or earnings). Since the firm does not earn interest from the cash position, there are holding cost for the cash reserve if it is positive, but also interest (cost) in case it is negative. The cash reserve can be increased or decreased by the management at each decision epoch which implies transfer costs. What is the optimal cash balance policy?
The state x of the system is here the current cash reserve. The action $a = f(x)$ is either the new cash reserve or the amount of money which is transferred from the cash reserve to assets. The reward is a negative cost determined by the transfer cost and the holding or understocking cost. This example is treated in Sections 2.6.2 as a finite horizon problem and in Section 7.6.2 as an infinite horizon problem. ◆

Example 1.1.3 (Mean-Variance Problem). Consider a small investor who acts on a given financial market. Her aim is to choose among all portfolios which yield at least a certain expected return (benchmark) after N periods, the one with smallest portfolio variance. What is the optimal investment strategy?
This is an optimization problem with an additional constraint. As in the first example the state x of the system is the available capital. The action $a = f(x)$ is the investment decision. When we assume that there are d different assets available, then $a = (a_1, \dots, a_d) \in \mathbb{R}^d$ and a_k gives the amount of money which is invested in asset k. The aim is to solve

$$(MV) \quad \begin{cases} \operatorname{Var}_{x_0}^\pi [X_N] \to \min \\ \\ \mathbb{E}_{x_0}^\pi [X_N] \ge \mu \end{cases}$$

where the minimization is over all policies $\pi = (f_0, \dots, f_{N-1})$. In order to get rid of the constraint and to define the one-stage reward there is some work needed. This problem is investigated intensively in Section 4.6 and with partial observation in Section 6.2. ◆

Example 1.1.4 (Dividend Problem in Risk Theory). Imagine we consider the risk reserve of an insurance company which earns some premia on the one hand but has to pay out possible claims on the other hand. At the beginning

of each period the insurer can decide upon paying a dividend. A dividend can only be paid when the risk reserve at that time point is positive. Once the risk reserve got negative we say that the company is ruined and has to stop its business. Which dividend pay-out policy maximizes the expected discounted dividends until ruin?
The state x of the system is here the current risk reserve. The action $a = f(x)$ is the dividend which is paid out where $a \le x$. The one-stage reward is the dividend which is paid. This problem has to be dealt with as one with infinite horizon since the time horizon is not fixed in advance. This example is treated in Section 9.2. It can be shown that the optimal policy is stationary and has a certain structure which is called *band-policy*. ♦

Example 1.1.5 (Bandit Problem). Suppose we have two slot machines with unknown success probability θ_1 and θ_2. At each stage we have to choose one of the arms. We receive one Euro if the arm wins, else no cash flow appears. How should we play in order to maximize our expected total reward over N trials?
This problem is a *Partially Observable Markov Decision Problem* since the success probabilities are not known. Hence the state of the system must here be interpreted as the available information of the decision maker. This information can be represented as the number of successes and failures at both arms up to this time point. Here $x = (m_1, n_1, m_2, n_2) \in \mathbb{N}_0^4$ denotes the number of successes m_i and failures n_i at arm i. An estimate for the win probability at arm i is then $\frac{m_i}{m_i+n_i}$. The action is obviously to choose one of the arms. The one-stage reward is the expected one-stage reward under the given information. This problem is treated in Section 5.5. Under some assumptions it can be shown that a so-called *index-policy* is optimal.
Bandit problems are generic problems which have a number of serious applications, for example medical trials of a new drug. ♦

Example 1.1.6 (Pricing of American Options). In order to find the fair price of an American option and its optimal exercise time, one has to solve an optimal stopping problem. In contrast to a European option the buyer of an American option can choose to exercise any time up to and including the expiration time. In Section 11.1 we show how such an optimal stopping problem can be solved in the framework of Markov Decision Processes. ♦

1.2 Organization of the Book

The book consists of eleven chapters which can be roughly grouped into four parts. The first part from Chapter 2 to 4 deals with the theory of Markov Decision Problems with finite time horizon, introduces the financial markets which are used later and provides some applications. The second part, which consists of Chapters 5 and 6, presents the theory of Partially Observable

Markov Decision Processes and provides some applications. Part III, which consists of Chapters 7, 8 and 9, investigates Markov Decision Problems with infinite time horizon, Piecewise Deterministic Markov Decision Processes, as well as applications. The last part – Chapters 10 and 11 – deals with stopping problems. Chapters with theory and applications alternate. The theory of Markov Decision Problems is presented in a self-contained way in Chapters 2, 5, 7, 8 and 10. Section 2.4 deals with conditions under which Markov Decision Problems satisfy the structure assumption. This part is slightly more advanced than the other material and might be skipped at first reading. Chapters 5 and 6 are not necessary for the understanding of the remaining chapters of the book (despite two examples in Chapters 10 and 11).

1.3 Notes and References

Historical Notes:
The first important books on Markov Decision Processes are Bellman (1957) (for a reprint see Bellman (2003)) and Howard (1960). The term 'Markov Decision Process' was coined by Bellman (1954). Shapley (1953) (for a reprint see Shapley (2003)) was the first study of Markov Decision Processes in the context of stochastic games. For more information on the origins of this research area see Puterman (1994) and Feinberg and Shwartz (2002). Later a more mathematical rigorous treatment of this theory appeared in Dubins and Savage (1965), Blackwell (1965), Shiryaev (1967) and Hinderer (1970). The fascinating book of Dubins and Savage (1965) deals with gambling models, however the underlying ideas are essentially the same. Blackwell (1965) introduces the model description which is used up to now. He was the first to give a rigorous treatment of discounted problems with general state spaces. Hinderer (1970) deals with general non-stationary models where reward functions and transition kernels may depend on the whole history of the underlying process. Another step towards generalizing the models are the books of Bertsekas and Shreve (1978) and Dynkin and Yushkevich (1979). There also the basic measurability questions are investigated.

Related Textbooks:
Nowadays a lot of excellent textbooks and handbooks on Markov Decision Processes exist and we are not able to give a complete list here. Thus we restrict to those books which we have frequently consulted or which are a reasonable addition and contain supplementary material.
By now, classical textbooks on Markov Decision Processes (besides the ones we have already mentioned in the 'Historical Notes') are Derman (1970), Ross (1970, 1983), Hordijk (1974), Whittle (1982, 1983), Schäl (1990), White (1993), Puterman (1994), Hernández-Lerma and Lasserre (1996), Filar and

Vrieze (1997), Bertsekas (2001, 2005) and Powell (2007). Schäl (1990) deals with discrete Markov Decision Processes and presents an interesting bridge between stochastic dynamic programming, gambling models and stopping problems. The book of Puterman (1994) is very comprehensive and treats also average reward criteria and semi-Markov decision models. The state and action spaces are mostly assumed to be discrete. Hernández-Lerma and Lasserre (1996) investigate Markov Decision Processes with Borel state and action spaces which have a semicontinuous-semicompact structure. Bertsekas (2001, 2005) considers deterministic and stochastic dynamic programs with a view towards applications and practical implementation issues. Moreover the Handbook of Markov Decision Processes, edited by Feinberg and Shwartz (2002), contains recent state-of-the-art contributions. Of particular interest in our context is the contribution of Schäl (2002) on 'Markov decision processes in finance and dynamic options'. Finally the focus of Powell (2007) is on *approximate dynamic programming* which helps to solve large-scale problems efficiently.
It is also worth mentioning that Kumar and Varaiya (1986) and Hernández-Lerma (1989) consider adaptive control problems or *Partially Observable Markov Decision Processes.* This research area is often not covered in textbooks. The same is true for Piecewise Deterministic Markov Decision Processes. Davis (1993) studies these optimization problems as control problems in continuous and in discrete time. Continuous-time Markov Decision Chains are an important subclass (see e.g. Guo and Hernández-Lerma (2009)).
The presentation of Markov Decision Processes is here restricted to the total reward criterion. Average-reward and sensitive discount criteria are treated in e.g. Hernández-Lerma and Lasserre (1999), whereas risk-sensitive control problems are considered in e.g. Whittle (1990) and Bielecki et al. (1999). Non-standard criteria for Markov Decision Processes are studied in e.g. Jaquette (1973), White (1988), Bouakiz and Sobel (1992), Huang and Kallenberg (1994), Feinberg and Shwartz (1994) and Boda et al. (2004). Markov Decision Processes with delayed dynamics are investigated in Bauer (2004) and Bauer and Rieder (2005).
Books with various applications of Markov Decision Processes in economics are Heyman and Sobel (2004a,b) and Stokey and Lucas (1989). Ingersoll (1987) and Huang and Litzenberger (1988) investigate intertemporal portfolio selection problems and Pliska (2000) considers consumption-investment problems in discrete-time financial markets (with finite state space). Also interesting is the reprint of Ziemba and Vickson (2006) on static and dynamic portfolio selection problems.

Complementary Textbooks:
Of course, Markov Decision Processes can be used in quite different applications. A very active area is the control of queues. References for this area are e.g. Kitaev and Rykov (1995), Sennott (1999), Tijms (2003) and Meyn

(2008). The latter book is very comprehensive and also deals with recent numerical approaches, e.g. fluid approximations.
Markov Decision Processes with constraints are systematically treated in Kallenberg (1983), Borkar (1991), Piunovskiy (1997) and Altman (1999). In particular, they also describe the convex analytic approach for solving Markov Decision Problems.

Stochastic control problems with diffusions are not treated in this book. In principle, the solution method is similar: Here one considers a local form of the Bellman equation which is called the *Hamilton–Jacobi–Bellman* equation. Since the *Hamilton–Jacobi–Bellman* equation is a partial differential equation, the mathematical problems are quite different. In particular, it is often not clear whether the optimal value function itself satisfies the *Hamilton–Jacobi–Bellman* equation because this would involve that the optimal value function is twice continuously differentiable. In discrete-time problems the optimal value function is always a solution of the Bellman equation (if the optimization problem is well-defined). Moreover, the study of controlled diffusion problems needs advanced knowledge in stochastic analysis and stochastic processes. Recent books on controlled diffusions are e.g. Fleming and Soner (1993), Yong and Zhou (1999), Kushner and Dupuis (2001) and Øksendal and Sulem (2005). The latter two books deal with the control of jump-diffusion processes. Yong and Zhou (1999) also use Pontryagin's optimality principle to solve stochastic control problems. The recent textbook of Seierstad (2009) treats stochastic control problems in discrete and continuous time. Books on portfolio optimization problems in continuous-time are e.g. Korn (1997) and Pham (2009).

Part I
Finite Horizon Optimization Problems and Financial Markets

Chapter 2
Theory of Finite Horizon Markov Decision Processes

In this chapter we will establish the theory of Markov Decision Processes with a finite time horizon and with general state and action spaces. Optimization problems of this kind can be solved by a backward induction algorithm. Since state and action space are arbitrary, we will impose a structure assumption on the problem in order to prove the validity of the backward induction and the existence of optimal policies. The chapter is organized as follows.
Section 2.1 provides the basic model data and the definition of policies. The precise mathematical model is then presented in Section 2.2 along with a sufficient integrability assumption which implies a well-defined problem. The solution technique for these problems is explained in Section 2.3. Under structure assumptions on the model it will be shown that Markov Decision Problems can be solved recursively by the so-called *Bellman equation*. The next section summarizes a number of important special cases in which the structure assumption is satisfied. Conditions on the model data are given such that the value functions are upper semicontinuous, continuous, measurable, increasing, concave or convex respectively. Also the monotonicity of the optimal policy under some conditions is established. This is an essential property for computations. Finally the important concept of upper bounding functions is introduced in this section. Whenever an upper bounding function for a Markov Decision Model exists, the integrability assumption is satisfied. This concept will be very fruitful when dealing with infinite horizon Markov Decision Problems in Chapter 7. In Section 2.5 the important case of stationary Markov Decision Models is investigated. The notion 'stationary' indicates that the model data does not depend on the time index. The relevant theory is here adopted from the non-stationary case. Finally Section 2.6 highlights the application of the developed theory by investigating three simple examples. The first example is a special card game, the second one a cash balance problem and the last one deals with the classical stochastic LQ-problems. The last section contains some notes and references.

N. Bäuerle and U. Rieder, *Markov Decision Processes with Applications to Finance*, Universitext, DOI 10.1007/978-3-642-18324-9_2,

2.1 Markov Decision Models

After having discussed the scope of Markov Decision Models informally in Chapter 1 we will now give a precise definition of a Markov Decision Model. This can be done by defining the ingredients or input data of the model in mathematical terms.

Definition 2.1.1. A (non-stationary) *Markov Decision Model* with planning horizon $N \in \mathbb{N}$ consists of a set of data $(E, A, D_n, Q_n, r_n, g_N)$ with the following meaning for $n = 0, 1, \dots, N-1$:

- E is the *state space*, endowed with a σ-algebra $\mathfrak{E}$. The elements (states) are denoted by $x \in E$.
- A is the *action space*, endowed with a σ-algebra $\mathfrak{A}$. The elements (actions) are denoted by $a \in A$.
- $D_n \subset E \times A$ is a measurable subset of $E \times A$ and denotes the set of possible state-action combinations at time n. We assume that D_n contains the graph of a measurable mapping $f_n : E \to A$, i.e. $(x, f_n(x)) \in D_n$ for all $x \in E$. For $x \in E$, the set $D_n(x) = \{a \in A \mid (x, a) \in D_n\}$ is the set of *admissible actions* in state x at time n.
- Q_n is a stochastic transition kernel from D_n to E, i.e. for any fixed pair $(x, a) \in D_n$, the mapping $B \mapsto Q_n(B|x, a)$ is a probability measure on $\mathfrak{E}$ and $(x, a) \mapsto Q_n(B|x, a)$ is measurable for all $B \in \mathfrak{E}$. The quantity $Q_n(B|x, a)$ gives the probability that the next state at time $n+1$ is in B if the current state is x and action a is taken at time n. Q_n describes the *transition law*.
- $r_n : D_n \to \mathbb{R}$ is a measurable function. $r_n(x, a)$ gives the (discounted) *one-stage reward* of the system at time n if the current state is x and action a is taken.
- $g_N : E \to \mathbb{R}$ is a measurable mapping. $g_N(x)$ gives the (discounted) *terminal reward* of the system at time N if the state is x.

Remark 2.1.2. a) In many applications the state and action spaces are Borel subsets of Polish spaces (i.e. complete, separable, metric spaces) or finite or countable sets. The σ-algebras $\mathfrak{E}$ and $\mathfrak{A}$ are then given by the σ-algebras $\mathcal{B}(E)$ and $\mathcal{B}(A)$ of all Borel subsets of E and A respectively. Often in applications E and A are subsets of $\mathbb{R}^d$ or $\mathbb{R}^d_+$.

b) If the one-stage reward function r'_n also depends on the next state, i.e. $r'_n = r'_n(x, a, x')$, then define

$$r_n(x, a) := \int r'_n(x, a, x') Q_n(dx'|x, a).$$

c) Often D_n and Q_n are independent of n and $r_n(x, a) := \beta^n r(x, a)$ and $g_N(x) := \beta^N g(x)$ for a (discount) factor $\beta \in (0, 1]$. In this case the Markov Decision Model is called *stationary* (see Section 2.5). $\Diamond$

The stochastic transition law of a Markov Decision Model is often given by a *transition* or *system function.* To make this more precise, suppose that $Z_1, Z_2, \dots, Z_N$ are random variables with values in a measurable space $(\mathcal{Z}, \mathfrak{Z})$. These random variables are called *disturbances.* Z_{n+1} influences the transition from the state at time n of the system to the state at time $n+1$. The distribution Q_n^Z of Z_{n+1} may depend on the current state and action at time n such that $Q_n^Z(\cdot|x,a)$ is a stochastic kernel for $(x,a) \in D_n$. The new state of the system at time $n+1$ can now be described by a *transition* or *system function* $T_n : D_n \times \mathcal{Z} \to E$ such that

$$x_{n+1} = T_n(x_n, a_n, z_{n+1}).$$

Thus, the transition law of the Markov Decision Model is here determined by T_n and Q_n^Z.

Theorem 2.1.3. *A* Markov Decision Model *is equivalently described by the set of data* $(E, A, D_n, \mathcal{Z}, T_n, Q_n^Z, r_n, g_N)$ *with the following meaning:*

- E, A, D_n, r_n, g_N *are as in Definition 2.1.1.*
- $\mathcal{Z}$ *is the* disturbance space, *endowed with a* σ*-algebra* $\mathfrak{Z}$.
- $Q_n^Z(B|x,a)$ *is a stochastic transition kernel for* $B \in \mathfrak{Z}$ *and* $(x,a) \in D_n$ *and* $Q_n^Z(B|x,a)$ *denotes the probability that* Z_{n+1} *is in* B *if the current state is* x *and action* a *is taken.*
- $T_n : D_n \times \mathcal{Z} \to E$ *is a measurable function and is called* transition *or* system function. $T_n(x,a,z)$ *gives the next state of the system at time* $n+1$ *if at time* n *the system is in state* x, *action* a *is taken and the disturbance* z *occurs at time* $n+1$.

Proof. Suppose first a Markov Decision Model as in Definition 2.1.1 is given. Obviously we can choose $\mathcal{Z} := E$, $T_n(x,a,z) := z$ and $Q_n^Z(B|x,a) := Q_n(B|x,a)$ for $B \in \mathfrak{E}$. Conversely, if the data $(E, A, D_n, \mathcal{Z}, T_n, Q_n^Z, r_n, g_N)$ is given then by setting

$$Q_n(B|x,a) := Q_n^Z\Big(\{z \in \mathcal{Z} \mid T_n(x,a,z) \in B\} \,\Big|\, x, a\Big), \quad B \in \mathfrak{E},$$

we obtain the stochastic kernel of the Markov Decision Model. □

Let us next consider the consumption problem of Example 1.1.1 and set it up as a Markov Decision Model.

Example 2.1.4 (Consumption Problem; continued). Let us consider the consumption problem of Example 1.1.1. We denote by Z_{n+1} the random return of our risky asset over period $[n, n+1)$. Further we suppose that $Z_1, \dots, Z_N$ are non-negative, independent random variables and we assume that the consumption is evaluated by utility functions $U_n : \mathbb{R}_+ \to \mathbb{R}$. The final capital is also evaluated by a utility function U_N. Thus we choose the following data:

- $E := \mathbb{R}_+$ where $x_n \in E$ denotes the wealth of the investor at time n,
- $A := \mathbb{R}_+$ where $a_n \in A$ denotes the wealth which is consumed at time n,
- $D_n(x) := [0, x]$ for all $x \in E$, i.e. we are not allowed to borrow money.
- $\mathcal{Z} := \mathbb{R}_+$ where z denotes the random return of the asset,
- $T_n(x_n, a_n, z_{n+1}) := (x_n - a_n)z_{n+1}$ is the transition function,
- $Q_n^Z(\cdot|x, a) :=$ distribution of Z_{n+1} (independent of (x, a)),
- $r_n(x, a) := U_n(a)$ is the one-stage reward,
- $g_N(x) := U_N(x)$. ♦

In what follows we assume that there is a fixed planning horizon $N \in \mathbb{N}$, i.e. N denotes the number of stages. Of course when we want to control a Markov Decision Process, due to its stochastic transitions, it is not reasonable to determine all actions at all time points at the beginning. Instead we have to react to random changes. Thus we have to choose a control at the beginning which takes into account the future time points and states.

Definition 2.1.5. a) A measurable mapping $f_n : E \to A$ with the property $f_n(x) \in D_n(x)$ for all $x \in E$, is called a *decision rule* at time n. We denote by F_n the set of all decision rules at time n.

b) A sequence of decision rules $\pi = (f_0, f_1, \dots, f_{N-1})$ with $f_n \in F_n$ is called an *N-stage policy* or *N-stage strategy.*

Note that $F_n \neq \emptyset$ since by Definition 2.1.1 D_n contains the graph of a measurable mapping $f_n : E \to A$.

Remark 2.1.6 (Randomized Policies). It is sometimes reasonable to allow for *randomized policies* or decision rules respectively. A *randomized Markov policy* $\pi = (f_0, f_1, \dots, f_{N-1})$ is given if $f_n(B|x)$ is a stochastic kernel with $f_n(D_n(x)|x) = 1$ for all $x \in E$. In order to apply such a policy we have to do a random experiment to determine the action. Randomized decision rules are related to *relaxed controls* or *Young measures* and are sometimes necessary to guarantee the existence of optimal policies (cf. Section 8.2). ◊

We consider a Markov Decision Model as an N-stage random experiment. Thus, in order to be mathematically precise we have to define the underlying probability space. The *canonical construction* is as follows. Define a measurable space $(\Omega, \mathcal{F})$ by

$$\Omega = E^{N+1}, \quad \mathcal{F} = \mathfrak{E} \otimes \dots \otimes \mathfrak{E}.$$

We denote $\omega = (x_0, x_1, \dots, x_N) \in \Omega$. The random variables $X_0, X_1, \dots, X_N$ are defined on the measurable space $(\Omega, \mathcal{F})$ by

$$X_n(\omega) = X_n((x_0, x_1, \dots, x_N)) = x_n,$$

being the n-th projection of ω. The random variable X_n represents the state of the system at time n and (X_n) is called *Markov Decision Process.* Suppose now that $\pi = (f_0, f_1, \dots, f_{N-1})$ is a fixed policy and $x \in E$ is a fixed initial state. According to the Theorem of Ionescu-Tulcea (see Appendix B) there exists a unique probability measure $\mathbb{P}_x^\pi$ on $(\Omega, \mathcal{F})$ with

(i) $\mathbb{P}_x^\pi(X_0 \in B) = \delta_x(B)$ for all $B \in \mathfrak{E}$.
(ii) $\mathbb{P}_x^\pi(X_{n+1} \in B | X_1, \dots, X_n) = \mathbb{P}_x^\pi(X_{n+1} \in B | X_n) = Q_n\big(B | X_n, f_n(X_n)\big)$.

Equation (ii) is the so-called *Markov property*, i.e. the sequence of random variables $X_0, X_1, \dots, X_n$ is a non-stationary Markov process with respect to $\mathbb{P}_x^\pi$. By $\mathbb{E}_x^\pi$ we denote the expectation with respect to $\mathbb{P}_x^\pi$. Moreover we denote by $\mathbb{P}_{nx}^\pi$ the conditional probability $\mathbb{P}_{nx}^\pi(\cdot) := \mathbb{P}^\pi(\cdot \mid X_n = x)$. $\mathbb{E}_{nx}^\pi$ is the corresponding expectation operator.

2.2 Finite Horizon Markov Decision Models

Now we have to impose an assumption which guarantees that all appearing expectations are well-defined. By $x^+ = \max\{0, x\}$ we denote the positive part of x.

Integrability Assumption ($\mathbf{A}_N$): For $n = 0, 1, \dots, N$

$$\delta_n^N(x) := \sup_\pi \mathbb{E}_{nx}^\pi \left[\sum_{k=n}^{N-1} r_k^+(X_k, f_k(X_k)) + g_N^+(X_N) \right] < \infty, \quad x \in E.$$

We assume that (A_N) holds for the N-stage Markov Decision Problems throughout the following chapters. Obviously Assumption (A_N) is satisfied if all r_n and g_N are bounded from above. The main results are even true under a weaker assumption than (A_N) (see Remark 2.3.14).

Example 2.2.1 (Consumption Problem; continued). In the consumption problem Assumption (A_N) is satisfied if we assume that the utility functions are increasing and concave and $\mathbb{E}\, Z_n < \infty$ for all n, because then r_n and g_N can be bounded by an affine-linear function $c_1 + c_2 x$ with $c_1, c_2 \ge 0$ and since $X_n \le x Z_1 \dots Z_n$ a.s. under every policy, the function δ_n^N satisfies

$$\begin{aligned}\delta_n^N(x) &= \sup_\pi \mathbb{E}_{nx}^\pi \left[\sum_{k=n}^{N-1} U_k^+(f_k(X_k)) + U_N^+(X_N) \right] \\ &\le N c_1 + x c_2 \sum_{k=n}^{N} \mathbb{E}\, Z_1 \dots \mathbb{E}\, Z_k < \infty, \quad x > 0\end{aligned}$$

which implies the statement. ♦

We can now introduce the expected discounted reward of a policy and the N-stage optimization problem we are interested in. For $n = 0, 1, \dots, N$ and a policy $\pi = (f_0, \dots, f_{N-1})$ let $V_{n\pi}(x)$ be defined by

$$V_{n\pi}(x) := \mathbb{E}^{\pi}_{nx}\left[\sum_{k=n}^{N-1} r_k(X_k, f_k(X_k)) + g_N(X_N)\right], \quad x \in E.$$

$V_{n\pi}(x)$ is the *expected total reward at time n over the remaining stages n to N* if we use policy π and start in state $x \in E$ at time n. The *value function* V_n is defined by

$$V_n(x) := \sup_{\pi} V_{n\pi}(x), \quad x \in E.$$

$V_n(x)$ is the *maximal expected total reward at time n over the remaining stages n to N if we start in state $x \in E$ at time n.* The functions $V_{n\pi}$ and V_n are well-defined since

$$V_{n\pi}(x) \le V_n(x) \le \delta_n^N(x) < \infty, \quad x \in E.$$

Note that $V_{N\pi}(x) = V_N(x) = g_N(x)$ and that $V_{n\pi}$ depends only on $(f_n, \dots, f_{N-1})$. Moreover, it is in general not true that V_n is measurable. This causes theoretical inconveniences. Some further assumptions are needed to imply this (see Section 2.4).

A policy $\pi \in F_0 \times \dots \times F_{N-1}$ is called *optimal* for the N-stage Markov Decision Model if $V_{0\pi}(x) = V_0(x)$ for all $x \in E$.

Until now we have considered *Markov policies.* One could ask why the decision rules are only functions of the current state and do not depend on the complete history? Let us now introduce the *sets of histories* which are denoted by

$$H_0 := E,$$
$$H_n := H_{n-1} \times A \times E.$$

An element $h_n = (x_0, a_0, x_1, \dots, x_n) \in H_n$ is called *history up to time n.*

Definition 2.2.2. a) A measurable mapping $f_n : H_n \to A$ with the property $f_n(h_n) \in D_n(x_n)$ for all $h_n \in H_n$ is called a *history-dependent decision rule* at stage n.

b) A sequence $\pi = (f_0, f_1, \dots, f_{N-1})$ where f_n is a history-dependent decision rule at stage n, is called a *history-dependent N-stage policy.* We denote by Π_N the set of all history-dependent N-stage policies.

Let $\pi \in \Pi_N$ be a history-dependent policy. Then $V_{n\pi}(h_n)$ is defined as the conditional expectation of the total reward in $[n, N]$, given the history $h_n \in H_n$. The following theorem states that history-dependent policies do not improve the maximal expected rewards. For a proof see Hinderer (1970), Theorem 18.4.

Theorem 2.2.3. *For $n = 0, \dots, N$ it holds:*

$$V_n(x_n) = \sup_{\pi \in \Pi_N} V_{n\pi}(h_n), \quad h_n = (x_0, a_0, x_1, \dots, x_n).$$

Though we are in general satisfied with the value function $V_0(x)$, it turns out that on the way computing $V_0(x)$ we also have to determine the value function $V_n(x)$. This is a standard feature of many multistage optimization techniques and explained in the next section.

2.3 The Bellman Equation

For a fixed policy $\pi \in F_0 \times \dots \times F_{N-1}$ we can compute the expected discounted rewards recursively by the so-called *reward iteration*. But first we introduce some important operators which simplify the notation. In what follows let us denote by

$$\mathbb{M}(E) := \{v : E \to [-\infty, \infty) \mid v \text{ is measurable}\}.$$

Due to our assumptions we have $V_{n\pi} \in \mathbb{M}(E)$ for all π and n.

Definition 2.3.1. We define the following operators for $n = 0, 1, \dots, N-1$.

a) For $v \in \mathbb{M}(E)$ define

$$(L_n v)(x, a) := r_n(x, a) + \int v(x') Q_n(dx'|x, a), \quad (x, a) \in D_n$$

whenever the integral exists.

b) For $v \in \mathbb{M}(E)$ and $f \in F_n$ define

$$(\mathcal{T}_{nf} v)(x) := (L_n v)(x, f(x)), \quad x \in E.$$

c) For $v \in \mathbb{M}(E)$ define

$$(\mathcal{T}_n v)(x) := \sup_{a \in D_n(x)} (L_n v)(x, a), \quad x \in E.$$

$\mathcal{T}_n$ is called the *maximal reward operator at time n*.

Remark 2.3.2. a) We have the following relation between the operators

$$\mathcal{T}_n v = \sup_{f \in F_n} \mathcal{T}_{nf} v.$$

b) It holds that $\mathcal{T}_{nf} v \in I\!M(E)$ for all $v \in I\!M(E)$, but $\mathcal{T}_n v$ does not belong to $I\!M(E)$ in general.

c) If a Markov Decision Model with disturbances (Z_n) is given as in Theorem 2.1.3, then $L_n v$ can be written as

$$(L_n v)(x,a) = r_n(x,a) + \mathbb{E}\Big[v\big(T_n(x,a,Z_{n+1})\big)\Big].$$

This representation is often more convenient. ◊

Notation: In what follows we will skip the brackets $(\mathcal{T}_n v)(x)$ around the operators and simply write $\mathcal{T}_n v(x)$ in order to ease notation. When we have a sequence of operators like $\mathcal{T}_n \mathcal{T}_{n+1} v$ then the order of application is given by $(\mathcal{T}_n(\mathcal{T}_{n+1} v))$, i.e. the inner operator is applied first. The same convention applies to the other operators.

The operators have the following important properties.

Lemma 2.3.3. *All three operators are monotone, i.e. for $v, w \in I\!M(E)$ with $v(x) \le w(x)$ for all $x \in E$ it holds:*

a) $L_n v(x,a) \le L_n w(x,a)$ for all $(x,a) \in D_n$,
b) $\mathcal{T}_{nf} v(x) \le \mathcal{T}_{nf} w(x)$ for all $x \in E, f \in F_n$,
c) $\mathcal{T}_n v(x) \le \mathcal{T}_n w(x)$ for all $x \in E$.

Proof. Let $v(x) \le w(x)$ for all $x \in E$. Then

$$\int v(x') Q_n(dx'|x,a) \le \int w(x') Q_n(dx'|x,a).$$

Thus, $L_n v(x,a) \le L_n w(x,a)$ which implies the first and second statement. Taking the supremum over all $a \in D_n(x)$ implies the third statement. □

The operators $\mathcal{T}_{nf}$ can now be used to compute the value of a policy recursively.

Theorem 2.3.4 (Reward Iteration). *Let $\pi = (f_0, \dots, f_{N-1})$ be an N-stage policy. For $n = 0, 1, \dots, N-1$ it holds:*

a) $V_{N\pi} = g_N$ and $V_{n\pi} = \mathcal{T}_{nf_n} V_{n+1,\pi}$,
b) $V_{n\pi} = \mathcal{T}_{nf_n} \dots \mathcal{T}_{N-1 f_{N-1}} g_N$.

Proof. a) For $x \in E$ we have

$$\begin{aligned}
V_{n\pi}(x) &= \mathbb{E}_{nx}^{\pi}\left[\sum_{k=n}^{N-1} r_k\big(X_k, f_k(X_k)\big) + g_N(X_N)\right] \\
&= \mathbb{E}_{nx}^{\pi}[r_n\big(x, f_n(x)\big)] + \mathbb{E}_{nx}^{\pi}\left[\sum_{k=n+1}^{N-1} r_k\big(X_k, f_k(X_k)\big) + g_N(X_N)\right] \\
&= r_n\big(x, f_n(x)\big) \\
&\quad + \mathbb{E}_{nx}^{\pi}\left[\mathbb{E}_{nx}^{\pi}\left[\sum_{k=n+1}^{N-1} r_k\big(X_k, f_k(X_k)\big) + g_N(X_N) \,\Big|\, X_{n+1}\right]\right] \\
&= r_n\big(x, f_n(x)\big) \\
&\quad + \int \mathbb{E}_{n+1,x'}^{\pi}\left[\sum_{k=n+1}^{N-1} r_k\big(X_k, f_k(X_k)\big) + g_N(X_N)\right] Q_n\big(dx'|x, f_n(x)\big) \\
&= r_n\big(x, f_n(x)\big) + \int V_{n+1,\pi}(x') Q_n\big(dx'|x, f_n(x)\big)
\end{aligned}$$

where we have used the properties of $\mathbb{P}_{xn}^{\pi}$ in the fourth equality.
b) Follows from a) by induction. □

Example 2.3.5 (Consumption Problem; continued). We revisit again Example 2.1.4. First note that for $f_n \in F_n$ the $\mathcal{T}_{nf_n}$ operator in this example reads

$$\mathcal{T}_{nf_n} v(x) = U_n\big(f_n(x)\big) + \mathbb{E}\, v\big((x - f_n(x)) Z_{n+1}\big).$$

Now let us assume that $U_n(x) := \log x$ for all n and $g_N(x) := \log x$. Moreover, we assume that the return distribution is independent of n and has finite expectation $\mathbb{E}\, Z$. Then (A_N) is satisfied as we have shown in Example 2.2.1. If we choose the N-stage policy $\pi = (f_0, \ldots, f_{N-1})$ with $f_n(x) = cx$ and $c \in [0,1]$, i.e. we always consume a constant fraction of the wealth, then the Reward Iteration in Theorem 2.3.4 implies by induction on N that

$$V_{0\pi}(x) = (N+1)\log x + N \log c + \frac{(N+1)N}{2}\Big(\log(1-c) + \mathbb{E}\log Z\Big).$$

Hence $\pi^* = (f_0^*, \ldots, f_{N-1}^*)$ with $f_n^*(x) = c^* x$ and $c^* = \frac{2}{N+3}$ maximizes the expected log-utility (among all linear consumption policies). ♦

The next definition will be crucial for the solution of Markov Decision Problems.

Definition 2.3.6. Let $v \in I\!M(E)$. A decision rule $f \in F_n$ is called a *maximizer of* v at time n if $\mathcal{T}_{nf} v = \mathcal{T}_n v$, i.e. for all $x \in E$, $f(x)$ is a maximum point of the mapping $a \mapsto L_n v(x,a)$, $a \in D_n(x)$.

Theorem 2.3.8 below gives the key solution method for Markov Decision Problems. They can be solved by successive application of the $\mathcal{T}_n$-operators. As mentioned earlier it is in general not true that $\mathcal{T}_n v \in I\!M(E)$ for $v \in I\!M(E)$. However, it can be shown that V_n is analytically measurable and the sequence (V_n) satisfies the so-called *Bellman equation*

$$\begin{aligned} V_N &= g_N, \\ V_n &= \mathcal{T}_n V_{n+1}, \quad n = 0, 1, \dots, N-1, \end{aligned}$$

see e.g. Bertsekas and Shreve (1978). In the next theorem we show that whenever a solution of the Bellman equation exists together with a sequence of maximizers, then this yields the solution of our optimization problem.

Theorem 2.3.7 (Verification Theorem). *Let $(v_n) \subset I\!M(E)$ be a solution of the Bellman equation. Then it holds:*

a) $v_n \geq V_n$ for $n = 0, 1, \dots, N$.
b) If f_n^* is a maximizer of v_{n+1} for $n = 0, 1, \dots, N-1$, then $v_n = V_n$ and the policy $\pi^* = (f_0^*, f_1^*, \dots, f_{N-1}^*)$ is optimal for the N-stage Markov Decision Problem.

Proof. a) For $n = N$ we have $v_N = g_N = V_N$. Suppose $v_{n+1} \geq V_{n+1}$, then for all $\pi = (f_0, \dots, f_{N-1})$

$$v_n = \mathcal{T}_n v_{n+1} \geq \mathcal{T}_n V_{n+1} \geq \mathcal{T}_{n,f_n} V_{n+1,\pi} = V_{n\pi}.$$

Taking the supremum over all policies π yields $v_n \geq V_n$.

b) We show recursively that $v_n = V_n = V_{n\pi^*}$. For $n = N$ this is obvious. Suppose the statement is true for $n+1$, then

$$V_n \leq v_n = \mathcal{T}_{nf_n^*} v_{n+1} = \mathcal{T}_{nf_n^*} V_{n+1} = V_{n\pi^*} \leq V_n$$

and the theorem holds. □

The Verification Theorem is similar to statements which are usually delivered for stochastic control problems in continuous time. It is sufficient for applications where a solution of the Bellman equation is obvious and the existence of maximizers easy (e.g. if state and action spaces are finite). In general the existence of an optimal policy is not guaranteed. We have to make further assumptions about the structure of the problem to ensure this. In what follows we will first make a structure assumption to state our main theorem. Important cases where this assumption is satisfied are then discussed in Section 2.4. Also note that the value of an optimization problem is always unique whereas an optimal policy may not be unique.

Structure Assumption (SA$_N$): *There exist sets $I\!M_n \subset I\!M(E)$ and $\Delta_n \subset F_n$ such that for all $n = 0, 1, \dots, N-1$:*

(i) $g_N \in I\!M_N$.
(ii) *If $v \in I\!M_{n+1}$ then $\mathcal{T}_n v$ is well-defined and $\mathcal{T}_n v \in I\!M_n$.*
(iii) *For all $v \in I\!M_{n+1}$ there exists a maximizer f_n of v with $f_n \in \Delta_n$.*

Often $I\!M_n$ is independent of n and it is possible to choose $\Delta_n = F_n \cap \Delta$ for a set $\Delta \subset \{f : E \to A \text{ measurable}\}$, i.e all value functions and all maximizers have the same structural properties.

The next theorem is the main result of this section. It shows how Markov Decision Problems can be solved recursively by solving N (one-stage) optimization problems.

Theorem 2.3.8 (Structure Theorem). *Let (SA$_N$) be satisfied. Then it holds:*

a) $V_n \in I\!M_n$ and the sequence (V_n) satisfies the Bellman equation, *i.e. for $n = 0, 1, \dots, N-1$*

$$V_N(x) = g_N(x),$$
$$V_n(x) = \sup_{a \in D_n(x)} \left\{ r_n(x,a) + \int V_{n+1}(x') Q_n(dx'|x,a) \right\}, \quad x \in E.$$

b) $V_n = \mathcal{T}_n \mathcal{T}_{n+1} \dots \mathcal{T}_{N-1} g_N$.
c) For $n = 0, 1, \dots, N-1$ there exist maximizers f_n of V_{n+1} with $f_n \in \Delta_n$, and every sequence of maximizers f_n^* of V_{n+1} defines an optimal policy $(f_0^*, f_1^*, \dots, f_{N-1}^*)$ for the N-stage Markov Decision Problem.

Proof. Since b) follows directly from a) it suffices to prove a) and c). We show by induction on $n = N-1, \dots, 0$ that $V_n \in I\!M_n$ and that

$$V_{n\pi^*} = \mathcal{T}_n V_{n+1} = V_n$$

where $\pi^* = (f_0^*, \dots, f_{N-1}^*)$ is the policy generated by the maximizers of $V_1, \dots, V_N$ and $f_n^* \in \Delta_n$. We know $V_N = g_N \in I\!M_N$ by (SA$_N$) (i). Now suppose that the statement is true for $N-1, \dots, n+1$. Since $V_k \in I\!M_k$ for $k = N, \dots, n+1$, the maximizers $f_n^*, \dots, f_{N-1}^*$ exist and we obtain with the reward iteration and the induction hypothesis (note that $f_0^*, \dots, f_{n-1}^*$ are not relevant for the following equation)

$$V_{n\pi^*} = \mathcal{T}_{nf_n^*} V_{n+1,\pi^*} = \mathcal{T}_{nf_n^*} V_{n+1} = \mathcal{T}_n V_{n+1}.$$

Hence $V_n \geq \mathcal{T}_n V_{n+1}$. On the other hand for an arbitrary policy π

$$V_{n\pi} = \mathcal{T}_{nf_n} V_{n+1,\pi} \leq \mathcal{T}_{nf_n} V_{n+1} \leq \mathcal{T}_n V_{n+1}$$

where we have used the order preserving property of $\mathcal{T}_{nf_n}$. Taking the supremum over all policies yields $V_n \leq \mathcal{T}_n V_{n+1}$. Altogether it follows that

$$V_{n\pi^*} = \mathcal{T}_n V_{n+1} = V_n$$

and in view of (SA$_N$), $V_n \in I\!M_n$. □

From this result we conclude directly the following corollary.

Corollary 2.3.9. *Let (SA$_N$) be satisfied. If $n \leq m \leq N$ then it holds:*

$$V_n(x) = \sup_\pi \mathbb{E}^\pi_{nx} \Big[\sum_{k=n}^{m-1} r_k\big(X_k, f_k(X_k)\big) + V_m(X_m)\Big], \quad x \in E.$$

Theorem 2.3.8 implies the following recursive algorithm to solve Markov Decision Problems:

Backward Induction Algorithm.

1. Set $n := N$ and for $x \in E$:
$$V_N(x) := g_N(x).$$
2. Set $n := n - 1$ and compute for all $x \in E$
$$V_n(x) = \sup_{a \in D_n(x)} \Big\{ r_n(x,a) + \int V_{n+1}(x') Q_n(dx'|x,a) \Big\}.$$
Compute a maximizer f_n^* of V_{n+1}.
3. If $n = 0$, then the value function V_0 is computed and the optimal policy π^* is given by $\pi^* = (f_0^*, \ldots, f_{N-1}^*)$. Otherwise, go to step 2.

Theorem 2.3.8 tells us that the maximizers yield an optimal strategy. However the reverse statement is not true: optimal strategies do not necessarily contain only maximizers. This is shown by the following example.

Example 2.3.10. Let $N = 2$ be the planning horizon and state and action spaces be given by $S = \{0,1\} = A = D_n(x)$ for all $x \in E$. The transition probabilities are given by $Q_n(\{x'\}|x,a) = 1$ if $a = x'$ and zero otherwise (see Figure 2.1). The reward functions are given by $r_n(x,a) = a$ for $(x,a) \in D_n$ and $g_2(x) = x$. The optimal policy is easily computed to be $\pi^* = (f_0^*, f_1^*)$ with $f_0^*(x) = 1$ and $f_1^*(x) = 1$ for all $x \in E$. However, it is easy to see that $\pi = (f_0, f_1)$ with $f_0(x) \equiv 1$, and $f_1(0) = 0,\ f_1(1) = 1$ yields the same expected total reward and is thus optimal, too. Obviously the reason is that under an optimal policy state 0 will not be visited after time 1. ◇

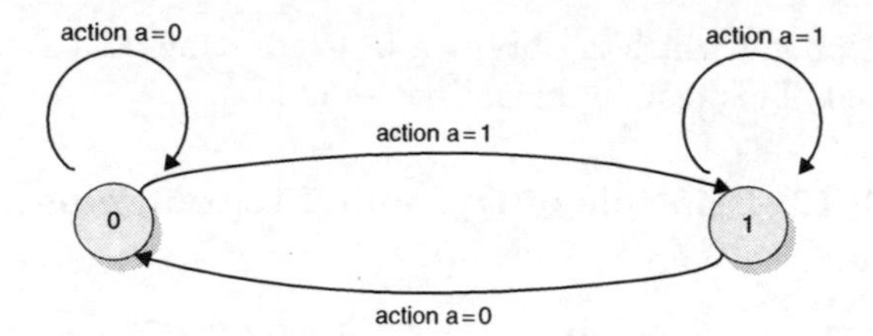

Fig. 2.1 Transition probabilities of Example 2.3.10.

Example 2.3.11 (Consumption Problem; continued). Let us now solve the consumption problem of Example 2.3.5. First suppose that the Structure Assumption (SA_N) is satisfied and we can apply Theorem 2.3.8. Thus, we obtain $V_N(x) = \log x$ and

$$\begin{aligned} V_{N-1}(x) = \mathcal{T}_{N-1}V_N(x) &= \sup_{a\in[0,x]} \Big\{ \log a + \log(x-a) + \mathbb{E} \log Z \Big\} \\ &= 2\log x + 2\log 0.5 + \mathbb{E}\log Z \end{aligned}$$

where the maximizer is given by $f^*_{N-1}(x) = 0.5x$. Now by induction we obtain

$$V_n(x) = (N-n+1)\log x + d_n, \quad 0 \le n \le N$$

where $d_N = 0$ and

$$d_n = d_{n+1} + (N-n)\,\mathbb{E}\log Z - \log\big(N-n+1\big) + (N-n)\log\Big(\frac{N-n}{N-n+1}\Big),$$

and we obtain the maximizer $f^*_n(x) = \frac{1}{N-n+1}x$. Thus, the optimal fraction which is consumed is independent of the wealth and increases over time. Finally it remains to show that (SA_N) is satisfied. But this can now be easily verified by choosing

$$\mathbb{M}_n := \{v \in \mathbb{M}(E) \mid v(x) = b\log x + d \text{ for constants } b, d \in \mathbb{R}\}$$

$$\Delta_n := \{f \in F_n \mid f(x) = cx \text{ for } c \in \mathbb{R}\}.$$

Indeed, the necessary calculations are pretty much the same as we have performed before. ♦

In order to obtain a good guess about how $\mathbb{M}_n$ and Δ_n look like it is reasonable to compute the first steps of the Backward Induction Algorithm and investigate the structure of the value functions.

The solution method in Theorem 2.3.8 relies on a very simple but general observation which is called the *Principle of Dynamic Programming.* Informally it says that whenever we have an optimal policy π^* over a certain horizon N and consider the process now only on a subinterval of $[0, N]$, then the corresponding policy which is obtained by restricting π^* to this subinterval is again optimal. This can be formalized as follows.

Theorem 2.3.12 (Principle of Dynamic Programming). *Let* (SA_N) *be satisfied. Then it holds for* $n \le m \le N$*:*

$$V_{n\pi^*}(x) = V_n(x) \quad \Rightarrow \quad V_{m\pi^*} = V_m \;\; \mathbb{P}^{\pi^*}_{nx} - a.s.,$$

i.e. if $(f_n^*, \dots, f_{N-1}^*)$ *is optimal for the time period* $[n, N]$ *then* $(f_m^*, \dots, f_{N-1}^*)$ *is optimal for* $[m, N]$.

Proof. It follows from the Reward Iteration (Theorem 2.3.4) and the definition of V_m that

$$\begin{aligned} V_n(x) &= V_{n\pi^*}(x) = \mathcal{T}_{nf_n^*} \dots \mathcal{T}_{m-1,f_{m-1}^*} V_{m\pi^*}(x) \\ &= \mathbb{E}^{\pi^*}_{nx}\Big[\sum_{k=n}^{m-1} r_k\big(X_k, f_k^*(X_k)\big) + V_{m\pi^*}(X_m)\Big] \\ &\le \mathbb{E}^{\pi^*}_{nx}\Big[\sum_{k=n}^{m-1} r_k\big(X_k, f_k^*(X_k)\big) + V_m(X_m)\Big] \le V_n(x) \end{aligned}$$

where we have used Corollary 2.3.9 for the last inequality. This implies that we have indeed equality and that $\mathbb{E}^{\pi^*}_{nx}\big[V_m(X_m) - V_{m\pi^*}(X_m)\big] = 0$ which means that $V_{m\pi^*} = V_m \; \mathbb{P}^{\pi^*}_{nx} - a.s.$ □

Sometimes Markov Decision Models are such that the state space contains an absorbing subset ('cemetery' subset) which will never be left once it is reached and where we obtain no reward. Let us call this set $G \subset E$. Obviously such a set is not very interesting and can in principle be neglected when formulating the Markov Decision Model. However this leads at least for some stages to a substochastic transition law. This is explained in the next example.

Example 2.3.13 (Absorbing Markov Decision Model). Suppose that a Markov Decision Model is given, where E is the state space, $\emptyset \neq G \subset E$ and A is the action space. We suppose that $Q_n(G|x, a) = 1$ for $x \in G$, i.e. G will never be left, once it is entered. Moreover $r_n(x, a) = 0$ and $g_N(x) = 0$ for all $x \in G$ and $a \in D_n(x)$. Then it is sufficient to consider the following *substochastic* Markov Decision Model $(\tilde{E}, \tilde{A}, \tilde{D}_n, \tilde{Q}_n, \tilde{r}_n, \tilde{g}_N)$ with

- $\tilde{E} := E \setminus G$,
- $\tilde{A} := A$,
- $\tilde{D}_n(x) := D_n(x)$ for $x \in \tilde{E}$ and $\tilde{D}_n := \{(x, a) \mid x \in \tilde{E}, a \in \tilde{D}_n(x)\}$,
- $\tilde{Q}_n(B|x, a) := Q_n(B|x, a)$ for $(x, a) \in \tilde{D}_n$ and $B \in \mathfrak{E}, B \subset \tilde{E}$,
- $\tilde{r}_n(x, a) := r_n(x, a)$ for $(x, a) \in \tilde{D}$,

- $\tilde{g}_N(x) := g_N(x)$ for $x \in \tilde{E}$.

Typically $\tilde{Q}_n(\tilde{E}|x,a) < 1$ for some $x \in E$, since G is deleted from the state space and may be entered with positive probability. Moreover, by definition we have $V_N(x) = 0$ for $x \in G$. It is now easy to see by induction using Theorem 2.3.8 that $V_n(x) = 0$ for all $x \in G$. This implies that it holds for all $x \in \tilde{E}$:

$$\begin{aligned} V_n(x) &= \sup_{a \in D_n(x)} \Big\{ r_n(x,a) + \int V_{n+1}(x') Q_n(dx'|x,a) \Big\} \\ &= \sup_{a \in \tilde{D}_n(x)} \Big\{ \tilde{r}_n(x,a) + \int_{\tilde{E}} V_{n+1}(x') \tilde{Q}_n(dx'|x,a) \Big\} \end{aligned}$$

where we have used that V_{n+1} vanishes on G. As a consequence, we can restrict ourselves to consider the Bellman equation on $\tilde{E}$. For an application see Section 2.6.1. ♦

Remark 2.3.14. a) The statements in Theorem 2.3.8 remain valid when we replace the Assumption (A_N) by the following weaker assumption: Assume that for $n = 0, 1, \dots, N$

$$\sum_{k=n}^{N-1} r_k^+(X_k, f_k(X_k)) + g_N^+(X_N)$$

is P_{nx}^π-integrable for all π and $x \in E$. However, in this case we might have $V_n(x) = +\infty$ for some $x \in E$.

b) It is well known that the Bellman equation holds under much weaker assumptions than in Theorem 2.3.8 (see Hinderer (1970), Theorem 14.4). In particular, if the reward functions r_n and g_N are non-negative (and without any further assumptions), the value functions V_n satisfy the Bellman equation. ◊

Remark 2.3.15 (Minimizing Cost). Instead of one-stage rewards r_n and a terminal reward g_N, sometimes problems are given where we have a one-stage cost c_n and a terminal cost h_N. In this case we want to minimize

$$\mathbb{E}_{nx}^\pi \left[\sum_{k=n}^{N-1} c_k\big(X_k, f_k(X_k)\big) + h_N(X_N) \right], \quad x \in E$$

for $\pi = (f_0, \dots, f_{N-1})$. But this problem can be transformed into a reward maximization problem by setting $r_n(x,a) := -c_n(x,a)$, $g_N(x) := -h_N(x)$. Thus, all the statements so far remain valid. We will use the same notation $V_{n\pi}$ and V_n for the cost functions under policy π and the minimal cost function. Moreover, the minimal cost operator $\mathcal{T}_n$ has the form

$$(\mathcal{T}_n v)(x) = \inf_{a \in D_n(x)} \Big\{ c_n(x,a) + \int v(x') Q_n(dx'|x,a) \Big\}.$$

In this case V_n is also called *cost-to-go function.* ◊

2.4 Structured Markov Decision Models

In this section we give sufficient conditions under which assumptions (A_N) and (SA_N) in the preceding section are satisfied and thus imply the validity of the Bellman equation and the existence of optimal policies. For (SA_N) we will identify conditions which imply that special sets $\mathbb{M}_n$ and Δ_n can be chosen. Of course it is interesting to choose the sets $\mathbb{M}_n$ and Δ_n as small as possible. The smaller the sets, the more information we have about the value functions and the optimal policy. On the other hand, small sets imply that we have to prove a lot of properties of $\mathcal{T}_n v$ if $v \in \mathbb{M}_{n+1}$.
Let us first consider the Integrability Assumption (A_N). It is fulfilled when the Markov Decision Model has a so-called upper bounding function.

Definition 2.4.1. A measurable function $b : E \to \mathbb{R}_+$ is called an *upper bounding function* for the Markov Decision Model if there exist $c_r, c_g, \alpha_b \in \mathbb{R}_+$ such that for all $n = 0, 1, \dots, N-1$:

(i) $r_n^+(x,a) \le c_r b(x)$ for all $(x,a) \in D_n$,
(ii) $g_N^+(x) \le c_g b(x)$ for all $x \in E$,
(iii) $\int b(x') Q_n(dx'|x,a) \le \alpha_b b(x)$ for all $(x,a) \in D_n$.

If r_n and g_N are bounded from above, then obviously $b \equiv 1$ is an upper bounding function.
Let b be an upper bounding function for the Markov Decision Model. For $v \in \mathbb{M}(E)$ we denote the *weighted supremum norm* by

$$\|v\|_b := \sup_{x \in E} \frac{|v(x)|}{b(x)}$$

(with the convention $\frac{0}{0} := 0$) and define the set

$$\mathbb{B}_b := \{ v \in \mathbb{M}(E) \mid \|v\|_b < \infty \}.$$

Equivalently $\mathbb{B}_b$ can be written as

$$\mathbb{B}_b := \{ v \in \mathbb{M}(E) \mid |v(x)| \le c b(x) \text{ for all } x \in E \text{ and for some } c \in \mathbb{R}_+ \}.$$

The concept of upper bounding functions is of particular interest for Markov Decision Models with infinite time horizon (cp. Chapter 7). The next result is fundamental for many applications.

Proposition 2.4.2. *If the Markov Decision Model has an upper bounding function b, then $\delta_n^N \in \mathbb{B}_b$ and the Integrability Assumption (A_N) is satisfied.*

Proof. Since $\delta_n^N \geq 0$ we have to show that $\delta_n^N(x) \leq cb(x)$ for some $c \in \mathbb{R}_+$. From the properties of an upper bounding function it follows that

$$\mathbb{E}_x^\pi \left[r_k^+(X_k, f_k(X_k)) \mid X_{k-1} \right] = \int r_k^+(x', f_k(x')) Q_k(dx' \mid X_{k-1}, f_{k-1}(X_{k-1}))$$
$$\leq c_r \alpha_b b(X_{k-1})$$

and by iteration we obtain

$$\mathbb{E}_x^\pi \left[r_k^+(X_k, f_k(X_k)) \right] \leq c_r \alpha_b^k b(x), \quad x \in E.$$

Analogously we get

$$\mathbb{E}_x^\pi \left[g_N^+(X_N) \right] \leq c_g \alpha_b^N b(x), \quad x \in E$$

and the result follows. □

For the rest of Section 2.4 we assume that E and A are Borel spaces (see Remark 2.1.2 a)). Also D_n is assumed to be a Borel subset of $E \times A$. Further we suppose that our Markov Decision Model has an upper bounding function b and we introduce the set

$$\mathbb{B}_b^+ := \{ v \in \mathbb{M}(E) \mid \|v^+\|_b < \infty \}.$$

Equivalently $\mathbb{B}_b^+$ can be written as

$$\mathbb{B}_b^+ := \{ v \in \mathbb{M}(E) \mid v^+(x) \leq cb(x) \text{ for all } x \in E \text{ and for some } c \in \mathbb{R}_+ \}.$$

2.4.1 Semicontinuous Markov Decision Models

In so-called *semicontinuous* Markov Decision Models the Structure Assumption (SA_N) is fulfilled with $\mathbb{M}_n$ being a subset of upper semicontinuous functions. This is a consequence of the following results (for the definition of upper semicontinuity and properties of set-valued functions the reader is referred to the Appendix A).

Proposition 2.4.3. *Let $v \in \mathbb{B}_b^+$ be upper semicontinuous. Suppose the following assumptions are satisfied:*

(i) *$D_n(x)$ is compact for all $x \in E$ and $x \mapsto D_n(x)$ is upper semicontinuous,*

(ii) $(x,a) \mapsto L_n v(x,a)$ *is upper semicontinuous on* D_n.

Then $\mathcal{T}_n v$ *is upper semicontinuous and there exists a maximizer* f_n *of* v.

Remark 2.4.4. Condition (i) in Proposition 2.4.3 can be replaced by the following condition: For all $x \in E$ the level sets $\{a \in D_n(x) \mid L_n v(x,a) \geq c\}$ are compact for all $c \in \mathbb{R}$ and the set-valued mapping

$$x \mapsto \{a \in D_n(x) \mid L_n v(x,a) = \mathcal{T}_n v(x)\}$$

is upper semicontinuous. ◊

Proof. To ease notation let us define

$$w(x,a) := L_n v(x,a), \quad w^*(x) := \mathcal{T}_n v(x)$$

and $D(x) := D_n(x)$. For $x_0 \in E$ select a sequence $(x_n) \subset E$ converging to x_0 such that the limit of $w^*(x_n)$ exists. We have to show that

$$\lim_{n\to\infty} w^*(x_n) \leq w^*(x_0).$$

Since $a \mapsto w(x,a)$ is upper semicontinuous on the compact set $D(x)$, it attains its supremum on $D(x)$ (see Theorem A.1.2). Let $a_n \in D(x_n)$ be a maximum point of $a \mapsto w(x_n,a)$ on $D(x_n)$. By the upper semicontinuity of $x \mapsto D(x)$ there is a subsequence (a_{n_k}) of (a_n) converging to some $a_0 \in D(x_0)$. The upper semicontinuity of w implies

$$\lim_{n\to\infty} w^*(x_n) = \lim_{k\to\infty} w^*(x_{n_k}) = \lim_{k\to\infty} w(x_{n_k}, a_{n_k}) \leq w(x_0,a_0) \leq w^*(x_0),$$

i.e. w^* is upper semicontinuous.
Since w and w^* are measurable, it follows easily that

$$D^* := \{(x,a) \in D \mid w(x,a) = w^*(x)\}$$

is a Borel subset of $E \times A$ and each $D^*(x)$ is compact since

$$D^*(x) := \{a \in D(x) \mid w(x,a) \geq w^*(x)\}.$$

Then, applying the selection theorem of Kuratowski and Ryll-Nardzewski (see Appendix, Theorem A.2.3), we obtain a Borel measurable selector f for D^*. This is the desired maximizer of v. □

Remark 2.4.5. If $A \subset \mathbb{R}$ then there exists a smallest and a largest maximizer of $v \in I\!B_b^+$. Note that the set

$$D_n^*(x) := \{a \in D_n(x) \mid L_n v(x,a) = T_n v(x)\}$$

is compact for $x \in E$. Then $\max D_n^*(x)$ and $\min D_n^*(x)$ are maximizers of v by the Selection Theorem A.2.3. ◊

A set of sufficient conditions on the data of a Markov Decision Model in order to assure that (SA$_N$) is satisfied with $I\!M_n$ being the set of upper semicontinuous functions $v \in I\!B_b^+$ and with $\Delta_n := F_n$ is given below. Consequently under these assumptions Theorem 2.3.8 is valid. The proof follows immediately from Proposition 2.4.3.

Theorem 2.4.6. *Suppose the Markov Decision Model has an upper bounding function b and for all $n = 0, 1, \dots, N-1$ it holds:*

(i) *$D_n(x)$ is compact for all $x \in E$ and $x \mapsto D_n(x)$ is upper semicontinuous,*
(ii) *$(x,a) \mapsto \int v(x')Q_n(dx'|x,a)$ is upper semicontinuous for all upper semicontinuous $v \in I\!B_b^+$,*
(iii) *$(x,a) \mapsto r_n(x,a)$ is upper semicontinuous,*
(iv) *$x \mapsto g_N(x)$ is upper semicontinuous.*

Then the sets $I\!M_n := \{v \in I\!B_b^+ \mid v$ is upper semicontinuous$\}$ and $\Delta_n := F_n$ satisfy the Structure Assumption (SA$_N$). In particular, $V_n \in I\!M_n$ and there exists a maximizer $f_n^* \in F_n$ of V_{n+1}. The policy $(f_0^*, \dots, f_{N-1}^*)$ is optimal.

Instead of checking condition (ii) of Theorem 2.4.6 directly, we can alternatively use the following Lemma:

Lemma 2.4.7. *Let b be a continuous upper bounding function. Then the following statements are equivalent:*

(i) *$(x,a) \mapsto \int v(x')Q(dx'|x,a)$ is upper semicontinuous for all upper semicontinuous $v \in I\!B_b^+$.*
(ii) *$(x,a) \mapsto \int b(x')Q(dx'|x,a)$ is continuous, and $(x,a) \mapsto \int v(x')Q(dx'|x,a)$ is continuous and bounded for all continuous and bounded v on E.*

A stochastic kernel Q with the last property is called *weakly continuous.*

Proof. The proof that (ii) implies (i) is as follows: Let $v \in I\!B_b^+$ be upper semicontinuous. Then we have $v - cb \le 0$ for some $c \in \mathbb{R}_+$ and $x \mapsto v(x) - cb(x)$ is upper semicontinuous. According to Lemma A.1.3 this implies the existence of a sequence $(\tilde{v}_k)$ of continuous and bounded functions such that $\tilde{v}_k \downarrow v - cb$. Due to our assumption the function $(x,a) \mapsto \int \tilde{v}_k(x')Q(dx'|x,a)$ is now continuous and bounded. Moreover, monotone convergence implies that

$$\int \tilde{v}_k(x')Q(dx'|x,a) \downarrow \int (v - cb)(x')Q(dx'|x,a) \quad \text{for } k \to \infty.$$

Thus, we can again conclude with Lemma A.1.3 that the limit is upper semicontinuous. However in view of our assumption this implies $(x,a) \mapsto \int v(x')Q(dx'|x,a)$ is upper semicontinuous.
Next we prove that (i) implies (ii): Since b and $-b$ are in $\mathbb{B}_b^+$, we get

$$(x,a) \mapsto \int b(x')Q(dx'|x,a)$$

is continuous. If v is bounded and continuous, then the functions $v - \|v\|$ and $-v - \|v\|$ belong to $\mathbb{B}_b^+$ and are upper semicontinuous. Hence the function $(x,a) \mapsto \int v(x')Q(dx'|x,a)$ is continuous. □

2.4.2 Continuous Markov Decision Models

Next we investigate when the Structure Assumption (SA_N) is satisfied with $\mathbb{M}_n$ being a subset of continuous functions.

Proposition 2.4.8. *Let $v \in \mathbb{B}_b^+$ be continuous. Suppose the following assumptions are satisfied:*

(i) *$D_n(x)$ is compact for all $x \in E$ and $x \mapsto D_n(x)$ is continuous,*
(ii) *$(x,a) \mapsto L_n v(x,a)$ is continuous on D_n.*

Then $\mathcal{T}_n v$ is continuous and there exists a maximizer $f_n \in F_n$ of v. If v has a unique maximizer $f_n \in F_n$ at time n, then f_n is continuous.

Proof. We use the same notation as in the proof of Proposition 2.4.3. In view of Proposition 2.4.3 it is sufficient to show that w^* is lower semicontinuous, i.e. that $w^*(x_0) \le \lim_{n\to\infty} w^*(x_n)$ for each sequence $(x_n) \subset E$ which converges to $x_0 \in E$ and for which $\lim_{n\to\infty} w^*(x_n)$ exists. We know by assumption that $w^*(x_0) = w(x_0, a_0)$ for some $a_0 \in D(x_0)$. Since $x \mapsto D(x)$ is continuous, there exists a subsequence (x_{n_k}) of (x_n) and a sequence of points $a_{n_k} \in D(x_{n_k})$ converging to a_0. Hence we have $(x_{n_k}, a_{n_k}) \to (x_0, a_0)$. It follows from the continuity of w that

$$w^*(x_0) = w(x_0, a_0) = \lim_{k\to\infty} w(x_{n_k}, a_{n_k}) \le \lim_{k\to\infty} w^*(x_{n_k}) = \lim_{n\to\infty} w^*(x_n).$$

Since $x \mapsto D(x)$ is upper semicontinuous, D is closed and it follows that

$$D^* := \{(x,a) \in D \mid w(x,a) = w^*(x)\}$$

is a closed subset of D. Then we obtain that $x \mapsto D^*(x)$ is also upper semicontinuous. Thus, if v has a unique maximizer f_n, i.e. $D_n^*(x) = \{f_n(x)\}$ for all $x \in E$, then f_n must be continuous. □

Remark 2.4.9. If $A \subset \mathbb{R}$, then the smallest (largest) maximizer of v is lower semicontinuous (upper semicontinuous). This can be seen as follows: If $A \subset \mathbb{R}$, then $x \mapsto f(x) := \max D^*(x)$ is the largest maximizer. Choose $x_0, x_n \in E$ such that $x_n \to x_0$ and the limit of $(f(x_n))$ exists. Since $a_n := f(x_n) \in D^*(x_n)$ and $x \mapsto D^*(x)$ is upper semicontinuous, (a_n) has an accumulation point in $D^*(x_0)$ which must be $\lim_{n\to\infty} f(x_n)$. It follows that

$$\lim_{n\to\infty} f(x_n) \le \max D^*(x_0) = f(x_0),$$

i.e. the largest maximizer is upper semicontinuous. In the same way it can be shown that the smallest maximizer is lower semicontinuous. ◊

It is now rather straightforward that the following conditions on the data of a Markov Decision Model imply by using Proposition 2.4.8 that (SA$_N$) is satisfied and that Theorem 2.3.8 is valid.

Theorem 2.4.10. *Suppose a Markov Decision Model with upper bounding function b is given and for all $n = 0, 1, \dots, N-1$ it holds:*

(i) *$D_n(x)$ is compact for all $x \in E$ and $x \mapsto D_n(x)$ is continuous,*
(ii) *$(x,a) \mapsto \int v(x')Q_n(dx'|x,a)$ is continuous for all continuous $v \in \mathbb{B}_b^+$,*
(iii) *$(x,a) \mapsto r_n(x,a)$ is continuous,*
(iv) *$x \mapsto g_N(x)$ is continuous.*

Then the sets $\mathbb{M}_n := \{v \in \mathbb{B}_b^+ \mid v \text{ is continuous}\}$ and $\Delta_n := F_n$ satisfy the Structure Assumption (SA$_N$). If the maximizer of V_n is unique, then Δ_n can be chosen as the set of continuous functions.

2.4.3 Measurable Markov Decision Models

Sometimes the Structure Assumption (SA$_N$) can be fulfilled with $\mathbb{M}_n = \mathbb{B}_b^+$. For this case the following result is useful.

Proposition 2.4.11. *Let $v \in \mathbb{B}_b^+$ and suppose the following assumptions are satisfied:*

(i) *$D_n(x)$ is compact for all $x \in E$,*
(ii) *$a \mapsto L_n v(x,a)$ is upper semicontinuous on $D_n(x)$ for all $x \in E$.*

Then $\mathcal{T}_n v$ is measurable and there exists a maximizer $f_n \in F_n$ of v.

Proof. We use the same notation as in the proof of Proposition 2.4.3. Let $\alpha \in \mathbb{R}$. It is sufficient to prove that $\{x \in E \mid w^*(x) \ge \alpha\}$ is a Borel set. But

$$\begin{aligned}\{x \in E \mid w^*(x) \ge \alpha\} &= \{x \in E \mid w(x,a) \ge \alpha \text{ for some } a \in D(x)\} \\ &= \text{proj}_E\{(x,a) \in D \mid w(x,a) \ge \alpha\}.\end{aligned}$$

This set is Borel by a result of Kunugui and Novikov (see Himmelberg et al. (1976)), since D is Borel with compact values (i.e. compact vertical sections) and $\{(x,a) \in D \mid w(x,a) \ge \alpha\}$ is a Borel subset of D with closed (and therefore compact) values. Actually, Kunugui and Novikov prove that the projection of a Borel subset of $E \times A$ with compact values is a Borel subset of E. The existence of a maximizer can be shown in the same way as in the proof of Proposition 2.4.3. □

Remark 2.4.12. a) If A is countable and $D_n(x)$ is finite for all $x \in E$, then both assumptions (i) and (ii) of Proposition 2.4.11 are fulfilled.
b) Condition (i) in Proposition 2.4.11 can be replaced by: For all $x \in E$ the level set $\{a \in D_n(x) \mid L_n v(x,a) \ge c\}$ is compact for all $c \in \mathbb{R}$. ◊

The following theorem follows directly from Proposition 2.4.11. In particular the main result (Theorem 2.3.8) holds under the assumptions of Theorem 2.4.13.

Theorem 2.4.13. *Suppose a Markov Decision Model with upper bounding function b is given and for all $n = 0, 1, \dots, N-1$ it holds:*

(i) *$D_n(x)$ is compact for all $x \in E$,*
(ii) *$a \mapsto \int v(x') Q_n(dx'|x,a)$ is upper semicontinuous for all $v \in \mathbb{B}_b^+$ and for all $x \in E$,*
(iii) *$a \mapsto r_n(x,a)$ is upper semicontinuous for all $x \in E$.*

Then the sets $\mathbb{M}_n := \mathbb{B}_b^+$ and $\Delta_n := F_n$ satisfy the Structure Assumption (SA_N).

In a more general framework one can choose $\mathbb{M}_n$ as the set of upper semianalytic functions (see Bertsekas and Shreve (1978)). But of course, one wants to choose $\mathbb{M}_n$ and Δ_n as small as possible.

2.4.4 Monotone and Convex Markov Decision Models

Structural properties (e.g. monotonicity, concavity and convexity) for the value functions and also for the maximizers are important for applications.

Results like these also simplify numerical solutions. In order to ease the exposition we assume now that $E \subset \mathbb{R}^d$ and $A \subset \mathbb{R}^m$ endowed with the usual preorder $\le$ of componentwise comparison e.g. $x \le y$ for $x, y \in \mathbb{R}^d$ if $x_k \le y_k$ for $k = 1, \dots, d$.

Theorem 2.4.14. *Suppose a Markov Decision Model with upper bounding function b is given and for all $n = 0, 1, \dots, N-1$ it holds:*

(i) *$D_n(\cdot)$ is increasing, i.e. $x \le x'$ implies $D_n(x) \subset D_n(x')$,*
(ii) *the stochastic kernels $Q_n(\cdot|x,a)$ are stochastically monotone for all $a \in D_n(x)$, i.e. the mapping $x \mapsto \int v(x')Q_n(dx'|x,a)$ is increasing for all increasing $v \in \mathbb{B}_b^+$ and for all $a \in D_n(x)$,*
(iii) *$x \mapsto r_n(x,a)$ is increasing for all a,*
(iv) *g_N is increasing on E,*
(v) *for all increasing $v \in \mathbb{B}_b^+$ there exists a maximizer $f_n \in \Delta_n$ of v.*

Then the sets $\mathbb{M}_n := \{v \in \mathbb{B}_b^+ \mid v \text{ is increasing}\}$ and Δ_n satisfy the Structure Assumption (SA_N).

Proof. Obviously condition (iv) shows that $g_N \in \mathbb{M}_N$. Let now $v \in \mathbb{M}_{n+1}$. Then conditions (ii) and (iii) imply that $x \mapsto L_n v(x,a)$ is increasing for all a. In view of (i) we obtain $\mathcal{T}_n v \in \mathbb{M}_n$. Condition (v) is equivalent to condition (iii) of (SA_N). Thus, the statement is shown. □

It is more complicated to identify situations in which the maximizers are increasing. For this property we need the following definition.

Definition 2.4.15. A set $D \subset E \times A$ is called *completely monotone* if for all points $(x,a'), (x',a) \in D$ with $x \le x'$ and $a \le a'$ it follows that $(x,a), (x',a') \in D$.

An important special case where D is completely monotone is given if $D(x)$ is independent of x. If $A = \mathbb{R}$ and $D(x) = [\underline{d}(x), \bar{d}(x)]$. Then D is completely monotone if and only if $\underline{d} : E \to \mathbb{R}$ and $\bar{d} : E \to \mathbb{R}$ are increasing.

For a definition and properties of *supermodular* functions see Appendix A.3.

Proposition 2.4.16. *Let $v \in \mathbb{B}_b^+$ and suppose the following assumptions are satisfied where $D_n^*(x) := \{a \in D_n(x) \mid L_n v(x,a) = \mathcal{T}_n v(x)\}$ for $x \in E$:*

(i) *D_n is completely monotone,*
(ii) *$L_n v$ is supermodular on D_n,*
(iii) *there exists a largest maximizer f_n^* of v i.e. for all $x \in E$ it holds: $f_n^*(x) \ge a$ for all $a \in D_n^*(x)$ which are comparable with $f_n^*(x)$.*

Then f_n^* is weakly increasing, i.e. $x \le x'$ implies $f_n^*(x) \le f_n^*(x')$, whenever $f_n^*(x)$ and $f_n^*(x')$ are comparable.

Proof. Suppose that f_n^* is not increasing, i.e. there exist $x, x' \in E$ with $x \le x'$ and $f_n^*(x) =: a > a' := f_n^*(x')$. Due to our assumptions (i) and (ii) we know that $(x, a'), (x', a) \in D_n$ and

$$\big(L_n v(x, a') - L_n v(x, a)\big) + \big(L_n v(x', a) - L_n v(x', a')\big) \ge 0.$$

Since a is a maximum point of $b \mapsto L_n v(x, b)$ and a' is a maximum point of $b \mapsto L_n v(x', b)$ the expressions in brackets are non-positive. Thus, we must have $L_n v(x', a') = L_n v(x', a)$ which means in particular that a is also a maximum point of $b \mapsto L_n v(x', b)$. But this contradicts the definition of f_n^* as the largest maximizer of v, and the statement follows. □

Remark 2.4.17. a) A similar result holds for the smallest maximizer of v.
b) If we reverse the relation on the state or the action space we obtain conditions for weakly decreasing maximizers. ◊

If our Markov Decision Model fulfills all assumptions of Theorem 2.4.14 and Proposition 2.4.16, then

$$\mathbb{M}_n := \{v \in \mathbb{B}_b^+ \mid v \text{ is increasing}\}$$
$$\Delta_n := \{f \in F_n \mid f \text{ is weakly increasing}\}$$

satisfy the Structure Assumption (SA_N).

Of particular interest are concave or convex value functions.

Proposition 2.4.18. *Let $v \in \mathbb{B}_b^+$ and suppose the following assumptions are satisfied:*

(i) *D_n is convex in $E \times A$,*
(ii) *$L_n v(x, a)$ is concave on D_n.*

Then $\mathcal{T}_n v$ is concave on E.

Proof. First note that $\mathcal{T}_n v(x) < \infty$ for all $x \in E$ and that E is convex. Let $x, x' \in E$ and $\alpha \in (0, 1)$. For $\varepsilon > 0$ there exist $a \in D_n(x)$ and $a' \in D_n(x')$ with

$$L_n v(x, a) \ge \mathcal{T}_n v(x) - \varepsilon,$$
$$L_n v(x', a') \ge \mathcal{T}_n v(x') - \varepsilon.$$

The convexity of D_n implies

$$\alpha(x, a) + (1 - \alpha)(x', a') \in D_n$$

which means that $\alpha a + (1 - \alpha)a' \in D_n(\alpha x + (1 - \alpha)x')$. Hence by (ii)

$$\begin{aligned}\mathcal{T}_n v(\alpha x + (1-\alpha)x') &\geq L_n v(\alpha x + (1-\alpha)x', \alpha a + (1-\alpha)a') \\ &\geq \alpha L_n v(x,a) + (1-\alpha) L_n v(x',a') \\ &\geq \alpha \mathcal{T}_n v(x) + (1-\alpha)\mathcal{T}_n v(x') - \varepsilon.\end{aligned}$$

This is true for all $\varepsilon > 0$, and the statement follows. □

Proposition 2.4.18 now directly implies that the following conditions on the data of the Markov Decision Model guarantee that (SA_N) is satisfied with the set $\mathbb{M}_n$ being a subset of concave functions.

Theorem 2.4.19. *Suppose a Markov Decision Model with upper bounding function b is given and for all $n = 0, 1, \ldots, N-1$ it holds:*

(i) D_n *is convex in* $E \times A$,
(ii) *the mapping* $(x,a) \mapsto \int v(x') Q_n(dx'|x,a)$ *is concave for all concave* $v \in \mathbb{B}_b^+$,
(iii) $(x,a) \mapsto r_n(x,a)$ *is concave,*
(iv) g_N *is concave on* E,
(iv) *for all concave* $v \in \mathbb{B}_b^+$ *there exists a maximizer* $f_n \in \Delta_n$ *of* v.

Then the sets $\mathbb{M}_n := \{v \in \mathbb{B}_b^+ \mid v \text{ is concave}\}$ *and* Δ_n *satisfy the Structure Assumption (SA_N).*

Remark 2.4.20. If $A = \mathbb{R}$ and $D(x) = [\underline{d}(x), \bar{d}(x)]$ then D is convex in $E \times A$ if and only if E is convex, $\underline{d} : E \to \mathbb{R}$ is convex and $\bar{d} : E \to \mathbb{R}$ is concave. ◇

Proposition 2.4.21. *Let $v \in \mathbb{B}_b^+$ and suppose that the following assumptions are satisfied:*

(i) E *is convex and* $D_n := E \times A$,
(ii) $x \mapsto L_n v(x,a)$ *is convex for all* $a \in A$.

Then $\mathcal{T}_n v$ is convex on E. If moreover A is a polytope and $a \mapsto L_n v(x,a)$ is convex for all $x \in E$, then there exists a so-called bang-bang *maximizer f_n^* of v at time n, i.e. $f_n^*(x)$ is a vertex of A for all $x \in E$.*

Proof. The first statement follows from the fact that the supremum of an arbitrary number of convex functions is again convex (because $\mathcal{T}_n v < \infty$). Now if A is a polytope, the convex function $a \mapsto L_n v(x,a)$ attains its supremum in a vertex and the set of all maximum points of $a \mapsto L_n v(x,a)$ which are vertices, is finite for all $x \in E$. Then, by applying the Selection Theorem A.2.3, we obtain the second statement. □

Proposition 2.4.21 now directly implies that the following conditions on the data of the Markov Decision Model guarantee that (SA_N) is satisfied with the set $\mathbb{M}_n$ being a subset of convex functions.

Theorem 2.4.22. *Suppose a Markov Decision Model with upper bounding function b is given and for all $n = 0, 1, \dots, N-1$ it holds:*

(i) *E is convex and $D_n := E \times A$,*
(ii) *for all convex $v \in \mathbb{B}_b^+$, $x \mapsto \int v(x')Q_n(dx'|x,a)$ is convex for all $a \in A$,*
(iii) *$x \mapsto r_n(x,a)$ is convex for all $a \in A$,*
(iv) *g_N is convex,*
(v) *for all convex $v \in \mathbb{B}_b^+$ there exists a maximizer $f_n \in \Delta_n$ of v.*

Then the sets $\mathbb{M}_n := \{v \in \mathbb{B}_b^+ \mid v \text{ is convex}\}$ and Δ_n satisfy the Structure Assumption (SA_N).

2.4.5 Comparison of Markov Decision Models

When the value functions of a Markov Decision Model have one of the properties of the last section (e.g. monotonicity, convexity, concavity), then it is possible to discuss the qualitative influence of the transition kernel on the value function. More precisely, we want to know in which direction the value function changes, if the transition kernel $Q_n(\cdot|x,a)$ is replaced by $\tilde{Q}_n(\cdot|x,a)$. To this end, denote

$$\begin{aligned} \mathbb{M}_{st} &:= \{v \in \mathbb{B}_b^+ \mid v \text{ is increasing}\} \\ \mathbb{M}_{cv} &:= \{v \in \mathbb{B}_b^+ \mid v \text{ is concave}\} \\ \mathbb{M}_{cx} &:= \{v \in \mathbb{B}_b^+ \mid v \text{ is convex}\}. \end{aligned}$$

We denote by $(\tilde{V}_n)$ the value functions of the Markov Decision Model with transition kernels $\tilde{Q}_n$. In the following theorem we use stochastic orders for the transition kernels (see Appendix B.3 for details).

Theorem 2.4.23. *Suppose a Markov Decision Model with upper bounding function b is given which satisfies the Structure Assumption (SA_N) with the set $\mathbb{M}_n^\diamond$ where $\diamond \in \{st, cv, cx\}$. If for all $n = 0, 1, \dots, N-1$*

$$Q_n(\cdot|x,a) \le_\diamond \tilde{Q}_n(\cdot|x,a), \quad \textit{for all } (x,a) \in D$$

then $V_n \le \tilde{V}_n$ for $n = 0, 1, \dots, N-1$.

The proof follows directly from the properties of the stochastic orders (see Appendix B.3).

2.5 Stationary Markov Decision Models

In this section we consider stationary Markov Decision Models, i.e. the data does not depend on n and is given by (E, A, D, Q, r_n, g_N) with $r_n := \beta^n r$, $g_N := \beta^N g$ and $\beta \in (0, 1]$.

We denote by F the set of all decision rules $f : E \to A$ with $f(x) \in D(x)$ for $x \in E$. Then F^N is the set of all N-stage policies $\pi = (f_0, \ldots, f_{N-1})$.

The expected discounted reward over n stages under a policy $\pi \in F^n$ is given by

$$J_{n\pi}(x) := \mathbb{E}_x^\pi \left[\sum_{k=0}^{n-1} \beta^k r\big(X_k, f_k(X_k)\big) + \beta^n g(X_n) \right], \quad x \in E$$

when the system starts in state $x \in E$. The maximal expected discounted reward over n stages is defined by

$$J_0(x) := g(x)$$
$$J_n(x) := \sup_{\pi \in F^n} J_{n\pi}(x), \quad x \in E, \ 1 \leq n \leq N.$$

In order to obtain a well-defined stochastic optimization problem we need the following integrability assumption (see Section 2.2):

Assumption ($\mathbf{A}_N$): *For $x \in E$*

$$\delta_N(x) := \sup_\pi \mathbb{E}_x^\pi \left[\sum_{k=0}^{N-1} \beta^k r^+(X_k, f_k(X_k)) + \beta^N g^+(X_N) \right] < \infty.$$

Remark 2.5.1. Since $\delta_0 := g^+ \leq \delta_{n-1} \leq \delta_n \leq \delta_N$ and

$$\delta_n^N = \beta^n \delta_{N-n}, \quad n = 0, 1 \ldots, N,$$

the Integrability Assumption (A_N) is equivalent to the integrability assumption in Section 2.2 when we have a stationary Markov Decision Model. ◊

As explained in Section 2.4 it is convenient to show that (A_N) is satisfied by proving the existence of an upper bounding function. The definition of an upper bounding function for a stationary Markov Decision Model is as follows.

Definition 2.5.2. A measurable mapping $b : E \to \mathbb{R}_+$ is called an *upper bounding function* for the stationary Markov Decision Model if there exist $c_r, c_g, \alpha_b \in \mathbb{R}_+$, such that:

(i) $r^+(x,a) \le c_r b(x)$ for all $(x,a) \in D$,
(ii) $g^+(x) \le c_g b(x)$ for all $x \in E$,
(iii) $\int b(x')Q(dx'|x,a) \le \alpha_b b(x)$ for all $(x,a) \in D$.

If the stationary Markov Decision Model has an upper bounding function b, then we have $\delta_N \in \mathbb{B}_b$ and the Integrability Assumption (A_N) is satisfied (cf. Proposition 2.4.2). Obviously every stationary model is a special non-stationary model. We obtain the following relation between the value functions J_n and V_n:

$$V_n(x) = \beta^n J_{N-n}(x), \quad x \in E,\ n = 0, 1, \dots, N.$$

But on the other hand, every non-stationary Markov Decision Model can be formulated as a stationary one. The idea is to extend the state space by including the time parameter.

As in Definition 2.3.1 we introduce the following operators for $v \in \mathbb{M}(E)$:

$$\begin{aligned} Lv(x,a) &:= r(x,a) + \beta \int v(x')Q(dx'|x,a), \quad (x,a) \in D, \\ \mathcal{T}_f v(x) &:= Lv\big(x, f(x)\big), \quad x \in E \\ \mathcal{T} v(x) &:= \sup_{a \in D(x)} Lv(x,a), \quad x \in E. \end{aligned}$$

$\mathcal{T}$ is called the *maximal reward operator*. The reward iteration reads now as follows.

Theorem 2.5.3 (Reward Iteration). *For $\pi = (f_0, \dots, f_{n-1})$ it holds:*

$$J_{n\pi} = \mathcal{T}_{f_0} \dots \mathcal{T}_{f_{n-1}} g.$$

The Structure Assumption (SA_N) has to be modified for stationary Markov Decision Models.

Structure Assumption ($\mathbf{SA}_N$): *There exist sets $\mathbb{M} \subset \mathbb{M}(E)$ and $\Delta \subset F$ such that:*

(i) $g \in \mathbb{M}$.
(ii) *If $v \in \mathbb{M}$ then $\mathcal{T}v(x)$ is well-defined and $\mathcal{T}v \in \mathbb{M}$.*
(iii) *For all $v \in \mathbb{M}$ there exists a maximizer $f \in \Delta$ of v, i.e.*

$$\mathcal{T}_f v(x) = \mathcal{T}v(x), \quad x \in E.$$

The main Theorem 2.3.8 about the recursive computation of the optimal value functions has now the following form.

Theorem 2.5.4 (Structure Theorem). *Let (SA_N) be satisfied.*

a) Then $J_n \in I\!M$ and the Bellman equation $J_n = \mathcal{T} J_{n-1}$ holds, i.e.

$$J_0(x) = g(x)$$
$$J_n(x) = \sup_{a \in D(x)} \Big\{ r(x,a) + \beta \int J_{n-1}(x') Q(dx'|x,a) \Big\}, \qquad x \in E.$$

Moreover, $J_n = \mathcal{T}^n g$.

b) For $n = 1, \dots, N$ there exist maximizers f_n^* of J_{n-1} with $f_n^* \in \Delta$, and every sequence of maximizers f_n^* of J_{n-1} defines an optimal policy $(f_N^*, \dots, f_1^*)$ for the stationary N-stage Markov Decision Model.

In many examples we will see that the Structure Assumption is naturally fulfilled. For some conditions which imply (SA_N) see Section 2.4. The simplest case arises when both E and A are finite. Here the Structure Assumption (SA_N) is satisfied with $I\!M := \{v : E \to \mathbb{R}\}$ because every function is measurable and maximizers exist. Moreover, the transition kernel has a discrete density and we denote

$$q(x'|x,a) := Q(\{x'\}|x,a)$$

for $x, x' \in E$ and $a \in D(x)$.
Analogously to the non-stationary case, Theorem 2.5.4 gives a recursive algorithm to solve Markov Decision Problems. Due to the stationarity of the data however, it is not necessary to formulate the algorithm as a backward algorithm.

Forward Induction Algorithm.

1. Set $n := 0$ and for $x \in E$:

$$J_0(x) := g(x).$$

2. Set $n := n + 1$ and compute for all $x \in E$

$$J_n(x) = \sup_{a \in D(x)} \Big\{ r(x,a) + \int J_{n-1}(x') Q(dx'|x,a) \Big\}.$$

   Compute a maximizer f_n^* of J_{n-1}.
3. If $n = N$, then the value function J_N is computed and the optimal policy π^* is given by $\pi^* = (f_N^*, \dots, f_1^*)$. Otherwise, go to step 2.

The *induction algorithm* computes the n-stage value functions and the optimal decision rules recursively over the stages, beginning with the terminal reward function. We illustrate this procedure with the following numerical example which is known as Howard's toymaker in the literature.

Example 2.5.5 (Howard's Toymaker). Suppose a Markov Decision Model is given by the following data. The planning horizon is $N = 4$. The state space consists of two states $E = \{1, 2\}$ as well as the action space $A = \{1, 2\}$. We have no restriction on the actions, i.e. $D(x) = A$. The reward is discounted by a factor $\beta \in (0, 1)$ and the one-stage reward is given by $r(1, 1) = 6$, $r(2, 1) = -3$, $r(1, 2) = 4$, $r(2, 2) = -5$. The terminal reward is $g(1) = 105$, $g(2) = 100$. The transition probabilities are denoted by the following matrices (see Figure 2.2)

$$q(\cdot|\cdot, 1) = \begin{pmatrix} 0.5 \; 0.5 \\ \\ 0.4 \; 0.6 \end{pmatrix} \quad q(\cdot|\cdot, 2) = \begin{pmatrix} 0.8 \; 0.2 \\ \\ 0.7 \; 0.3 \end{pmatrix}.$$

Note that $q(\cdot|\cdot, 1)$ gives the transition probabilities if action $a = 1$ is chosen and $q(\cdot|\cdot, 2)$ gives the transition probabilities if action $a = 2$ is chosen.

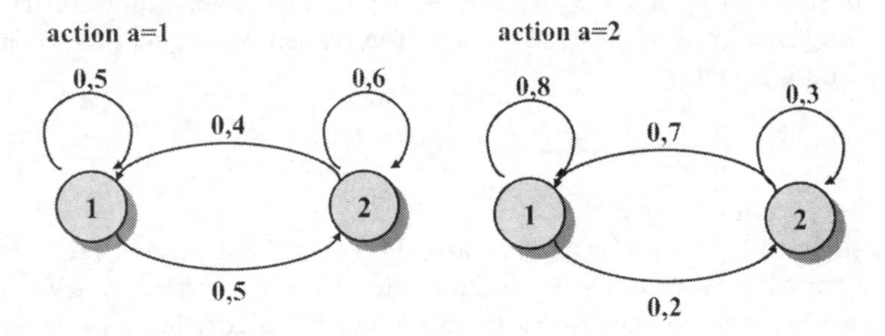

Fig. 2.2 Transition probabilities of Howard's toymaker.

The result of the computation with $\beta = 0.98$ is shown in Table 2.1. By $D_n^*(x)$ we denote the set

$$D_n^*(x) := \{a \in D(x) \mid a \text{ maximizes } b \mapsto r(x, b) + \beta \sum_{x' \in E} q(x'|x, b) J_{n-1}(x')\}.$$

In particular, the value function of the 4-stage problem is given by $J_4(1) = 106.303$ and $J_4(2) = 96.326$.
Moreover, if $\Delta_0 := J_0(1) - J_0(2) \geq 0$ then the following *Turnpike Theorem* can be shown:

- If $\beta < \frac{20}{29}$, then there exists an $N^* = N^*(\beta, \Delta_0) \in \mathbb{N}$ such that $D_n^*(1) = D_n^*(2) = \{1\}$ for $n \geq N^*$ and $2 \in D_n^*(1) = D_n^*(2)$ for $n < N^*$.
- If $\beta > \frac{20}{29}$, then there exists an $N^* = N^*(\beta, \Delta_0) \in \mathbb{N}$ such that $D_n^*(1) = D_n^*(2) = \{2\}$ for $n \geq N^*$ and $1 \in D_n^*(1) = D_n^*(2)$ for $n < N^*$.

n	$J_n(1)$	$J_n(2)$	$D_n^*(1)$	$D_n^*(2)$
0	105	100		
1	106.45	96.96	1	1
2	106.461	96.531	2	2
3	106.385	96.412	2	2
4	106.303	96.326	2	2

Table 2.1 Computational results of the Backward induction algorithm.

- If $\beta = \frac{20}{29}$, then

$$\begin{array}{lll} D_n^*(1) = D_n^*(2) = \{1\} & \text{for all } n \in \mathbb{N} & \text{if } \Delta_0 < \frac{29}{3} \\ D_n^*(1) = D_n^*(2) = A & \text{for all } n \in \mathbb{N} & \text{if } \Delta_0 = \frac{29}{3} \\ D_n^*(1) = D_n^*(2) = \{2\} & \text{for all } n \in \mathbb{N} & \text{if } \Delta_0 > \frac{29}{3}. \end{array}$$ ♦

The following Markov Decision Model with random discounting has important applications in finance and insurance.

Example 2.5.6 (Random Discounting). Suppose a stationary Markov Decision Model $(E, A, D, Q, \beta^n r, \beta^N g)$ is given. Sometimes the discount factors for the stages vary randomly. Here we assume that the (non-negative) discount factors (β_n) form a Markov process, given by a transition kernel $Q^\beta(B|y)$ and β_0 is given. We suppose that (β_n) is independent of the state process. (Also the more general case where (β_n) depends on the state process can be dealt with.) Then we are interested in finding the maximal expected discounted reward over all policies $\pi \in F^N$, i.e. we want to maximize the expression

$$\mathbb{E}_x^\pi \left[\sum_{n=0}^{N-1} \Big(\prod_{k=0}^{n-1} \beta_k \Big) r(X_n, f_n(\beta_0, \ldots, \beta_{n-1}, X_n)) + \Big(\prod_{k=0}^{N-1} \beta_k \Big) g(X_N) \right], \quad x \in E.$$

Of course we assume that the Markov process (β_n) can be observed by the decision maker and thus the decision rules are allowed to depend on it. This problem can again be solved via a standard Markov Decision Model by extending the state space E. Let us define:

- $\tilde{E} := E \times [0, \infty) \times [0, \infty)$ where $(x, \beta, \delta) \in \tilde{E}$ denotes the state x, the new discount factor β and the product δ of the discount factors realized so far,
- $\tilde{A} = A$,
- $\tilde{D}(x, \beta, \delta) = D(x)$ for all $(x, \beta, \delta) \in E$,
- $\tilde{Q}(B \times B_1 \times B_2 | x, \beta, \delta, a) = Q(B|x, a) \otimes Q^\beta(B_1|\beta) \otimes 1_{B_2}(\delta \cdot \beta)$ for all $(x, \beta, \delta) \in \tilde{E}$ and suitable measurable sets B, B_1, B_2,
- $\tilde{r}\big((x, \beta, \delta), a\big) = \delta r(x, a)$ for all $(x, \beta, \delta) \in \tilde{E}$,
- $\tilde{g}(x, \beta, \delta) = \delta g(x)$.

Then the maximal reward operator for $v \in I\!M(E)$ is given by

$$\mathcal{T}v(x,\beta,\delta) = \sup_{a\in D(x)} \left\{ \delta r(x,a) + \int Q(dx'|x,a) \int Q^{\beta}(dy'|\beta) v(x',y',\delta\beta) \right\}$$

and the solution of the preceding problem with random discounting is given by the value

$$J_N(x,\beta_0,1) = \mathcal{T}^N \tilde{g}(x,\beta_0,1).$$

Note that the optimal decision rule at time n depends on x_n, β_n and the product of the discount factors $\delta_n = \prod_{k=0}^{n-1} \beta_k$. ♦

2.6 Applications and Examples

In this section we collect some applications of the theory presented so far. Further examples with a focus on financial optimization problems can be found in Chapter 4.

2.6.1 Red-and-Black Card Game

Consider the following simple card game: The dealer uncovers successively the cards of a well-shuffled deck which initially contains b_0 black and r_0 red cards. The player can at any time stop the uncovering of the cards. If the next card at the stack is black (red), the player wins (loses) 1 Euro. If the player does not stop the dealer, then the colour of the last card is decisive. When the player says stop right at the beginning then the probability of winning 1 Euro is obviously $\frac{b_0}{b_0+r_0}$ and her expected gain will be $\frac{b_0-r_0}{b_0+r_0}$. The same holds true when the player waits until the last card. Is there a strategy which yields a larger expected gain? The answer is no! We will prove this by formulating the problem as a stationary Markov Decision Model. The state of the system is the number of cards in the stack, thus

$$E := \{x = (b,r) \in \mathbb{N}_0^2 \mid b \le b_0, r \le r_0\}.$$

The state $(0,0)$ will be absorbing, thus in view of Example 2.3.13 we define $G := \{(0,0)\}$ as the set of absorbing states. Once, we have entered G the reward is zero and we stay in G. For $x \in E$ and $x \notin \{(0,1),(1,0)\}$ we have $D(x) = A = \{0,1\}$ with the interpretation that $a = 0$ means 'go ahead' and $a = 1$ means 'stop'. Since the player has to take the last card if she had not stopped before we have $D\big((0,1)\big) = D\big((1,0)\big) = \{1\}$. The transition probabilities are given by

$$\begin{aligned} q\big((b,r-1) \mid (b,r),0\big) &:= \frac{r}{r+b}, \quad r \ge 1, b \ge 0 \\ q\big((b-1,r) \mid (b,r),0\big) &:= \frac{b}{r+b}, \quad r \ge 0, b \ge 1 \\ q\big((0,0) \mid (b,r),1\big) &:= 1, \quad (b,r) \in E. \\ q\big((0,0) \mid (0,0),a\big) &:= 1, \quad a \in A. \end{aligned} \tag{2.1}$$

The one-stage reward is given by the expected reward (according to Remark 2.1.2),

$$r\big((b,r),1\big) := \frac{b-r}{b+r} \quad \text{for } (b,r) \in E \setminus G,$$

and the reward is zero otherwise. Finally we define

$$g(b,r) := \frac{b-r}{b+r} \quad \text{for } (b,r) \in E \setminus G$$

and $g((0,0)) = 0$. We summarize now the data of the stationary Markov Decision Model.

- $E := \{x = (b,r) \in \mathbb{N}_0^2 \mid b \le b_0, r \le r_0\}$ where $x = (b,r)$ denotes the number of black and red cards in the stack,
- $A := \{0,1\}$ where $a = 0$ means 'go ahead' and $a = 1$ means 'stop',
- $D(x) = A$ for $x \notin \{(0,1),(1,0)\}$ and $D\big((0,1)\big) = D\big((1,0)\big) = \{1\}$,
- the transition probabilities are given by equation (2.1),
- the one-stage reward is $r\big(x,1\big) = \frac{b-r}{b+r}$ for $x = (b,r) \in E \setminus G$ and 0 otherwise,
- $g(x) = \frac{b-r}{b+r}$ for $x = (b,r) \in E \setminus G$, $g((0,0)) = 0$,
- $N := r_0 + b_0$ and $\beta := 1$.

Since E and A are finite, (A_N) and also the Structure Assumption (SA_N) is clearly satisfied with

$$I\!M := \{v : E \to \mathbb{R} \mid v(x) = 0 \text{ for } x \in G\} \quad \text{and} \quad \Delta := F.$$

In particular we immediately know that an optimal policy exists. The maximal reward operator is given by

$$\mathcal{T}v(b,r) := \max\left\{\frac{b-r}{b+r}, \frac{r}{r+b}v(r-1,b) + \frac{b}{r+b}v(r,b-1)\right\} \quad \text{for } b+r \ge 2,$$

$$\mathcal{T}v(1,0) := 1,$$

$$\mathcal{T}v(0,1) := -1,$$

$$\mathcal{T}v(0,0) := 0.$$

It is not difficult to see that $g = \mathcal{T}g$. For $x = (b,r) \in E$ with $r + b \ge 2$ the computation is as follows:

$$\begin{aligned}\mathcal{T}g(b,r) &= \max\left\{\frac{b-r}{b+r},\ \frac{r}{r+b}g(r-1,b)+\frac{b}{r+b}g(r,b-1)\right\}\\ &= \max\left\{\frac{b-r}{b+r},\ \frac{r}{r+b}\cdot\frac{b-r+1}{r+b-1}+\frac{b}{r+b}\cdot\frac{b-r-1}{r+b-1}\right\}\\ &= \max\left\{\frac{b-r}{b+r},\ \frac{b-r}{b+r}\right\} = g(b,r).\end{aligned}$$

Since both expressions for $a = 0$ and $a = 1$ are identical, every $f \in F$ is a maximizer of g. Applying Theorem 2.5.4 we obtain that $J_n = \mathcal{T}^n g = g$ and we can formulate the solution of the card game.

Theorem 2.6.1. *The maximal value of the card game is given by*

$$J_{r_0+b_0}(b_0,r_0) = g(b_0,r_0) = \frac{b_0-r_0}{b_0+r_0},$$

and every strategy is optimal.

Thus, there is no strategy which yields a higher expected reward than the trivial ones discussed above. Note that the game is fair (i.e. $J_N(b_0,r_0) = 0$) if and only if $r_0 = b_0$.

2.6.2 A Cash Balance Problem

The cash balance problem involves the decision about the optimal cash level of a firm over a finite number of periods. The aim is to use the firm's liquid assets efficiently. There is a random stochastic change in the cash reserve each period (which can be both positive and negative). Since the firm does not earn interest from the cash position, there are holding cost or *opportunity cost* for the cash reserve if it is positive. But also in case the cash reserve is negative the firm incurs an out-of-pocket expense and has to pay interest. The cash reserve can be increased or decreased by the management at the beginning of each period which implies transfer costs. To keep the example simple we assume that the random changes in the cash flow are given by independent and identically distributed random variables (Z_n) with finite expectation. The transfer cost are linear. More precisely, let us define a function $c : \mathbb{R} \to \mathbb{R}_+$ by

$$c(z) := c_u z^+ + c_d z^-$$

where $c_u, c_d > 0$. The transfer cost are then given by $c(z)$ if the amount z is transferred. The cost $L(x)$ have to be paid at the beginning of a period for cash level x. We assume that

- $L : \mathbb{R} \to \mathbb{R}_+$, $L(0) = 0$,

- $x \mapsto L(x)$ is convex,
- $\lim_{|x|\to\infty} \frac{L(x)}{|x|} = \infty$.

This problem can be formulated as a Markov Decision Model with disturbances (Z_n) and with state space $E := \mathbb{R}$, where the state $x \in E$ is the current cash level. At the beginning of each period we have to decide upon the new cash level $a \in A := \mathbb{R}$. All actions are admissible, i.e. $D(x) := A$. The reward is then given as the negative cost $r(x,a) := -c(a-x) - L(a)$ of transfer and holding cost. The transition function is given by

$$T(x,a,z) := a - z$$

where z is a realization of the stochastic cash change Z_{n+1}. There is no terminal reward, i.e. $g \equiv 0$ and cost are discounted by a factor $\beta \in (0,1]$. The planning horizon N is given. We summarize the data of the stationary Markov Decision Model:

- $E := \mathbb{R}$ where $x \in E$ denotes the cash level,
- $A := \mathbb{R}$ where $a \in A$ denotes the new cash level after transfer,
- $D(x) := A$,
- $\mathcal{Z} := \mathbb{R}$ where $z \in \mathcal{Z}$ denotes the cash change,
- $T(x,a,z) := a - z$,
- $Q^Z(\cdot|x,a) :=$ distribution of Z_{n+1} (independent of (x,a)),
- $r(x,a) := -c(a-x) - L(a)$,
- $g \equiv 0$,
- $\beta \in (0,1]$.

Obviously the reward is bounded from above, i.e. $b \equiv 1$ is an upper bounding function. In what follows we will treat this problem as one of minimizing cost which seems to be more natural. The minimal cost operator is given by:

$$\mathcal{T}v(x) := \min \left\{ \begin{array}{ll} \inf_{a>x} \Big\{(a-x)c_u + L(a) + \beta \mathbb{E}\, v(a-Z)\Big\}, & (2.2) \\ L(x) + \beta \mathbb{E}\, v(x-Z), & (2.3) \\ \inf_{a<x} \Big\{(x-a)c_d + L(a) + \beta \mathbb{E}\, v(a-Z)\Big\} & (2.4) \end{array} \right.$$

where $Z := Z_1$. We will next check the Structure Assumption (SA$_N$). Thus, we first have to find a reasonable set $\mathbb{M}$. Looking at $\mathcal{T}v$ we choose the Ansatz:

$$\mathbb{M} := \{v : E \to \mathbb{R}_+ \mid v \text{ is convex and } v(x) \le c(-x) + d \text{ for some } d \in \mathbb{R}_+\}.$$

Moreover, we will see below that the set of minimizers is of a special form. Obviously $0 \in \mathbb{M}$. Now let $v \in \mathbb{M}$ and define the functions

$$h_u(a) := (a - x)c_u + L(a) + \beta \mathbb{E}\, v(a - Z),$$
$$h_d(a) := (x - a)c_d + L(a) + \beta \mathbb{E}\, v(a - Z).$$

By the definition of $\mathbb{M}$ both functions are finite on $\mathbb{R}$, since for $a \in A$ we obtain

$$\mathbb{E}\, v(a - Z) \le d + \mathbb{E}\, c(Z - a) \le d + \mathbb{E}\, |a - Z|(c_u + c_d) < \infty.$$

Also both functions are convex and $\lim_{|a|\to\infty} h_u(a) = \lim_{|a|\to\infty} h_d(a) = \infty$. Thus, both have a well-defined finite minimum point. Moreover, the convexity implies that the right- and left-hand side derivative at each point exist. A minimum point is characterized by a non-negative right-hand side derivative and a non-positive left-hand side derivative. Thus, we define

$$S_- := \inf\Big\{a \in \mathbb{R} \mid \frac{\partial^+}{\partial a} h_u(a) \ge 0\Big\},$$
$$S_+ := \sup\Big\{a \in \mathbb{R} \mid \frac{\partial^-}{\partial a} h_d(a) \le 0\Big\},$$

where $\frac{\partial^+}{\partial a} h$ and $\frac{\partial^-}{\partial a} h$ denote the right- and left-hand side derivative respectively. Since $h_u(a) - h_d(a) = (a - x)(c_u + c_d)$ is increasing in a, we have $S_- \le S_+$. It is important to note that S_- and S_+ do not depend on x. In order to determine a minimum point of $\mathcal{T} v$ we distinguish three cases:

(i) $x < S_-$: In this case the infimum of (2.4) is obtained if we plug in $a = x$ and thus the values of (2.4) and (2.3) are equal. However, the infimum of (2.2) is attained in $a = S_-$ and is less or equal to the value of (2.3) since $h_u(S_-) \le h_u(x) = L(x) + \beta \mathbb{E}\, v(x - Z)$.

(ii) $S_- \le x \le S_+$: Here the minimum values of the three expressions are equal and $a = x$ is the global minimum point.

(iii) $S_+ < x$: This case is analogous to the first one and the global minimum is attained in $a = S_+$.

Hence we have shown that a minimizer f^* exists and is of the form

$$f^*(x) = \begin{cases} S_- & \text{if } x < S_- \\ x & \text{if } S_- \le x \le S_+ \\ S_+ & \text{if } x > S_+. \end{cases} \tag{2.5}$$

This means that if the cash level is below S_-, sell enough securities to bring the cash level up to S_-. If the cash level is between the limits do nothing, and if the cash level is above S_+, buy securities to reduce the cash level to this critical level. Note that S_+ and S_- depend on v. As a consequence we define

$$\Delta := \Big\{ f \in F \;\Big|\; \text{there exist } S_-, S_+ \in \mathbb{R} \text{ with } S_- \le S_+ \text{ and } f \text{ is of the form (2.5)} \Big\}.$$

Inserting the minimizer gives

$$\mathcal{T}v(x) = \begin{cases} (S_- - x)c_u + L(S_-) + \beta \mathbb{E}\, v(S_- - Z) & \text{if } x < S_- \\ L(x) + \beta \mathbb{E}\, v(x - Z) & \text{if } S_- \le x \le S_+ \\ (x - S_+)c_d + L(S_+) + \beta \mathbb{E}\, v(S_+ - Z) & \text{if } x > S_+. \end{cases}$$

It is not difficult to verify that this function is again in $\mathbb{M}$. First there exists $d \in \mathbb{R}_+$ such that $\mathcal{T}v(x) \le d + c(-x)$. The convexity of $\mathcal{T}v$ on the intervals $(-\infty, S_-)$, (S_-, S_+), (S_+, ∞) is also obvious. It remains to investigate the points S_- and S_+. Here we have to show that the left-hand side derivative is less than or equal to the right-hand side derivative. Due to the definition of S_- and S_+ we obtain

$$c_u + \frac{\partial^+}{\partial x}\Big(L(x) + \beta \mathbb{E}\, v(x - Z) \Big)\Big|_{x = S_-} \ge 0$$

$$-c_d + \frac{\partial^-}{\partial x}\Big(L(x) + \beta \mathbb{E}\, v(x - Z) \Big)\Big|_{x = S_+} \le 0$$

since S_- and S_+ are the minimum points. This observation yields $\mathcal{T}v \in \mathbb{M}$. Thus, the Structure Assumption (SA$_N$) is satisfied for $\mathbb{M}$ and Δ. Theorem 2.5.4 can be applied to the cash balance problem and we obtain the following result.

Theorem 2.6.2. *a) There exist critical levels S_{n-} and S_{n+} such that for $n = 1, \dots, N$*

$$J_n(x) = \begin{cases} (S_{n-} - x)c_u + L(S_{n-}) + \beta \mathbb{E}\, J_{n-1}(S_{n-} - Z) & \text{if } x < S_{n-} \\ L(x) + \beta \mathbb{E}\, J_{n-1}(x - Z) & \text{if } S_{n-} \le x \le S_{n+} \\ (x - S_{n+})c_d + L(S_{n+}) + \beta \mathbb{E}\, J_{n-1}(S_{n+} - Z) & \text{if } x > S_{n+}. \end{cases}$$

with $J_0 \equiv 0$.
b) The optimal cash balance policy is given by $(f_N^*, \dots, f_1^*)$ where f_n^* is

$$f_n^*(x) := \begin{cases} S_{n-} & \text{if } x < S_{n-} \\ x & \text{if } S_{n-} \le x \le S_{n+} \\ S_{n+} & \text{if } x > S_{n+}. \end{cases} \tag{2.6}$$

Note that the critical levels which determine the transfer regions (sell, buy, do nothing) depend on n. Obviously the transfer cost imply that it is unlikely that many transfers occur. Hence the problem is sometimes also called *smoothing problem.*

2.6.3 Stochastic Linear-Quadratic Problems

A famous class of control problems with various different applications are linear-quadratic problems (LQ-problems). The name stems from the linear state transition function and the quadratic cost function. In what follows we suppose that $E := \mathbb{R}^m$ is the state space of the underlying system and $D_n(x) := A := \mathbb{R}^d$, i.e. all actions are admissible. The state transition functions are linear in state and action with random coefficient matrices $A_1, B_1, \dots, A_N, B_N$ with suitable dimensions, i.e. the system transition functions are given by

$$T_n(x, a, A_{n+1}, B_{n+1}) := A_{n+1}x + B_{n+1}a.$$

Thus, the disturbance in $[n, n+1)$ is given by $Z_{n+1} := (A_{n+1}, B_{n+1})$. The distribution of Z_{n+1} is influenced neither by the state nor by the action, and the random matrices $Z_1, Z_2, \dots$ are supposed to be independent but not necessarily identically distributed and have finite expectation and covariance. Moreover, we assume that $\mathbb{E}\left[B_{n+1}^\top Q B_{n+1}\right]$ is positive definite for all symmetric positive definite Q. Obviously we obtain a non-stationary problem. The one-stage reward is a negative cost function

$$r_n(x, a) := -x^\top Q_n x$$

and the terminal reward is

$$g_N(x, a) := -x^\top Q_N x$$

with deterministic, symmetric and positive definite matrices $Q_0, Q_1, \dots, Q_N$. There is no discounting. The aim is to minimize

$$\mathbb{E}_x^\pi \left[\sum_{k=0}^N X_k^\top Q_k X_k \right]$$

over all N-stage policies π. Thus, the aim is here to keep the state of the system close to zero. We summarize the data of the Markov Decision Model with disturbances (Z_n) as follows.

- $E := \mathbb{R}^m$ where $x \in E$ denotes the system state,
- $A := \mathbb{R}^d = D_n(x)$ where $a \in A$ denotes the action,
- $\mathcal{Z} := \mathbb{R}^{(m,m)} \times \mathbb{R}^{(m,d)}$ where $Z = (A, B)$ with values in $\mathcal{Z}$ denotes the random transition coefficients of the linear system,

- $T_n(x,a,A,B) := Ax + Ba$,
- $Q^Z(\cdot|x,a)$:= distribution of $Z_{n+1} := (A_{n+1}, B_{n+1})$ (independent of (x,a)),
- $r_n(x,a) := -x^\top Q_n x$,
- $g_N(x,a) := -x^\top Q_N x$,
- $\beta := 1$.

We have $r \leq 0$ and $b \equiv 1$ is an upper bounding function. Thus, (A_N) is satisfied. We will treat this problem as a cost minimization problem, i.e. we suppose that V_n is the minimal cost in the period $[n,N]$. For the calculation below we assume that all expectations exist. There are various applications of this regulation problem in engineering, but it will turn out that problems of this type are also important for example for quadratic hedging or mean-variance problems. The minimal cost operator is given by

$$\mathcal{T}_n v(x) = \inf_{a\in\mathbb{R}^d} \left\{x^\top Q_n x + \mathbb{E}\, v\big(A_{n+1}x + B_{n+1}a\big)\right\}.$$

We will next check the Structure Assumption (SA_N). It is reasonable to assume that $I\!M_n$ is given by

$$I\!M_n := \{v : \mathbb{R}^m \to \mathbb{R}_+ \mid v(x) = x^\top Qx \text{ with } Q \text{ symmetric, positive definite}\}.$$

It will also turn out that the sets $\Delta_n := \Delta \cap F_n$ can be chosen as the set of all linear functions, i.e.

$$\Delta := \{f : E \to A \mid f(x) = Cx \text{ for some } C \in \mathbb{R}^{(d,m)}\}.$$

Let us start with (SA_N)(i): Obviously $x^\top Q_N x \in I\!M_N$. Now let $v(x) = x^\top Qx \in I\!M_{n+1}$. We try to solve the following optimization problem

$$\begin{aligned}\mathcal{T}_n v(x) &= \inf_{a\in\mathbb{R}^d} \left\{x^\top Q_n x + \mathbb{E}\, v\big(A_{n+1}x + B_{n+1}a\big)\right\}\\ &= \inf_{a\in\mathbb{R}^d} \Big\{x^\top Q_n x + x^\top\, \mathbb{E}\left[A_{n+1}^\top Q A_{n+1}\right]x + 2x^\top\, \mathbb{E}\left[A_{n+1}^\top Q B_{n+1}\right]a\\ &\qquad + a^\top\, \mathbb{E}\left[B_{n+1}^\top Q B_{n+1}\right]a\Big\}.\end{aligned}$$

Since Q is positive definite, we have by assumption that $\mathbb{E}\left[B_{n+1}^\top Q B_{n+1}\right]$ is also positive definite and thus regular and the function in brackets is convex in a (for fixed $x \in E$). Differentiating with respect to a and setting the derivative equal to zero, we obtain that the unique minimum point is given by

$$f_n^*(x) = -\Big(\mathbb{E}\left[B_{n+1}^\top Q B_{n+1}\right]\Big)^{-1}\mathbb{E}\left[B_{n+1}^\top Q A_{n+1}\right]x.$$

Inserting the minimum point into the equation for $\mathcal{T}_n v$ yields

$$\mathcal{T}_n v(x) = x^\top \Big(Q_n + \mathbb{E}\big[A_{n+1}^\top Q A_{n+1}\big] - \mathbb{E}\big[A_{n+1}^\top Q B_{n+1}\big] \big(\mathbb{E}\big[B_{n+1}^\top Q B_{n+1}\big]\big)^{-1} \mathbb{E}\left[B_{n+1}^\top Q A_{n+1}\right]\Big) x = x^\top \tilde{Q} x$$

where $\tilde{Q}$ is defined as the expression in the brackets. Note that $\tilde{Q}$ is symmetric and since $x'\tilde{Q}x = \mathcal{T}_n v(x) \geq x^\top Q_n x$, it is also positive definite. Thus $\mathcal{T} v \in I\!M_n$ and the Structure Assumption (SA$_N$) is satisfied for $I\!M_n$ and $\Delta_n = \Delta \cap F_n$. Now we can apply Theorem 2.3.8 to solve the stochastic LQ-problem.

Theorem 2.6.3. *a) Let the matrices $\tilde{Q}_n$ be recursively defined by*

$$\begin{aligned}
\tilde{Q}_N &:= Q_N \\
\tilde{Q}_n &:= Q_n + \mathbb{E}\left[A_{n+1}^\top \tilde{Q}_{n+1} A_{n+1}\right] \\
&\quad - \mathbb{E}\big[A_{n+1}^\top \tilde{Q}_{n+1} B_{n+1}\big] \big(\mathbb{E}\big[B_{n+1}^\top \tilde{Q}_{n+1} B_{n+1}\big]\big)^{-1} \mathbb{E}\big[B_{n+1}^\top \tilde{Q}_{n+1} A_{n+1}\big].
\end{aligned}$$

Then $\tilde{Q}_n$ are symmetric, positive semidefinite and $V_n(x) = x^\top \tilde{Q}_n x$ for $x \in E$.
b) The optimal policy $(f_0^*, \dots, f_{N-1}^*)$ is given by

$$f_n^*(x) := -\Big(\mathbb{E}\left[B_{n+1}^\top \tilde{Q}_{n+1} B_{n+1}\right]\Big)^{-1} \mathbb{E}\left[B_{n+1}^\top \tilde{Q}_{n+1} A_{n+1}\right] x.$$

Note that the optimal decision rule is a linear function of the state and the coefficient matrix can be computed off-line. The minimal cost function is quadratic.
Our formulation of the stochastic LQ-problem can be generalized in different ways without leaving the LQ-framework. For example the transition function can be extended to

$$T_n(x, a, A_{n+1}, B_{n+1}, C_{n+1}) := A_{n+1}x + B_{n+1}a + C_{n+1}$$

where C_n are vectors with random entries. Thus, the stochastic disturbance variable is extended to $Z_n := (A_n, B_n, C_n)$ with the usual independence assumptions. Moreover, the cost function can be generalized to

$$\mathbb{E}_x^\pi \left[\sum_{k=0}^{N} (X_k - b_k)^\top Q_k (X_k - b_k) + \sum_{k=0}^{N-1} f_k(X_k)^\top \hat{Q}_k f_k(X_k) \right]$$

where $\hat{Q}_k$ are deterministic, symmetric positive semidefinite matrices and b_k are deterministic vectors. In this formulation the control itself is penalized and the distance of the state process to the benchmarks b_k has to be kept small. Note that in both generalizations the value functions remain of linear-quadratic form.

2.7 Exercises

Exercise 2.7.1 (Howard's Toymaker). Consider Howard's toymaker of Example 2.5.5. Show the stated Turnpike Theorem by conducting the following steps:

a) For a function $v : E \to \mathbb{R}$ let $\Delta v := v(1) - v(2)$ and show for all $a \in A$ that $Lv(1,a) - Lv(2,a) = 9 + 0.1\beta\Delta v$.
b) Show that $\Delta J_n = 9\sum_{k=0}^{n-1}(0.1\beta)^k + (0.1\beta)^n \Delta J_0$ for $n \in \mathbb{N}$.

Exercise 2.7.2. Suppose a stationary Markov Decision Model with planning horizon N is given which satisfies (A_N) and (SA_N). We define $J_n = \mathcal{T}^n g$ and $\hat{J}_n = \mathcal{T}^n \hat{g}$ for two terminal reward functions $g, \hat{g} \in I\!M$. Show:

a) For all $k = 1, \dots, N$ it holds:

$$J_N - \hat{J}_k \le \beta^N \sup_x \big(g(x) - \hat{g}(x)\big) + \sup_x \big(\hat{J}_k(x) - \hat{J}_{k-1}(x)\big) \sum_{j=1}^{N-k} \beta^j.$$

b) For all $k = 1, \dots, N$ it holds:

$$J_N - \hat{J}_k \ge \beta^N \inf_x \big(g(x) - \hat{g}(x)\big) + \inf_x \big(\hat{J}_k(x) - \hat{J}_{k-1}(x)\big) \sum_{j=1}^{N-k} \beta^j.$$

c) The bounds for J_N in a) and b) are decreasing in k.

Exercise 2.7.3 (Card Game). Consider the following variant of the red-and-black card game. Suppose we have a deck of 52 cards which is turned over and cards are uncovered one by one. The player has to say 'stop' when she thinks that the next card is the ace of spades. Which strategy maximizes the probability of a correct guess? This example is taken from Ross (1983).

Exercise 2.7.4 (Casino Game). Imagine you enter a casino and are allowed to play N times the same game. The probability of winning one game is $p \in (0,1)$ and the games are independent. You have an initial wealth $x > 0$ and are allowed to stake any amount in the interval $[0, x]$. When you win, you obtains twice your stake otherwise it is lost. The aim is to maximize the expected wealth $\mathbb{E}_x^\pi[X_N]$ after N games.

a) Set this up as a Markov Decision Model.
b) Find an upper bounding function and show that (SA_N) can be satisfied.
c) Determine an optimal strategy for the cases $p < \frac{1}{2}$, $p = \frac{1}{2}$ and $p > \frac{1}{2}$.
d) What changes if you want to maximize $\mathbb{E}_x^\pi[U(X_N)]$ where $U : \mathbb{R}_+ \to \mathbb{R}$ is a strictly increasing and strictly concave utility function?

Exercise 2.7.5 (LQ-problem). Consider the following special LQ-problem (see Bertsekas (2005) Section 4.1 for more details): The transition function is given by

$$T_n(x, a, z) := A_{n+1}x + B_{n+1}a + z$$

where $x \in \mathbb{R}^m, a \in \mathbb{R}^d$ and A_n, B_n are deterministic matrices of appropriate dimension. The disturbances $Z_1, Z_2, \ldots$ are independent and identically distributed with finite expectation and covariance matrix. The cost

$$\mathbb{E}_x^\pi \left[\sum_{k=0}^{N} X_k^\top Q_k X_k \right]$$

have to be minimized where the Q_n are positive definite.

a) Show that (A_N) and (SA_N) are satisfied.

b) Show that the minimal cost-to-go function is given by

$$V_0(x) = x^\top \tilde{Q}_0 x + \sum_{k=1}^{N} \mathbb{E}[Z_k^\top \tilde{Q}_k Z_k], \quad x \in \mathbb{R}^m$$

where

$$\begin{aligned}
\tilde{Q}_N &:= Q_N \\
\tilde{Q}_n &:= Q_n + A_{n+1}^\top \tilde{Q}_{n+1} A_{n+1} \\
&\quad - A_{n+1}^\top \tilde{Q}_{n+1} B_{n+1} \big(B_{n+1}^\top \tilde{Q}_{n+1} B_{n+1}\big)^{-1} B_{n+1}^\top \tilde{Q}_{n+1} A_{n+1}
\end{aligned}$$

and the optimal policy $(f_0^*, \ldots, f_{N-1}^*)$ is given by

$$f_n^*(x) = -\Big(B_{n+1}^\top \tilde{Q}_{n+1} B_{n+1}\Big)^{-1} B_{n+1}^\top \tilde{Q}_{n+1} A_{n+1} x.$$

c) Let now $A_k = A$, $B_k = B$ and $Q_k = Q$ for all k and consider the so-called *discrete Riccati equation*

$$\begin{aligned}
\tilde{Q}_N &:= Q \\
\tilde{Q}_n &:= Q + A^\top \tilde{Q}_{n+1} A - A^\top \tilde{Q}_{n+1} B \big(B^\top \tilde{Q}_{n+1} B\big)^{-1} B^\top \tilde{Q}_{n+1} A.
\end{aligned}$$

Moreover, assume that the matrix

$$[B, AB, A^2B, \ldots, A^{N-1}B]$$

has full rank. Show that there exists a positive definite matrix $\tilde{Q}$ such that $\lim_{n\to\infty} \tilde{Q}_n = \tilde{Q}$. Moreover, $\tilde{Q}$ is the unique solution of

$$\tilde{Q} = Q + A^\top \tilde{Q} A - A^\top \tilde{Q} B \big(B^\top \tilde{Q} B\big)^{-1} B^\top \tilde{Q} A$$

within the class of positive semidefinite matrices.

Remark: The convergence of $\tilde{Q}_k$ in the case of stochastic coefficients is more delicate.

Exercise 2.7.6 (Binary Markov Decision Model). Suppose a stationary Markov Decision Model is given with $A = \{0,1\}$. Such a model is called a *binary* Markov Decision Model. We suppose that the reward functions r and g are bounded. Define $r(x,1) = r_1(x)$ and $r(x,0) = r_0(x)$ for $x \in E$. For $v : E \to \mathbb{R}$ measurable and bounded and $a \in A$ we denote

$$(Q_a v)(x) := \int v(y)Q(dy|x,a), \quad x \in E.$$

a) Show that (A_N) and (SA_N) are satisfied.
b) Show that the value function satisfies

$$J_n = \max\{r_0 + \beta Q_0 J_{n-1}, r_1 + \beta Q_1 J_{n-1}\} =: \max\{L_0 J_{n-1}, L_1 J_{n-1}\}.$$

c) If we denote $d_n(x) = L_1 J_{n-1}(x) - L_0 J_{n-1}(x), n \in \mathbb{N}, x \in E$ show that

$$d_{n+1} = L_1 L_0 J_{n-1} - L_0 L_1 J_{n-1} + \beta Q_1 d_n^+ - \beta Q_0 d_n^-.$$

Exercise 2.7.7 (Replacement Problem). A machine is in use over several periods. The state of the machine is randomly deteriorating and the reward which is obtained depends on the state of the machine. When should the machine be replaced by a new one? The new machine costs a fix amount $K \geq 0$.
We assume that the evolution of the state of the machine is a Markov process with state space $E = \mathbb{R}_+$ and transition kernel Q where $Q([x,\infty)|x) = 1$. A large state x refers to a worse condition/quality of the machine. The reward is $r(x)$ if the state of the machine is x. We assume that the measurable function $r : E \to \mathbb{R}$ is bounded and for the terminal reward $g = r$. In what follows we use the abbreviation

$$(Qv)(x) := \int v(x')Q(dx'|x), \quad (Q_0 v)(x) := \int v(x')Q(dx'|0).$$

Note that $(Q_0 v)(x)$ does not depend on x!

a) Show that (A_N) and (SA_N) are satisfied.
b) Show that the maximal reward operator is given by

$$\mathcal{T}v(x) = r(x) + \max\{\beta(Qv)(x), -K + \beta(Q_0 v)(x)\}, \quad x \geq 0.$$

c) Let $d_n(x) := -K + \beta(Q_0 J_{n-1})(x) - \beta(QJ_{n-1})(x), n \in \mathbb{N}, x \in E$. Show that

$$d_{n+1} = -(1-\beta)K - \beta Qr - \beta Q d_n^- + c_n$$

where $c_n := \beta Q_0 J_n - \beta^2 Q_0 J_{n-1}$ is independent of $x \in E$.
d) If r is decreasing and the transition kernel Q is stochastically monotone prove that a maximizer f_n^* of J_{n-1} is of the form

$$f_n^*(x) = \begin{cases} \text{replace} & \text{if } x \geq x_n^* \\ \text{not replace} & \text{if } x < x_n^* \end{cases}$$

for $x_n^* \in \mathbb{R}_+$. The value x_n^* is called the threshold or control limit.

Exercise 2.7.8 (Terminating Markov Decision Model). Suppose a stationary Markov Decision Model with $\beta = 1$ is given with the following properties:

- There exists a set $G \subset E$ such that $r(x,a) = 0$ and $Q(\{x\}|x,a) = 1$ for all $x \in G$ and $a \in D(x)$.
- For all $x \in E$ there exists an $N(x) \leq N$ such that $\mathbb{P}_x^\pi(X_{N(x)} \in G) = 1$ for all policies π.

Define $J(x) := J_{N(x)}(x)$ for all $x \in E$. Such a Markov Decision Model is called *terminating*.

a) Show that $J(x) = g(x)$ for $x \in G$ and $J(x) = \mathcal{T}J(x)$ for $x \notin G$.
b) If $f \in F$ satisfies $\mathcal{T}_f J(x) = \mathcal{T}J(x)$ for $x \notin G$ and $f(x) \in D(x)$ arbitrary for $x \in G$, then the stationary policy $(f, f, \ldots, f) \in F^N$ is optimal.
c) Show that the Red-and-Black card game of Section 2.6.1 is a terminating Markov Decision Model.

Exercise 2.7.9. You are leaving your office late in the evening when it suddenly starts raining and you realize that you have lost your umbrella somewhere during the day. The umbrella can only be at a finite number of places, labelled $1, \ldots, m$. The probability that it is at place i is p_i with $\sum_{i=1}^m p_i = 1$. The distance between two places is given by d_{ij} where $i, j \in \{0, 1, \ldots, m\}$ and label 0 is your office. In which sequence do you visit the places in order to minimize the expected length of the journey until you find your umbrella? Set this up as a Markov Decision Problem or as a terminating Markov Decision Model and write a computer program to solve it.

2.8 Remarks and References

In this chapter we consider Markov Decision Models with Borel state and action spaces and unbounded reward functions. In order to increase the readability and to reduce the mathematical framework (e.g. measurability and existence problems) we introduce the Structure Assumption (SA$_N$) and the notion of an (upper) bounding function in Section 2.4. This framework is very useful for applications in finance (see Chapter 4) where the state spaces are often uncountable subsets of Euclidean spaces and the utility functions are unbounded. A similar approach is also used in Schäl (1990) and Puterman (1994).

Section 2.4: Semi-continuous Markov Decision Models have been investigated e.g. in Bertsekas and Shreve (1978), Dynkin and Yushkevich (1979) and Hernández-Lerma and Lasserre (1996). Properties of the value functions like *increasing*, *concave*, *convex* and combinations thereof were first rigorously studied in Hinderer (1985). For a recent paper see Smith and McCardle (2002). Moreover, Hinderer (2005) and Müller (1997) discuss Lipschitz-continuity of the value functions. The fact that supermodular functions are important for obtaining monotone maximizers was first discussed by Topkis (1978). Altman and Stidham (1995) investigate so-called *binary* Markov Decision Models with two actions (e.g. in replacement problems) and derive general conditions for the existence of threshold policies. The comparison results for Markov Decision Problems can been found in Müller and Stoyan (2002), see also Bäuerle and Rieder (1997).

Section 2.6: The 'Red-and-Black' card game was presented by Connelly (1974) under the name 'Say red'. It can also be solved by martingale arguments. Other interesting game problems and examples can be found in Ross (1970, 1983). Heyman and Sobel (2004a,b) consider various stochastic optimization problems in Operations Research, in particular cash balance models, inventory and queueing problems. For a recent extension of the cash balance problem, where the cash changes depend on an underlying Markov process, see Hinderer and Waldmann (2001). Stochastic LQ-problems have been investigated by many authors. For a comprehensive treatment see e.g. Bertsekas (2001).

Chapter 3
The Financial Markets

In this chapter we introduce the financial markets which will appear in our applications. In Section 3.1 a financial market in discrete time is presented. A prominent example is the binomial model. However, we do not restrict to finite probability spaces in general. We will define portfolio strategies and characterize the absence of arbitrage in this market. In later chapters we will often restrict to Markov asset price processes in order to be able to use the Markov Decision Process framework. In Section 3.2 a special financial market in continuous time is considered which is driven by jumps only. More precisely the asset dynamics follow so-called *Piecewise Deterministic Markov Processes.* Though portfolio strategies are defined in continuous time here we will see in Section 9.3 that portfolio optimization problems in this market can be solved with the help of Markov Decision Processes. In Section 3.3 we will briefly investigate the relation of the discrete-time financial market to the standard Black-Scholes-Merton model as a widely used benchmark model in mathematical finance. Indeed if the parameters in the discrete-time financial market are chosen appropriately, this market can be seen as an approximation of the Black-Scholes-Merton model or of even more general models. This observation serves as one justification for the importance of discrete-time models. Other justifications are that trading in continuous time is not possible or expensive in reality (because of transaction cost) and that continuous-time trading strategies are often pretty risky. In Section 3.4 utility functions and the concept of expected utility are introduced and discussed briefly. The last section contains some notes and references.

3.1 Asset Dynamics and Portfolio Strategies

We assume that asset prices are monitored in discrete time. The German stock index DAX for example is computed every second. But larger time periods of no trade may be applicable. Thus, we suppose that time is divided

N. Bäuerle and U. Rieder, *Markov Decision Processes with Applications to Finance*, Universitext, DOI 10.1007/978-3-642-18324-9_3,

into periods of length Δt and $t_n = n\Delta t$. The most common form of an asset price is a multiplicative model, i.e. if S_n is the price at time $t_n > 0$ then

$$S_{n+1} = S_n \tilde{R}_{n+1}.$$

The positive random variable $\tilde{R}_{n+1}$ defines the *relative price change* S_{n+1}/S_n between time t_n and t_{n+1}. For a riskless bond the relative price change S^0_{n+1}/S^0_n is chosen to be $1 + i_{n+1}$ with deterministic interest rate $i_{n+1} \in \mathbb{R}_+$.

There are two important special cases of the multiplicative model. The first one is the *binomial model* or *Cox-Ross-Rubinstein model*. Here it is assumed that there exists one bond with $i_{n+1} \equiv i$ and one stock with relative price changes which are independent and identically distributed and can take two values: either $\boldsymbol{u} > 0$ for up or $\boldsymbol{d} > 0$ for down where we assume that $\boldsymbol{d} < \boldsymbol{u}$. The probabilities for up and down are p and $1 - p$ respectively, i.e.

$$\mathbb{P}(\tilde{R}_n = \boldsymbol{u}) = p, \quad \mathbb{P}(\tilde{R}_n = \boldsymbol{d}) = 1 - p.$$

The stock price evolution is shown in Figure 3.1. If we want to model d assets simultaneously a *multinomial model* is adequate, i.e $\tilde{R}_n$ can take values in $\{\boldsymbol{u}, \boldsymbol{d}\}^d = \{z_1, \ldots, z_{2^d}\}$ and

$$\mathbb{P}(\tilde{R}_n = z_k) = p_k, \quad k = 1, \ldots, 2^d$$

with $\sum_{k=1}^{2^d} p_k = 1, \; p_k \geq 0$.

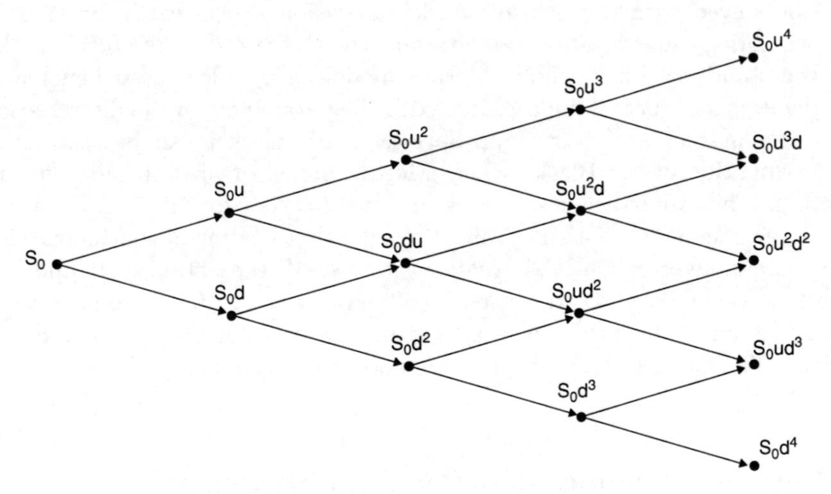

Fig. 3.1 Evolution of stock price in the binomial lattice model.

The second important special case is a discretization of the Black-Scholes-Merton model. In the standard Black-Scholes-Merton market model it is assumed that the stock price evolves according to

$$dS_t = S_t(\mu dt + \sigma dW_t)$$

where (W_t) is a Wiener process, $\mu \in \mathbb{R}$ and $\sigma > 0$ are given parameters. Thus, if we take a time step Δt we obtain

$$S_{n+1} = S_n \exp\Big((\mu - \frac{1}{2}\sigma^2)\Delta t + \sigma W_{\Delta t}\Big).$$

In this case $\tilde{R}_n$ has a *lognormal distribution* or $\log \tilde{R}_n$ is normally distributed. The multivariate version is thus $\log \tilde{R}_n \sim \mathcal{N}(a, \Sigma)$, i.e. $\log \tilde{R}_n$ (the *log* is taken componentwise) has a multivariate normal distribution. Note that in both special cases we have $\tilde{R}_n > 0$ and thus it is guaranteed that the stock price will stay positive with probability one.

In what follows we will consider an *N-period financial market* with d risky assets and one riskless bond. We assume that all random variables are defined on a probability space $(\Omega, \mathcal{F}, \mathbb{P})$ with filtration $(\mathcal{F}_n)$ and $\mathcal{F}_0 := \{\emptyset, \Omega\}$. The financial market is given by:

- A riskless bond with $S_0^0 \equiv 1$ and

  $$S_{n+1}^0 := S_n^0(1 + i_{n+1}), \quad n = 0, 1, \dots, N-1$$

  where i_{n+1} denotes the deterministic interest rate for the time period $[n, n+1)$. If the interest rate is constant, i.e. $i_n \equiv i$, then $S_n^0 = (1+i)^n$.
- There are d risky assets and the price process of asset k is given by $S_0^k = s_0^k$ and

  $$S_{n+1}^k = S_n^k \tilde{R}_{n+1}^k, \quad n = 0, 1, \dots, N-1.$$

  The processes (S_n^k) are assumed to be adapted with respect to the filtration $(\mathcal{F}_n)$ for all k. Moreover, we suppose that $\tilde{R}_{n+1}^k > 0$ $\mathbb{P}$-a.s. for all k and n and that s_0^k is deterministic. $\tilde{R}_{n+1}^k$ is the relative price change in the time interval $[n, n+1)$ for the risky asset k.

In what follows we denote $S_n := (S_n^1, \dots, S_n^d)$, $\tilde{R}_n := (\tilde{R}_n^1, \dots, \tilde{R}_n^d)$ and $\mathcal{F}_n^S := \sigma(S_0, \dots, S_n)$. Since (S_n) is $(\mathcal{F}_n)$-adapted it holds: $\mathcal{F}_n^S \subset \mathcal{F}_n$ for $n = 0, 1, \dots, N$. In most cases we will later assume that the random vectors $\tilde{R}_1, \dots, \tilde{R}_N$ are independent, however we do not impose this restrictive assumption now, because we will also consider some models where the independence assumption is not satisfied. Suppose now we are able to invest into this financial market.

Definition 3.1.1. A *portfolio* or a *trading strategy* is an $(\mathcal{F}_n)$-adapted stochastic process $\phi = (\phi_n^0, \phi_n)$ where $\phi_n^0 \in \mathbb{R}$ and $\phi_n = (\phi_n^1, \dots, \phi_n^d) \in \mathbb{R}^d$ for $n = 0, 1, \dots, N-1$. The quantity ϕ_n^k denotes the amount of money which is invested into asset k during the time interval $[n, n+1)$.

Remark 3.1.2. Note that in general we make no restriction on the sign of ϕ_n^k. In particular ϕ_n^k is allowed to be negative which in case $k = 0$ implies that a loan is taken and that the interest rate is the same for borrowing and lending. In case $\phi_n^k < 0$ for $k \neq 0$ this corresponds to short selling of asset k. ◊

The vector (ϕ_0^0, ϕ_0) is called the *initial portfolio* of the investor. The value of the initial portfolio is given by

$$X_0 := \sum_{k=0}^{d} \phi_0^k = \phi_0^0 + \phi_0 \cdot e$$

where $x \cdot y = \sum_{k=1}^{d} x_k y_k$ denotes the inner product of the vectors $x, y \in \mathbb{R}^d$ and $e := (1, \ldots, 1) \in \mathbb{R}^d$.

Let ϕ be a portfolio strategy and denote by X_{n-} the value of the portfolio at time n before trading. Then

$$X_n := X_{n-} := \sum_{k=0}^{d} \phi_{n-1}^k \tilde{R}_n^k = \phi_{n-1}^0 (1 + i_n) + \phi_{n-1} \cdot \tilde{R}_n.$$

The value of the portfolio at time n after trading is given by

$$X_{n+} := \sum_{k=0}^{d} \phi_n^k = \phi_n^0 + \phi_n \cdot e.$$

In what follows we sometimes write X_n^ϕ when we want to make the dependence on the portfolio strategy ϕ explicit.

Definition 3.1.3. A portfolio strategy ϕ is called *self-financing* if

$$X_{n-}^\phi = X_{n+}^\phi \quad \mathbb{P}\text{-a.s.}$$

for all $n = 1, \ldots, N-1$, i.e. the current wealth is just reassigned to the assets.

In the sections to come we will restrict to self-financing portfolio strategies and we will tacitly assume that whenever we consider a portfolio strategy that it is self-financing. The self-financing condition implies that the following equivalence holds for all $n = 1, \ldots, N-1$:

$$X_n = X_{n+}, \quad \Longleftrightarrow \quad \phi_n^0 + \phi_n \cdot e = \phi_{n-1}^0 (1 + i_n) + \phi_{n-1} \cdot \tilde{R}_n.$$

This equation can be used to derive an important recursive formula for the wealth evolution which will be used throughout the following chapter. This recursive equation is derived next (the self-financing condition is applied in the last equality):

$$
\begin{aligned}
X_{n+1} &= X_0 + \sum_{t=1}^{n+1} (X_t - X_{t-1}) \\
&= X_0 + \sum_{t=1}^{n+1} \Big(\phi_t^0 - \phi_{t-1}^0 + \phi_t \cdot e - \phi_{t-1} \cdot e \Big) \\
&= X_0 + \sum_{t=1}^{n+1} \Big(\phi_{t-1}^0 i_t + \phi_{t-1} \cdot (\tilde{R}_t - e) \Big).
\end{aligned}
$$

Since $\phi_n^0 = X_n - \phi_n \cdot e$, we can eliminate ϕ_n^0 to obtain

$$
\begin{aligned}
X_{n+1} &= X_n + (\phi_n^0 i_{n+1} + \phi_n \cdot (\tilde{R}_{n+1} - e)) \\
&= X_n (1 + i_{n+1}) + \sum_{k=1}^{d} \phi_n^k \Big(\tilde{R}_{n+1}^k - 1 - i_{n+1} \Big).
\end{aligned}
$$

When we introduce the so-called *relative risk process* (R_n) defined by $R_n := (R_n^1, \dots, R_n^d)$ and

$$
R_n^k := \frac{\tilde{R}_n^k}{1 + i_n} - 1, \quad k = 1, \dots, d,
$$

we obtain the important recursive formula

$$
X_{n+1} = (1 + i_{n+1}) \Big(X_n + \phi_n \cdot R_{n+1} \Big). \tag{3.1}
$$

The advantage of equation (3.1) is that only the investment into the stocks enters the equation and (ϕ_n) can be completed to a self-financing portfolio strategy. This formula will be used extensively throughout the next chapters. As usual we have to exclude arbitrage opportunities in the financial market. An arbitrage opportunity is formally defined as follows.

Definition 3.1.4. An *arbitrage opportunity* is a self-financing portfolio strategy $\phi = (\phi_n^0, \phi_n)$ with the following property: $X_0^\phi = 0$ and

$$
\mathbb{P}(X_N^\phi \geq 0) = 1 \quad \text{and} \quad \mathbb{P}(X_N^\phi > 0) > 0.
$$

Loosely speaking an arbitrage opportunity is a riskless investment strategy with the chance of a gain. In real markets such opportunities sometimes exist but once detected by traders they will soon disappear. The following theorem characterizes the absence of arbitrage opportunities. It shows that the market is free from arbitrage opportunities if and only if there is locally no arbitrage opportunity. This property is important when we reduce multiperiod optimization problems to one-period problems.

Theorem 3.1.5. *Consider an N-period financial market. The following two statements are equivalent:*

a) There are no arbitrage opportunities.
b) For $n = 0, 1, \dots, N-1$ and for all $\mathcal{F}_n$-measurable $\phi_n \in \mathbb{R}^d$ it holds:

$$\phi_n \cdot R_{n+1} \geq 0 \ \mathbb{P}\text{-a.s.} \quad \Rightarrow \quad \phi_n \cdot R_{n+1} = 0 \ \mathbb{P}\text{-a.s.}$$

Proof. Suppose first that there exists an $n \in \{1, \dots, N\}$ and an $\mathcal{F}_{n-1}$ measurable $\varphi \in \mathbb{R}^d$ such that

$$\varphi \cdot R_n \geq 0 \ \mathbb{P}\text{-a.s.} \quad \text{and} \quad \mathbb{P}(\varphi \cdot R_n > 0) > 0.$$

Then an arbitrage strategy ϕ can be constructed as follows:
Let $X_0^\phi = 0$ and define the portfolio strategy

$$\phi_k := \begin{cases} \varphi & \text{if } k = n-1, \\ 0 & \text{otherwise.} \end{cases}$$

Then $X_{n-1}^\phi = \phi_{n-1}^0 + \phi_{n-1} \cdot e = 0$ and $X_n^\phi = (1 + i_n)(\varphi \cdot R_n)$. Hence $\mathbb{P}(X_n^\phi \geq 0) = 1$ and $\mathbb{P}(X_n^\phi > 0) > 0$. Since $X_N^\phi = (1 + i_{n+1}) \cdots (1 + i_N) X_n^\phi$, the strategy ϕ is an arbitrage strategy.
In order to show the reverse implication suppose that there exists an arbitrage strategy $\phi = (\phi_n^0, \phi_n)$. Let

$$m := \min \{ n \mid \mathbb{P}(X_n^\phi \geq 0) = 1 \text{ and } \mathbb{P}(X_n^\phi > 0) > 0 \}.$$

Then $m \leq N$ by assumption. Define $\varphi := 1_{[X_{m-1}^\phi \leq 0]} \phi_{m-1}$. Then φ is $\mathcal{F}_{m-1}$-measurable and we can consider the portfolio strategy ψ which invests the same amounts as ϕ from time 0 to $m-2$ and at time $m-1$ the amount φ in the risky assets. It holds

$$X_m^\psi = (1 + i_m)(X_{m-1}^\phi + \varphi \cdot R_m).$$

Since $\mathbb{P}(X_m^\psi \geq 0) = 1$ it follows that $\varphi \cdot R_m \geq 0$ $\mathbb{P}$-a.s. By definition of m we have either $\mathbb{P}(X_{m-1}^\phi = 0) = 1$ or $\mathbb{P}(X_{m-1}^\phi < 0) > 0$. In the first case we obtain

$$\mathbb{P}(\varphi \cdot R_m > 0) = \mathbb{P}(X_m^\phi > 0) > 0$$

and in the second case we have $\mathbb{P}(\varphi \cdot R_m > 0) \geq \mathbb{P}(X_{m-1}^\phi < 0) > 0$. Altogether we have shown $\mathbb{P}(\varphi \cdot R_m \geq 0) = 1$ and $\mathbb{P}(\varphi \cdot R_m > 0) > 0$ which implies the statement. □

Remark 3.1.6. When the random variables $R_1, \dots, R_N$ are independent and $\mathcal{F}_n := \mathcal{F}_n^S := \sigma(S_0, \dots, S_n) = \sigma(R_1, \dots, R_n)$ we obtain the following even simpler characterization of the absence of arbitrage: For all $n = 0, 1, \dots, N-1$ and for all $\phi \in \mathbb{R}^d$ it holds:

$$\phi \cdot R_{n+1} \geq 0 \ \mathbb{P}\text{-a.s.} \quad \Rightarrow \quad \phi \cdot R_{n+1} = 0 \ \mathbb{P}\text{-a.s.} \tag{3.2}$$

Note that ϕ is deterministic here. It is not difficult to see that this statement is equivalent to Theorem 3.1.5 part b). The proof is as follows: Theorem 3.1.5 part b) implies (3.2) since every $\phi \in \mathbb{R}^d$ is $\mathcal{F}_n$-measurable. Now suppose (3.2) holds and there exists an $\mathcal{F}_n$-measurable $\phi_n \in \mathbb{R}^d$ such that

$$\phi_n \cdot R_{n+1} \geq 0 \ \mathbb{P}\text{-a.s.}$$

and $\mathbb{P}(\phi_n \cdot R_{n+1} > 0) > 0$. Define $\tilde{\phi}_n := \mathbb{E}[\phi_n] \in \mathbb{R}^d$. Then we obtain $\mathbb{E}[\phi_n \cdot R_{n+1} | R_{n+1}] = \tilde{\phi}_n \cdot R_{n+1} \geq 0$ $\mathbb{P}$-a.s. and $\mathbb{P}(\tilde{\phi}_n \cdot R_{n+1} > 0) > 0$ which is a contradiction. ◊

Example 3.1.7 (Binomial or Cox-Ross-Rubinstein Model). In the binomial financial market a necessary and sufficient condition for the absence of arbitrage opportunities is that the model parameters satisfy

$$\boldsymbol{d} < 1 + i < \boldsymbol{u}.$$

This can be seen as follows. First note that ϕR_1 can take the two values

$$\phi\Big(\frac{\boldsymbol{d}}{1+i} - 1\Big) \quad \text{and} \quad \phi\Big(\frac{\boldsymbol{u}}{1+i} - 1\Big).$$

Thus, if $\boldsymbol{d} < 1+i < \boldsymbol{u}$ then $\phi R_1 \geq 0$ implies $\phi \equiv 0$ and hence $\phi R_1 = 0$ holds. If $\boldsymbol{u} > \boldsymbol{d} \geq 1 + i$ we obtain for all $\phi > 0$ that $\phi R_1 \geq 0$ and $\mathbb{P}(\phi R_1 > 0) > 0$. Similarly if $1 + i \geq \boldsymbol{u} > \boldsymbol{d}$ we have for all $\phi < 0$ that $\phi R_1 \geq 0$ and $\mathbb{P}(\phi R_1 > 0) > 0$. Altogether the statement is shown. ♦

Remark 3.1.8. The absence of arbitrage opportunities is also equivalent to the existence of at least one equivalent martingale measure. This characterization is important when prices of contingent claims have to be computed (see Appendix C). ◊

3.2 Jump Markets in Continuous Time

In this section we consider a special class of continuous-time financial markets where asset dynamics follow a so-called *Piecewise Deterministic Markov Process*. The reason is twofold: First it can be shown that Lévy processes can be approximated arbitrarily close by such type of processes and second, optimization problems arising from these models can essentially be solved via discrete-time Markov Decision Models (see Chapter 8).
As before we suppose that we have d risky assets and one riskless bond. The evolution of the price processes is now given continuously over the time interval $[0,T]$. The definition of the price dynamics is as follows. Suppose now we have a probability space $(\Omega, \mathcal{F}, \mathbb{P})$ and on this space there is a homogeneous Poisson process (N_t) with rate $\lambda > 0$ and a sequence of independent and identically distributed random vectors (Y_n) with values in $(-1,\infty)^d$. The random vectors Y_n are assumed to have a distribution Q_Y and are independent of (N_t). Thus, we can define the $\mathbb{R}^d$-valued compound Poisson process

$$C_t := \sum_{n=1}^{N_t} Y_n$$

where the empty sum is supposed to be zero. By (C_t^k) we denote the k-th component i.e. $C_t^k := \sum_{n=1}^{N_t} Y_n^k$. The financial market is then given by:

- A riskless bond with deterministic price process (S_t^0) where

  $$S_t^0 := e^{\rho t}, \quad t \in [0,T]$$

  and $\rho \geq 0$ denotes the continuous interest rate.
- There are d risky assets and the price process of asset k is given by the stochastic differential equation:

  $$dS_t^k = S_{t-}^k \big(\mu_k dt + dC_t^k\big), \quad t \in [0,T]$$

  where $\mu_k \in \mathbb{R}$ is given and $S_0^k = s_0^k > 0$ is deterministic. Thus, we obtain

  $$S_t^k = s_0^k + \mu_k \int_0^t S_s^k ds + \int_0^t S_{s-}^k dC_s^k, \quad t \in [0,T].$$

We assume that the filtration $(\mathcal{F}_t)$ is generated by (C_t) or equivalently generated by the asset price processes (S_t^k) which implies that they are automatically adapted to the filtration.
In this financial market, the price processes are *Piecewise Deterministic Markov Processes*, i.e. they show a deterministic evolution between jumps and the jumps occur at random time points according to a Poisson process and have a random height. If we denote by $0 := T_0 < T_1 < T_2 < \dots$ the jump time points of the Poisson process then for $t \in [T_n, T_{n+1})$

$$S_t^k = S_{T_n}^k \exp\left(\mu_k(t - T_n)\right).$$

At the time of a jump we have

$$\Delta S_{T_n}^k = S_{T_n}^k - S_{T_n-}^k = S_{T_n-}^k Y_n^k.$$

Thus, Y_n^k gives the relative jump height of asset k at the n-th jump. Since $Y_n^k > -1$, our asset prices stay positive. At first glance it seems to be quite special that the asset prices can only have joint jumps at the time points $T_0 < T_1 < T_2 < \ldots$. However, note that the distribution of Y might well have probability mass on points $Y^k e_k$, where e_k is the k-th unit vector. In what follows we denote $S_t := (S_t^1, \ldots, S_t^d)$.
Of course we want to invest into this financial market. A decision about rearranging the portfolio has to be done before we know whether or not a jump in the asset prices will occur. In mathematical terms this means that the portfolio strategy has to be predictable (for a definition see Appendix B.2).

Definition 3.2.1. A *portfolio* or a *trading strategy* is an $(\mathcal{F}_t)$-predictable stochastic process $\phi = (\phi_t^0, \phi_t)$ where $\phi_t^0 \in \mathbb{R}$ and $\phi_t = (\phi_t^1, \ldots, \phi_t^d) \in \mathbb{R}^d$. The quantity ϕ_t^k denotes the amount of money which is invested in asset k at time t.

The wealth process (X_t^ϕ) determined by the trading strategy ϕ is then given by

$$X_t^\phi := \sum_{k=0}^{d} \phi_t^k = \phi_t^0 + \phi_t \cdot e$$

where

$$X_0 = x_0 := \phi_0^0 + \phi_0 \cdot e$$

is the initial wealth. The trading strategy ϕ is called *self-financing* when a change in the wealth at time t is only due to a change in an asset price and not due to additional money or consumption. Thus, the wealth at time t consists only of the initial wealth and the gains or losses which are accumulated over the interval $[0, t]$ by trading into the assets.

Definition 3.2.2. A portfolio strategy ϕ is called *self-financing* if

$$X_t^\phi = x_0 + \int_0^t \phi_s^0 \frac{dS_s^0}{S_s^0} + \sum_{k=1}^{d} \int_0^t \phi_s^k \frac{dS_s^k}{S_s^k}, \quad t \in [0, T].$$

In differential form the self-financing condition reads:

$$dX_t^\phi = \phi_t^0 \frac{dS_t^0}{S_t^0} + \sum_{k=1}^{d} \phi_t^k \frac{dS_t^k}{S_t^k} =: \phi_t^0 \frac{dS_t^0}{S_t^0} + \phi_t \frac{dS_t}{S_t}. \tag{3.3}$$

Inserting the wealth dynamics yields

$$dX_t = \Big(\phi_t^0 \rho + \phi_t \cdot \mu\Big) dt + \phi_t dC_t.$$

Using the fact that $X_t^\phi = \phi_t^0 + \phi_t \cdot e$ we can eliminate ϕ_t^0 in the differential equation and obtain

$$dX_t = \big(X_t \rho + \phi_t \cdot (\mu - \rho e)\big) dt + \phi_t dC_t. \tag{3.4}$$

Thus, between jumps, i.e. if $t \in [T_n, T_{n+1})$, the evolution of the wealth is given by

$$X_t = e^{\rho(t-T_n)} \Big(X_{T_n} + \int_{T_n}^t e^{-\rho(s-T_n)} \phi_s \cdot (\mu - \rho e) ds\Big). \tag{3.5}$$

At jumps we have

$$\Delta X_{T_n} = \phi_{T_n} \cdot Y_n. \tag{3.6}$$

In general we obtain the following explicit expression for the wealth process:

$$X_t = e^{\rho t} x_0 + \int_0^t \Big(e^{\rho(t-s)} \phi_s \cdot (\mu - \rho e)\Big) ds + \sum_{j=1}^{N_t} \phi_{T_j} \cdot Y_j \, e^{\rho(t-T_j)}.$$

For some considerations it is simpler to express the trading strategy in terms of the fraction of invested money. Let a trading strategy ϕ be given with $X_t^\phi > 0$ for all t and define

$$\pi_t^k := \frac{\phi_t^k}{X_{t-}}$$

which is the fraction of the money invested in asset k at time t. If we replace ϕ by π in the wealth equation (3.4) we obtain:

$$dX_t = X_{t-}\Big(\rho + \pi_t \cdot (\mu - \rho e) dt + \pi_t dC_t\Big). \tag{3.7}$$

On $t \in [T_n, T_{n+1})$, we obtain for an arbitrary trading strategy π

$$X_t = X_{T_n} \exp\Big(\int_{T_n}^t \big(\rho + \pi_s \cdot (\mu - \rho e)\big) ds\Big).$$

At the time of a jump we have

$$\Delta X_{T_n} = X_{T_n -} (\pi_{T_n} \cdot Y_n).$$

Altogether we obtain the following explicit expression for the wealth process

$$X_t = x_0 \exp\Big(\int_0^t (\rho + \pi_s \cdot (\mu - \rho e)) ds \Big) \prod_{j=1}^{N_t} \Big(1 + \pi_{T_j} \cdot Y_j \Big), \quad t \in [0, T].$$

3.3 Weak Convergence of Financial Markets

It is possible to approximate the asset price process of the Black-Scholes-Merton model by the asset price process in a binomial model. For this purpose we choose $\boldsymbol{u}, \boldsymbol{d}$ and p such that the first and second moments of the stock price in the binomial model match the respective moments in the Black-Scholes-Merton model. Recall that the stock price process in the Black-Scholes-Merton model is given by

$$dS_t = S_t(\mu dt + \sigma dW_t)$$

where (W_t) is a Wiener process, $\mu \in \mathbb{R}$ and $\sigma > 0$ are given parameters. Now we choose a small step length $h := \frac{1}{N}$ and equate the first and second moment of an increment of the stock price process in the Black-Scholes-Merton model over a time interval of length h with the first and second moment of the stock price increment in the binomial model. When we assume that the stock price process at the beginning of this time interval is s, we obtain the following *local consistency* conditions which have to be solved for $\boldsymbol{u}_N, \boldsymbol{d}_N$ and p_N :

$$s\mu h = p_N \Big(s(\boldsymbol{u}_N - 1) \Big) + (1 - p_N) \Big(s(\boldsymbol{d}_N - 1) \Big) \tag{3.8}$$

$$\sigma^2 s^2 h = p_N \Big(s(\boldsymbol{u}_N - 1) \Big)^2 + (1 - p_N) \Big(s(\boldsymbol{d}_N - 1) \Big)^2. \tag{3.9}$$

Obviously we have one degree of freedom for choosing the parameters $\boldsymbol{u}_N, \boldsymbol{d}_N$ and p_N. Thus, when we set $\boldsymbol{u}_N - 1 = 1 - \boldsymbol{d}_N$, we obtain

$$\boldsymbol{u}_N = 1 + \frac{\sigma}{\sqrt{N}}, \quad \boldsymbol{d}_N = 1 - \frac{\sigma}{\sqrt{N}}, \quad p_N = \frac{1}{2} + \frac{1}{2} \frac{\mu}{\sigma \sqrt{N}}.$$

This approach has been suggested in Cox and Rubinstein (1985). Let $(\tilde{R}_k^N)$ be a sequence of independent and identically distributed random variables which take the value $\boldsymbol{u}_N$ with probability p_N and $\boldsymbol{d}_N$ with probability $1 - p_N$. Then we can define the stochastic price processes

$$S_t^{(N)} := S_0 \prod_{k=1}^{\lfloor Nt \rfloor} \tilde{R}_k^N, \quad t \in [0, T]$$

which is a Semi-Markov chain with RCLL (right continuous with left-hand limits) sample paths. For large N this price process is now somehow close to the price process in the Black-Scholes-Merton model. More precisely, recall that weak convergence can be defined on metric spaces (see Definition B.1.2). The metric space which is used here is given by $D[0,T]$, the space of all RCLL functions on $[0,T]$ with the Skorohod topology. The next theorem follows essentially from Donsker's invariance principle (see e.g. Prigent (2003), chapter 3).

Theorem 3.3.1. *The sequence of stochastic processes* $(S_t^{(N)})$ *converges weakly to the stock price* (S_t) *in the Black-Scholes-Merton model for* $N \to \infty$.

Note that this particular choice of parameters $\boldsymbol{u}_N, \boldsymbol{d}_N$ and p_N establishes weak convergence however is not efficient for computational issues because the binomial tree is not recombining (for a discussion see Section 3.6).

3.4 Utility Functions and Expected Utility

There exists a well-established axiomatic theory on preference relations on sets of uncertain payoffs or lotteries. Under some conditions these preferences can be represented by a numerical function which assigns a real number to each lottery and thus induces a complete order on the lotteries. This numerical function is given by an expected utility. The axiomatic approach was pioneered by von Neumann and Morgenstern (1947). In general a utility function is a mapping $U : dom\, U \to \mathbb{R}$ which is applied to a random outcome of an investment. In particular, if we have two random variables X and Y, we can compare them by comparing $\mathbb{E}\, U(X)$ with $\mathbb{E}\, U(Y)$ where the larger value is preferred. This concept is closely related to stochastic orders (cf. Appendix B.3). The idea is that U is chosen by an individual and should to some extend also reflect her risk tolerance. A reasonable U should be increasing which means more money is better than less. Often it is also assumed that U is concave which means that for an individual the marginal utility of wealth is decreasing. For example if you own 100 Euro another 10 Euro of income is quite a lot. However if you own 1,000,000 Euro another 10 Euro is not worth much. If U is concave everywhere, then U or the individual who employs it, is said to be *risk averse*. This interpretation follows from the von Neumann-Morgenstern representation of the underlying preference relation. The Jensen inequality implies in particular that $\mathbb{E}\, U(X) \le U(\mathbb{E}\, X)$ if $\mathbb{E}\,|X| < \infty$ which means that a sure investment with the same expectation is always preferred by a risk averse investor.

Definition 3.4.1. A function $U : dom\, U \to \mathbb{R}$ is called a *utility function*, if U is strictly increasing, strictly concave and continuous on $dom\, U$.

If *dom* U is an open interval, then the concavity of U immediately implies that U is also continuous. If *dom* $U = [0, \infty)$, then U is continuous on $(0, \infty)$ and we suppose that U is also right-continuous in 0.
If an investor chooses $U(x) = x$ (which is not a utility function by definition since it is not strictly concave) then she is said to be *risk neutral* since no account for risk is made. The following utility functions are standard.

Example 3.4.2. a) *Logarithmic utility.* Here we have $U(x) = \log(x)$ and *dom* $U = (0, \infty)$. Note that the logarithm penalizes outcomes near zero heavily.

b) *Power utility.* Here we have $U(x) = \frac{1}{\gamma} x^\gamma$ and *dom* $U = [0, \infty)$ when $0 < \gamma < 1$. If $\gamma < 0$ we have *dom* $U = (0, \infty)$.

c) *Exponential utility.* Here we have $U(x) = -\frac{1}{\gamma} e^{-\gamma x}$ with $\gamma > 0$ and *dom* $U = \mathbb{R}$.

d) *Quadratic utility.* Here we have $U(x) = x - \gamma x^2$ for $\gamma > 0$ and *dom* $U = (-\infty, (2\gamma)^{-1})$. This function is only increasing for $x < (2\gamma)^{-1}$. ♦

Empirical and psychological experiments have revealed that the behaviour of many decision makers is in contrast to expected utility theory. In particular small probabilities are often overweighted. A popular example is the Allais Paradox where decision makers have to choose in two experiments between two lotteries. This has led to modified expected utility theory known as generalized expected utility or non-expected utility theory.
Let U be a utility function and X a random outcome with values in *dom* U. Due to the intermediate value theorem there exists a number $ceq(X) \in \mathbb{R}$ such that $\mathbb{E}\, U(X) = U(ceq(X))$. The value $ceq(X)$ is called *certainty equivalent* and $R(X) := \mathbb{E}\, X - ceq(X) > 0$ is called *risk premium.* If U is at least twice continuously differentiable we obtain with the Taylor series expansion:

$$U(ceq(X)) \approx U(\mathbb{E}\, X) + U'(\mathbb{E}\, X)\big(ceq(X) - \mathbb{E}\, X\big),$$
$$U(X) \approx U(\mathbb{E}\, X) + U'(\mathbb{E}\, X)(X - \mathbb{E}\, X) + \frac{1}{2} U''(\mathbb{E}\, X)(X - \mathbb{E}\, X)^2.$$

The last equation implies that

$$U(ceq(X)) = \mathbb{E}\, U(X) \approx U(\mathbb{E}\, X) + \frac{1}{2} U''(\mathbb{E}\, X) \operatorname{Var}(X).$$

Combining this with the first approximation and recalling the definition of the risk premium we obtain

$$R(X) \approx -\frac{1}{2} \frac{U''(\mathbb{E}\, X)}{U'(\mathbb{E}\, X)} \operatorname{Var}(X).$$

Hence the risk premium is equal to the variance multiplied by a coefficient which depends on the utility function. This coefficient determines the degree of risk aversion.

Definition 3.4.3. Let U be a utility function which is twice differentiable. Then the *Arrow-Pratt absolute risk aversion coefficient of U given level x* is defined by

$$\alpha_{AP}(x) := -\frac{U''(x)}{U'(x)}, \quad x \in dom\, U.$$

The function $\alpha_{AP}(x)$ shows how risk aversion changes with the level of wealth. A reasonable assumption is that $\alpha_{AP}(x)$ is decreasing, since more money increases the tendency to take a risk. The utility functions we have presented so far belong to certain classes of risk aversion.

Example 3.4.4. a) *Constant absolute risk aversion (CARA).* This class of utility functions is defined by $\alpha_{AP}(x) \equiv \alpha_{AP} > 0$. It consists of utility functions of the form

$$U(x) = a - be^{-\alpha x}$$

for $a \in \mathbb{R}, b, \alpha > 0$ and $dom\, U = \mathbb{R}$.

b) *Hyperbolic absolute risk aversion (HARA).* This class of utility functions is defined by $\alpha_{AP}(x) = (cx+d)^{-1}$. It consists of utility functions of the form

$$U(x) = \frac{1}{\gamma}(ax+b)^{\gamma}$$

for $\gamma < 1, \gamma \neq 1$ and $a > 0, b \geq 0$. If $0 < \gamma < 1$ then $dom\, U = [-\frac{b}{a}, \infty)$, otherwise $domU = (-\frac{b}{a}, \infty)$. All previously discussed utility functions may be seen as special cases of the HARA utility at least in a limiting sense (cf. Luenberger (1998)). ♦

3.5 Exercises

Exercise 3.5.1. Consider the financial market of Section 3.2. A self-financing trading strategy ϕ is an *arbitrage opportunity* if there exists a time point $t_0 \in (0,T]$ such that $\mathbb{P}(X_{t_0}^{\phi} \geq 0) = 1$ and $\mathbb{P}(X_{t_0} > 0) > 0$. A self-financing trading strategy ϕ is called bounded if $\sup_{t,\omega} |\phi_t^k(\omega)| < \infty$ for $k = 1, \ldots, d$. We suppose now that Q_Y has bounded support.
Show that there are no bounded arbitrage opportunities if and only if there exists a Q_Y-integrable function $h > 0$ such that for $k = 1, \ldots, d$

$$\int_{\mathbb{R}^d \setminus \{0\}} y_k h(y) Q_Y(dy) = \rho - \mu_k.$$

Hint: For a complete proof see Jacobsen (2006), Chapter 10.

Exercise 3.5.2 (St. Petersburg Paradox). Suppose a fair coin is tossed until a head appears. If a head appears on the n-th toss, the player receives 2^{n-1} Euro. Hence the distribution of the lottery is given by

$$\mathbb{P}(X = 2^{n-1}) = 2^{-n}, \quad n \in \mathbb{N}.$$

a) Show that $\mathbb{E}\, X = \infty$, i.e. the 'fair value' of the lottery is infinity.
b) Suppose now the player uses a utility function to evaluate the lottery. Compute the expected utility and the certainty equivalent for $U(x) = \sqrt{x}$ and $U(x) = \log x$.
   Hint: The certainty equivalents are 2.91 and 2 respectively and thus considerably smaller than ∞.

Exercise 3.5.3. We say an investor uses a *mean-variance criterion* when she prefers the lottery X to the lottery Y in case $\mathbb{E}\, X \geq \mathbb{E}\, Y$ and $Var(X) \leq Var(Y)$. When X and Y are normally distributed, the mean-variance criterion is equivalent to the expected utility approach. In order to prove this show that for $X \sim \mathcal{N}(\mu, \sigma^2)$ and $Y \sim \mathcal{N}(\nu, \tau^2)$ we have $\mathbb{E}\, U(X) \leq \mathbb{E}\, U(Y)$ for all utility functions U if and only if $\nu \geq \mu$ and $\tau^2 \leq \sigma^2$.
For a further treatment of mean-variance problems see Section 4.6.

3.6 Remarks and References

There are a lot of excellent textbooks on mathematical finance today. Discrete-time financial markets are investigated e.g. in Föllmer and Schied (2004), van der Hoek and Elliott (2006), Shreve (2004a) and Pliska (2000). A linear programming approach to general finite financial markets can be found in Cremer (1998). An introduction into financial markets driven by a Piecewise Deterministic Markov Process is given in Jacobsen (2006). General continuous-time financial markets are considered e.g. in Jeanblanc et al. (2009), Williams (2006), Cont and Tankov (2004), Shreve (2004b), Elliott and Kopp (2005), Musiela and Rutkowski (2005), Björk (2004), Bingham and Kiesel (2004), Duffie (1988, 2001) and Shiryaev (1999).
Details about the weak convergence of financial markets can be found in Prigent (2003). The particular choice of parameters in the binomial model given in Section 3.3 implies weak convergence to the Black-Scholes-Merton model, however this choice is not useful for computation of option prices since the tree is not recombining. A simple recombining tree is obtained when we choose

$$u_N = \exp\Big(\frac{\sigma}{\sqrt{N}}\Big), \quad d_N = \exp\Big(-\frac{\sigma}{\sqrt{N}}\Big), \quad p_N = \frac{1}{2} + \frac{1}{2}\frac{\mu - \frac{1}{2}\sigma^2}{\sigma\sqrt{N}}.$$

Similar parameters are proposed by Cox et al. (1979). For more computationally simple approximations consult Nelson and Ramaswamy (1990). For a different approach based on scenario tree generation see Pflug (2001).
For discussions about properties and applications of utility functions in finance we refer the reader to Föllmer and Schied (2004). For the Allais paradox see e.g. Levy (2006), Section 14. Kahneman and Tversky (1979) developed the so-called *Prospect Theory.* Daniel Kahneman received a Nobel prize in economics in 2002 for his insights from psychological research into economic sciences. Maurice Allais received the Nobel prize in economics in 1988 for his pioneering contributions to the theory of markets and efficient utilization of resources.

Chapter 4
Financial Optimization Problems

The theory of Markov Decision Processes which has been presented in Chapter 2 will now be applied to some selected dynamic optimization problems in finance. The basic underlying model is the financial market of Chapter 3. We will always assume that investors are small and cannot influence the asset price process. We begin in the first two sections with the classical problem of *maximizing the expected utility of terminal wealth*. In Section 4.1 we consider the general one-period model. It will turn out that the existence of an optimal portfolio strategy is equivalent to the absence of arbitrage in this market. Moreover, the one-period problem is the key building block for the multiperiod problems which are investigated in Section 4.2 and which can be solved with the theory of Markov Decision Processes. In this section we will also present some results for special utility functions and the relation to continuous-time models is highlighted. In Section 4.3 *consumption and investment problems* are treated and solved explicitly for special utility functions. The next section generalizes these models to include *regime switching*. Here a Markov chain is used to model the changing economic conditions which give rise to a changing return distribution. Under some simplifying assumptions this problem is solved and the influence of the environment is discussed. Section 4.5 deals with models with proportional *transaction cost*. For homogeneous utility functions it will turn out that the action space is separated into sell-, buy- and no-transaction regions which are defined by cones. The next section considers *dynamic mean-variance problems*. In contrast to utility functions the idea is now to measure the risk by the portfolio variance and to search among all portfolios which yield at least a certain expected return, the one with smallest portfolio variance. The challenge is here to reduce this problem to a Markov Decision Problem first. Essentially the task boils down to solving a linear-quadratic problem. In Section 4.7 the variance is replaced by the risk measure 'Average-Value-at-Risk'. In order to obtain an explicit solution in this *mean-risk model*, only the binomial model is considered here and the relation to the mean-variance problem is discussed. Section 4.8 deals with *index-tracking problems* and Section 4.9 investigates the

N. Bäuerle and U. Rieder, *Markov Decision Processes with Applications to Finance*, Universitext, DOI 10.1007/978-3-642-18324-9_4,

problem of *indifference pricing* in incomplete markets. Finally, the last but one section explains the relation to continuous-time models and introduces briefly the *approximating Markov chain approach.* The last section contains some remarks and references.

4.1 The One-Period Optimization Problem

In this section we investigate the one-period utility maximization problem. To ease notation we skip the time index on $i_1, \tilde{R}_1$ and R_1 in this section and write $i, \tilde{R}$ and R instead. We will see in Section 4.2 that problems of this type arise when we solve the multiperiod portfolio problem. We suppose that we have an investor with utility function $U : dom\, U \to \mathbb{R}$ and initial wealth $x > 0$. Let us denote by

$$\Delta(x) := \Big\{(\phi^0, \phi) \in \mathbb{R}^{1+d} \mid \phi^0(1+i) + \phi \cdot \tilde{R} \in dom\, U \;\; \mathbb{P}\text{-a.s.}, \phi^0 + \phi \cdot e \le x\Big\}$$

the set of admissible one-period portfolio strategies. A straightforward formulation of the one-period utility maximization problem is

$$\sup_{\phi \in \Delta(x)} \mathbb{E}\, U(\phi^0(1+i) + \phi \cdot \tilde{R}). \tag{4.1}$$

Hence the investor maximizes the expected utility of her investment under the constraint that the return is in the domain of the utility function with probability one and that the budget constraint $\phi^0 + \phi \cdot e \le x$ is satisfied. In order to have a well-defined model we make the following integrability assumption where $\|x\| := |x_1| + \ldots + |x_d|$ for $x \in \mathbb{R}^d$.

Assumption: $\mathbb{E}\, \|R\| < \infty$.

Obviously $\mathbb{E}\, \|R\| < \infty$ is equivalent to $\mathbb{E}\, \|\tilde{R}\| < \infty$. Since U is concave, the utility function can be bounded from above by an affine-linear function $c_u(1+x)$ with $c_u \in \mathbb{R}_+$. Thus the assumption implies that

$$\mathbb{E}\, U(\phi^0(1+i) + \phi \cdot \tilde{R}) \le c_u\big(1 + \phi^0(1+i) + \phi \cdot \mathbb{E}\, \|\tilde{R}\|\big) < \infty$$

for all $(\phi^0, \phi) \in \mathbb{R}^{1+d}$. The following observation is simple, but crucial: Suppose (ϕ^0, ϕ) is a portfolio with $\phi^0 + \phi \cdot e < x$. Then we can construct a new portfolio by investing the remaining wealth $x - \phi^0 - \phi \cdot e$ in the bond. This new portfolio is denoted by $(\varphi^0, \varphi) := (x - \phi \cdot e, \phi)$. Since utility functions are strictly increasing we obtain at once that

$$\mathbb{E}\, U(\varphi^0(1+i) + \varphi \cdot \tilde{R}) > \mathbb{E}\, U(\phi^0(1+i) + \phi \cdot \tilde{R}).$$

In view of problem (4.1) we can conclude that without loss of generality the budget constraint can be formulated as $\phi^0 + \phi \cdot e = x$ which allows to eliminate the bond component and to obtain an unconstrained problem. From our preceding considerations we know that we can write

$$\phi^0(1+i) + \phi \cdot \tilde{R} = X_1^\phi = (1+i)(x + \phi \cdot R).$$

Hence the admissible one-period investments in the risky assets are given by

$$D(x) := \{a \in \mathbb{R}^d \mid (1+i)(x + a \cdot R) \in dom\, U \;\; \mathbb{P}\text{-a.s.}\}.$$

For $a \in D(x)$ the amount invested in the bond is given by $x - a \cdot e$. Hence when we define

$$u(x,a) := \mathbb{E}\, U\big((1+i)(x + a \cdot R)\big)$$

an equivalent formulation of (4.1) is

$$v(x) := \sup_{a \in D(x)} u(x,a). \tag{4.2}$$

The following result shows that the absence of arbitrage opportunities is equivalent to the existence of an optimal solution for problem (4.2).

Theorem 4.1.1. *Let U be a utility function with dom $U = [0,\infty)$ or dom $U = (0,\infty)$. Then it holds:*

a) There are no arbitrage opportunities if and only if there exists a measurable function $f^* : dom\, U \to \mathbb{R}^d$ such that

$$u\big(x, f^*(x)\big) = v(x), \quad x \in dom\, U.$$

b) The function $v(x)$ is strictly increasing, strictly concave and continuous on dom U.

Remark 4.1.2. If *dom* $U = \mathbb{R}$ and U is bounded from above, then Theorem 4.1.1 also holds true. This is shown e.g. in Föllmer and Schied (2004), Theorem 3.5. This situation covers the exponential utility function. If U is not bounded from above on *dom* $U = \mathbb{R}$ then Theorem 4.1.1 is no longer true (see Example 7.3 in Rásonyi and Stettner (2005)). *dom* $U = (0,\infty)$ can be applied to the case of a logarithmic utility or a power utility with negative coefficient γ and *dom* $U = [0,\infty)$ applies to the case of a power utility with positive coefficient γ. ◊

Proof. We will only consider the case *dom* $U = (0,\infty)$. The case *dom* $U = [0,\infty)$ can be treated similarly.

We will first prove part a). Suppose that the market admits an arbitrage opportunity. Fix $x \in \mathit{dom}\, U$ and suppose $f^*(x)$ is an optimal solution of (4.2). Then according to Theorem 3.1.5 there exists a portfolio $a \in \mathbb{R}^d$ with $a \cdot R \geq 0$ $\mathbb{P}$-a.s. and $\mathbb{P}(a \cdot R > 0) > 0$. Thus $f^*(x) + a \in D(x)$ and

$$v(x) = u\big(x, f^*(x)\big) < u\big(x, f^*(x) + a\big)$$

which is a contradiction.
Now suppose that there are no arbitrage opportunities. Without loss of generality we suppose that $a \cdot R = 0$ $\mathbb{P}$-a.s. necessarily implies that $a = 0$. This property is called *non-redundancy.* Otherwise there exists at least one asset which can be perfectly replicated by a linear combination of the other assets. Thus, it can be eliminated from the market without losing a payoff profile.
Next we consider for a fixed initial wealth x the level sets

$$L(b) := \{a \in D(x) \mid u(x, a) \geq b\}, \quad b \in \mathbb{R}.$$

Since $a = 0$ is an element of $D(x)$ we have that at least $L\big(u(x, 0)\big) \neq \emptyset$. We will show that $L(b)$ is compact. First suppose that there exists an unbounded sequence $(a_n) \subset L(b)$. Choose a convergent subsequence $(a_{n_k}/\|a_{n_k}\|)$ which converges against a^*. Obviously $\|a^*\| = 1$ and since $a_{n_k} \in D(x)$ we obtain

$$a^* \cdot R = \lim_{k \to \infty} \frac{a_{n_k} \cdot R}{\|a_{n_k}\|} \geq \lim_{k \to \infty} \frac{-x}{\|a_{n_k}\|} = 0.$$

Moreover we must have $\mathbb{P}(a^* \cdot R > 0) > 0$ due to our non-redundancy assumption since $a^* \neq 0$. Thus, a^* constitutes an arbitrage opportunity which is a contradiction. Hence $L(b)$ must be bounded. Following the same proof it is also possible to show that $D(x)$ is bounded.
Next we show that the mapping $a \mapsto u(x, a)$ is upper semicontinuous. Since U is continuous the statement follows when we can apply Fatou's Lemma to interchange the limit and the expectation. But this can be done since

$$u(x, a) \leq c_u\big(1 + (1+i)(x + a \cdot \mathbb{E}\, R)\big) \leq c_u\big(1 + (1+i)(x + \max_{a \in \overline{D(x)}} a \cdot \mathbb{E}\, R)\big) < \infty$$

is an integrable upper bound (where $\overline{D(x)}$ is the closure of $D(x)$). Thus, $a \mapsto u(x, a)$ is upper semicontinuous and it follows from Lemma A.1.3 that $L(b)$ is closed. Altogether we have now shown that $L(b)$ is compact. By Proposition 2.4.11 together with Remark 2.4.12 there exists a measurable function $f^* : \mathit{dom}\, U \to \mathbb{R}^d$ such that

$$u\big(x, f^*(x)\big) = v(x), \quad x \in \mathit{dom}\, U.$$

Let us now prove part b). First we show that v is strictly increasing. Obviously $D(x) \subset D(x')$ if $x \leq x'$. Moreover, the mapping $x \mapsto u(x, a)$ is strictly increasing by our assumptions. Hence the statement follows from Theorem

2.4.14. The strict concavity follows from Proposition 2.4.18 since the set

$$\{(x,a) \mid a \in D(x),\ x > 0\}$$

is convex and $(x,a) \mapsto u(x,a)$ is strictly concave. The continuity follows from Proposition 2.4.8. □

Remark 4.1.3. Note that the optimal solution in Theorem 4.1.1 of the one-period utility maximization problem is unique if the financial market is *non-redundant*, i.e. there does not exist an asset which can be perfectly replicated by a linear combination of the other assets. ◊

4.2 Terminal Wealth Problems

At the beginning of this section we investigate the multiperiod extension of the utility maximization problem of Section 4.1. Suppose we have an investor with utility function $U : dom\,U \to \mathbb{R}$ with $dom\,U = [0,\infty)$ or $dom\,U = (0,\infty)$ and initial wealth $x > 0$. A financial market with d risky assets and one riskless bond is given (for a detailed description see Section 3.1). Here we assume that the random vectors $R_1,\dots,R_N$ are independent but not necessarily identically distributed. Moreover we assume that $(\mathcal{F}_n)$ is the filtration generated by the stock prices, i.e. $\mathcal{F}_n = \mathcal{F}_n^S$. The following assumption on the financial market is used throughout this section.

Assumption (FM):

(i) *There are no arbitrage opportunities.*
(ii) $\mathbb{E}\,\|R_n\| < \infty$ *for all* $n = 1,\dots,N$.

Our agent has to invest all the money into this market and is allowed to rearrange her portfolio over N stages. The aim is to maximize the expected utility of her terminal wealth. Recall that utility functions are strictly increasing, strictly concave and continuous by definition. According to (3.1) the wealth process (X_n) evolves as follows

$$X_{n+1} = \big(1 + i_{n+1}\big)\Big(X_n + \phi_n \cdot R_{n+1}\Big)$$

where $\phi = (\phi_n)$ is a portfolio strategy. The optimization problem is then

$$\begin{cases} \mathbb{E}_x\, U(X_N^\phi) \to \max \\ \phi \text{ is a portfolio strategy and } X_N^\phi \in dom\,U \ \mathbb{P}\text{-a.s.} \end{cases} \tag{4.3}$$

Since this problem has a Markovian structure it can be shown that the optimal portfolio strategy is among the set of all Markov portfolio strategies (ϕ_n) (see e.g. Theorem 2.2.3). Moreover ϕ_n depends only on the wealth X_n and not on the stock prices S_n which can be seen from the wealth equation. Thus, problem (4.3) can be solved via the following Markov Decision Model: The state space is $E := dom\, U$, where $x \in E$ is the available wealth. $A := \mathbb{R}^d$ is the action space and

$$D_n(x) := \Big\{ a \in \mathbb{R}^d \mid \big(1 + i_{n+1}\big)\big(x + a \cdot R_{n+1}\big) \in dom\, U \ \mathbb{P}\text{-a.s.} \Big\} \tag{4.4}$$

is the set of admissible actions. This restriction guarantees that a portfolio can be chosen for the remaining stages such that the final wealth X_N is with probability one in E. Though it is enough to have $X_N \in dom\, U$ $\mathbb{P}$-a.s., the absence of arbitrage implies that we have to require $X_n \in dom\, U$ $\mathbb{P}$-a.s. for all stages n. The transition function is given by

$$T_n(x, a, z) = \big(1 + i_{n+1}\big)\big(x + a \cdot z\big) \tag{4.5}$$

where $z \in [-1, \infty)^d$ denotes the relative risk in $[n, n+1)$. Since we do not have any intertemporal utility we choose $r_n \equiv 0$ and $g_N(x) = U(x)$. The data of our Markov Decision Model is summarized as follows:

- $E := dom\, U$ where x denotes the wealth,
- $A := \mathbb{R}^d$ where a is the amount of money invested in the risky assets,
- $D_n(x)$ is given by (4.4),
- $\mathcal{Z} := [-1, \infty)^d$ where z denotes the relative risk,
- $T_n(x, a, z)$ is given by (4.5),
- $Q_n^Z(\cdot|x, a) :=$ distribution of R_{n+1} (independent of (x, a)),
- $r_n \equiv 0$, and $g_N(x) := U(x)$.

Problem (4.3) can now equivalently be solved by the Markov Decision Model. The value functions are given by

$$V_n(x) = \sup_{\pi} \mathbb{E}_{nx}^{\pi}\, U(X_N) \tag{4.6}$$

where the supremum is taken over all policies π and $V_0(x)$ is the value of problem (4.3). Due to our assumption, the Markov Decision Model has an upper bounding function.

Proposition 4.2.1. *The function $b(x) := 1+x$ is an upper bounding function for the Markov Decision Model.*

Proof. We have to check conditions (i)–(iii) of Definition 2.4.1. Condition (i) is obvious. Condition (ii) follows since any concave function can be bounded from above by a linear affine function and (iii) holds since there exist constants $c, \alpha_b > 0$ such that

$$\int b(x')Q_n(dx'|x,a) = 1 + (1+i_{n+1})(x + a \cdot \mathbb{E}\, R_{n+1})$$
$$\leq 1 + (1+i_{n+1})(x + cx) \leq \alpha_b(1+x), \quad x \in E$$

for all $a \in D_n(x)$. The proof that $a \cdot \mathbb{E}\, R_{n+1} \leq cx$ is as follows: In view of the no arbitrage assumption the support of R_{n+1}^k contains elements $z_1^k \in (-1,0)$ and $z_2^k \in (0,\infty)$. Hence for all $a \in D(x)$ we have

$$-\frac{x}{z_2^k} \leq a^k \leq -\frac{x}{z_1^k}, \quad k = 1,\dots,d$$

which implies $a \cdot \mathbb{E}\, R_{n+1} \leq cx$. Note that $\mathbb{E}\, R_{n+1} < \infty$ due to Assumption (FM)(ii). □

We are now in a position to apply our main Theorem 2.3.8 and obtain the following statements.

Theorem 4.2.2. *For the multiperiod terminal wealth problem it holds:*

a) The value functions V_n are strictly increasing, strictly concave and continuous.

b) The value functions can be computed recursively by the Bellman equation

$$V_N(x) = U(x),$$
$$V_n(x) = \sup_{a \in D_n(x)} \mathbb{E}\, V_{n+1}\Big((1+i_{n+1})(x + a \cdot R_{n+1})\Big), \quad x \in E.$$

c) There exist maximizers f_n^* of V_{n+1}, and the portfolio strategy $(f_0^*, f_1^*, \dots, f_{N-1}^*)$ is optimal for the N-stage terminal wealth problem.

Proof. We show that the Structure Assumption (SA$_N$) is satisfied with

$$\mathbb{M}_n := \{v \in \mathbb{B}_b^+ \mid v \text{ is strictly increasing, strictly concave and continuous}\}$$

and $\Delta_n := F_n$. The statements then follow immediately from Theorem 2.3.8.

(i) $g_N = U \in \mathbb{M}_N$ holds by definition of utility functions.

(ii) Now suppose that $v \in \mathbb{M}_{n+1}$. Note that

$$\mathcal{T}_n v(x) = \sup_{a \in D_n(x)} \mathbb{E}\, v\Big((1+i_{n+1})(x + a \cdot R_{n+1})\Big), \quad x \in E.$$

Since v satisfies all properties of a utility function with *dom* $v = E$ and since the absence of arbitrage is a local property (see Theorem 3.1.5) we obtain $\mathcal{T}_n v \in \mathbb{M}_n$ from Theorem 4.1.1.

(iii) The existence of a maximizer follows again from Theorem 4.1.1. □

Remark 4.2.3. Theorem 4.2.2 can also be formulated in a similar way when $dom\, U = \mathbb{R}$. In this case $D_n(x) = \mathbb{R}$. We solve the terminal wealth problem with an exponential utility later in this section. ◊

Of course it is no problem to incorporate constraints on the portfolio strategy like short-selling constraints or constraints on the risk of the strategy. The only thing that changes is the set of admissible actions $D(x)$ which has to be defined accordingly.

There is one particular special case where the portfolio problem (4.3) has a very simple solution. This situation is formulated in the next theorem.

Theorem 4.2.4. *Let $\mathbb{E}\, R_n = 0$ for $n = 1, \dots, N$. Then it holds:*

a) The value functions are given by

$$V_n(x) = U\Big(x\, \frac{S_N^0}{S_n^0}\Big), \quad x \in E.$$

b) The optimal portfolio strategy $(f_0^*, f_1^*, \dots, f_{N-1}^*)$ is given by

$$f_n^*(x) \equiv 0,$$

i.e. the strategy 'invest all the money in the bond' is optimal.

Proof. Suppose that $\pi = (f_0, f_1, \dots, f_{N-1})$ is an arbitrary portfolio strategy. It is easy to verify that the discounted wealth process $\big(X_n/S_n^0\big)$ is a martingale under the assumptions of this theorem (for a definition of martingales see Section B.2): Obviously the expectation is well-defined and we obtain

$$\mathbb{E}^\pi\left[\frac{X_{n+1}}{S_{n+1}^0}\,\Big|\,\frac{X_n}{S_n^0}\right] = \mathbb{E}^\pi\left[(S_n^0)^{-1}\Big(X_n + f_n(X_n)\cdot R_{n+1}\Big)\,\Big|\,\frac{X_n}{S_n^0}\right] = \frac{X_n}{S_n^0}.$$

Thus, using Jensen's inequality it holds that

$$V_{n\pi}(x) = \mathbb{E}_{nx}^\pi\, U(X_N) \le U\Big(\mathbb{E}_{n,x}^\pi\left[\frac{X_N}{S_N^0}\right]S_N^0\Big) = U\Big(x\,\frac{S_N^0}{S_n^0}\Big)$$

which implies the statements. □

Remark 4.2.5. a) Since $\frac{S_{n+1}^k}{S_{n+1}^0} = \frac{S_n^k}{S_n^0}(1 + R_{n+1}^k)$, the condition $\mathbb{E}\, R_n^k = 0$ for all n means that the discounted stock price process $\big(S_n^k/S_n^0\big)$ is a martingale.

b) If short-selling is not allowed (i.e. $a \in \mathbb{R}^d_+$) then the portfolio strategy $(f_0^*, f_1^*, \ldots, f_{N-1}^*)$ with $f_n^*(x) \equiv 0$ is optimal if $\mathbb{E}\, R_n \le 0$ for $n = 1, \ldots, N$. This follows from the proof of Theorem 4.2.4. $\Diamond$

For some utility functions the portfolio optimization problem (4.6) can be solved rather explicitly. We summarize some of these results. Throughout we suppose that Assumption (FM) is valid.

Power Utility

Let us suppose that the utility function in problem (4.6) is given by

$$U(x) = \frac{1}{\gamma} x^\gamma, \quad x \in [0, \infty)$$

with $0 < \gamma < 1$. Here we have $E = [0, \infty)$. Since it will be more convenient to work with fractions of invested money instead of amounts we define the set of admissible fractions by

$$A_n := \{\alpha \in \mathbb{R}^d \mid 1 + \alpha \cdot R_{n+1} \ge 0 \ \mathbb{P}\text{-a.s.}\}$$

and the generic one-period optimization problem by

$$v_n := \sup_{\alpha \in A_n} \mathbb{E} \left(1 + \alpha \cdot R_{n+1}\right)^\gamma. \tag{4.7}$$

According to Theorem 4.1.1 this problem has a solution and we denote the optimal solution by $\alpha_n^* \in \mathbb{R}^d$.

Theorem 4.2.6. *Let U be the power utility with $0 < \gamma < 1$. Then it holds:*

a) The value functions are given by

$$V_n(x) = d_n x^\gamma, \quad x \ge 0$$

with

$$d_N = \frac{1}{\gamma} \text{ and } d_n = \frac{1}{\gamma} \prod_{k=n}^{N-1} (1 + i_{k+1})^\gamma v_k.$$

b) The optimal amounts which are invested in the stocks are given by

$$f_n^*(x) = \alpha_n^* \, x, \quad x \ge 0$$

where α_n^* is the optimal solution of (4.7). The optimal portfolio strategy is given by $(f_0^*, f_1^*, \ldots, f_{N-1}^*)$.

Note that the optimal portfolio strategy is myopic in the sense that the optimal fractions which are invested in the assets at time n depend only on the distribution of the relative risk R_{n+1}. Moreover, it is easy to see that the sequence d_n is decreasing in n.

Proof. In order to apply Theorem 2.3.8 we have to find sets $\mathbb{M}_n$ and Δ_n which satisfy the Structure Assumption (SA$_N$). We will first show that the sets can be chosen as

$$\mathbb{M}_n := \{v : E \to \mathbb{R}_+ \mid v(x) = bx^\gamma \text{ for } b > 0\},$$

$$\Delta_n := \{f \in F_n \mid f(x) = \alpha\, x \text{ for } \alpha \in \mathbb{R}^d\}.$$

We have to check the following conditions

(i) Obviously $g_N = U \in \mathbb{M}_N$.
(ii) Let $v \in \mathbb{M}_{n+1}$. Then we obtain

$$\begin{aligned} \mathcal{T}_n v(x) &= \sup_{a \in D_n(x)} \mathbb{E}\, v\Big((1 + i_{n+1})(x + a \cdot R_{n+1})\Big) \\ &= b(1 + i_{n+1})^\gamma \sup_{a \in D_n(x)} \mathbb{E}(x + a \cdot R_{n+1})^\gamma. \end{aligned}$$

If $x = 0$ then only $a = 0$ is admissible. Hence suppose $x > 0$. We use the transformation $a = \alpha x$ to obtain

$$\begin{aligned} \mathcal{T}_n v(x) &= b(1 + i_{n+1})^\gamma x^\gamma \sup_{\alpha \in A_n} \mathbb{E}(1 + \alpha \cdot R_{n+1})^\gamma \\ &= b(1 + i_{n+1})^\gamma x^\gamma v_n. \end{aligned}$$

Thus $\mathcal{T}_n v(x) = \tilde{b} x^\gamma \in \mathbb{M}_n$ with $\tilde{b} := b(1 + i_{n+1})^\gamma v_n > 0$.
(iii) For all $v \in \mathbb{M}_{n+1}$ there exists a maximizer f_n^* of v with $f_n^* \in \Delta_n$. This follows from Theorem 4.1.1 and (ii).

Hence we have shown that the Structure Assumption (SA$_N$) holds. The statement now follows by induction from Theorem 2.3.8. We obtain

$$\begin{aligned} V_n(x) &= \sup_{\alpha \in A_n} d_{n+1}(1 + i_{n+1})^\gamma x^\gamma\, \mathbb{E}\left(1 + \alpha \cdot R_{n+1}\right)^\gamma \\ &= d_{n+1}(1 + i_{n+1})^\gamma v_n x^\gamma. \end{aligned}$$

If we define $d_n := d_{n+1}(1 + i_{n+1})^\gamma v_n > 0$ then the statements in part a) follow and part b) can be concluded from the considerations in (ii). □

If the relative stock price changes $\tilde{R}_1, \tilde{R}_2, \ldots$ are identically distributed and $i_n = i$ for all n, then $\alpha_n^* \equiv \alpha^*$ is independent of n. In this case the Markov Decision Model is stationary. If we further assume that there is only one stock, then the optimal portfolio strategy can be characterized by a constant optimal stock to bond ratio (Merton-line) which is equal to $\frac{\alpha^*}{1-\alpha^*}$. If the current stock to bond ratio is larger, then the stock has to be sold and if it is smaller, then the stock has to be bought. This is illustrated in Figure 4.1.

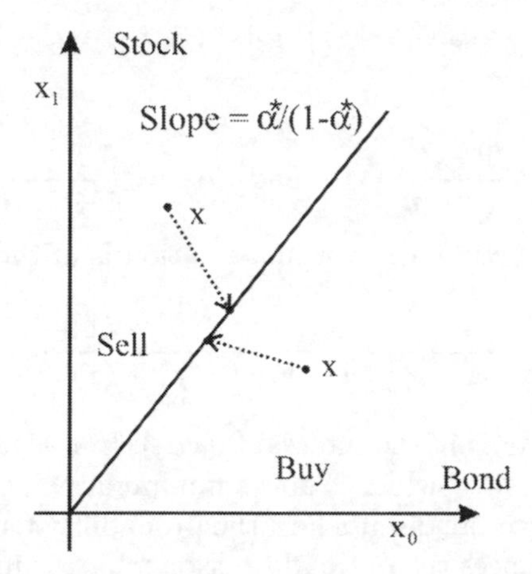

Fig. 4.1 Optimal stock to bond ratio (Merton-line).

Remark 4.2.7. In the case of a power utility function $U(x) = \frac{1}{\gamma}x^\gamma$ with $\gamma < 0$ we can proceed almost in the same way. In this case we do not need an integrability condition on the relative risk process, since the utility function is bounded from above by 0. Moreover, in this case *dom* $U = (0, \infty)$ and we define

$$A_n := \{\alpha \in \mathbb{R}^d \mid 1 + \alpha \cdot R_{n+1} > 0 \ \mathbb{P}\text{-a.s.}\}.$$

Theorem 4.2.6 can be stated in exactly the same way, the only difference is that v_n and α_n^* are now defined as the value and the optimal solution of

$$v_n := \inf_{\alpha \in A_n} \mathbb{E}\left(1 + \alpha \cdot R_{n+1}\right)^\gamma. \tag{4.8}$$

According to Theorem 4.1.1 this problem has a solution. ◇

Remark 4.2.8. Some authors also consider the power utility problem with parameter $\gamma > 1$ which reflects a risk-seeking investor. In this case the value functions are convex and it is reasonable to restrict to a compact set for the admissible portfolio strategy, e.g. by excluding short-sellings. In this case we have

$$D_n(x) = \{a \in \mathbb{R}_+^d \mid a \cdot e \le x\}.$$

Since a convex function attains its maximum at the boundary it is in this case optimal for the investor to concentrate her investment on one asset k, namely the one which maximizes $\mathbb{E}(1 + R_{n+1}^k)^\gamma$, i.e. she puts 'all her eggs in one basket'. ◇

Imagine we have one stock and the price process follows the *binomial model* as described in Section 3.1. The optimization problem (4.7) reduces to (we skip the constant $(1+i)^{-\gamma}$ in front of sup).

$$\sup_{\alpha_0 \le \alpha \le \alpha_1} \Big(1+i+\alpha(\boldsymbol{u}-1-i)\Big)^\gamma p + \Big(1+i+\alpha(\boldsymbol{d}-1-i)\Big)^\gamma (1-p)$$

where

$$\alpha_0 := \frac{1+i}{1+i-\boldsymbol{u}} < 0 \quad \text{and} \quad \alpha_1 := \frac{1+i}{1+i-\boldsymbol{d}} > 0.$$

If we define $\delta := (1-\gamma)^{-1}$, the optimal solution is of the form

$$\alpha^* := \frac{1+i}{(1+i-\boldsymbol{d})(\boldsymbol{u}-1-i)} \cdot \frac{(\boldsymbol{u}-1-i)^\delta p^\delta - (1+i-\boldsymbol{d})^\delta (1-p)^\delta}{(\boldsymbol{u}-1-i)^{\delta\gamma} p^\delta + (1+i-\boldsymbol{d})^{\delta\gamma}(1-p)^\delta}. \tag{4.9}$$

Note that this number is in the interval $[\alpha_0, \alpha_1]$. In the case $\gamma < 0$, we obtain the same expression for α^* which is now a minimum point of problem (4.8). In the binomial model we can discuss how the probability for an up movement of the stock price influences the optimal fraction which is invested in the stock.

Lemma 4.2.9. *Consider the binomial model with power utility and parameter $\gamma < 1, \gamma \neq 0$.*

a) The optimal fraction α^* which is invested in the stock is given in (4.9).
b) The function $\alpha^* = \alpha^*(p)$ is increasing in p.
c) If $p = \frac{1+i-\boldsymbol{d}}{\boldsymbol{u}-\boldsymbol{d}}$ then $\alpha^* = \alpha^*(p) = 0$.

Proof. Suppose first that $0 < \gamma < 1$. For our purpose it is convenient to discuss the maximum points of the mapping:

$$h(\alpha) := \big(1+i+\alpha(\boldsymbol{u}-1-i)\big)^\gamma + \lambda\big(1+i+\alpha(\boldsymbol{d}-1-i)\big)^\gamma$$

where $\lambda = \frac{1-p}{p}$. Differentiating with respect to α we obtain that the maximum is attained at (we denote $\delta := (1-\gamma)^{-1}$)

$$\alpha^*(\lambda) = (1+i)\frac{\lambda^{-\delta}(1+i-\boldsymbol{d})^{-\delta} - (\boldsymbol{u}-1-i)^{-\delta}}{(\boldsymbol{u}-1-i)^{-\delta\gamma} + (1+i-\boldsymbol{d})^{-\delta\gamma}\lambda^{-\delta}}$$

whenever this point is in the interval $[\alpha_0, \alpha_1]$. Rearranging terms we end up with formula (4.9). Since $h(\alpha)$ is concave, $\alpha^*(\lambda)$ is indeed a maximum point. The function $\lambda = \lambda(p) = \frac{1-p}{p}$ is decreasing in p, thus it remains to determine whether $\alpha^*(\lambda)$ is increasing or decreasing in λ. Differentiating $\alpha^*(\lambda)$ with respect to λ yields $\frac{\partial}{\partial\lambda}\alpha^*(\lambda) \le 0$ if and only if

$$(\boldsymbol{u}-1-r)^\gamma + (1+r-\boldsymbol{d}) \ge 0$$

which is true. Thus, $\alpha^*(p)$ is increasing in p. In particular it can now easily be verified that indeed $\alpha^* \in [\alpha_0, \alpha_1]$ by inserting $p = 0$ and $p = 1$ in equation (4.9). If $p = \frac{1+i-\boldsymbol{d}}{\boldsymbol{u}-\boldsymbol{d}}$, then $\mathbb{E}\, R_n = 0$ for all n and Theorem 4.2.4 applies. □

As a consequence we obtain from Lemma 4.2.9 that if $p < \frac{1+i-\boldsymbol{d}}{\boldsymbol{u}-\boldsymbol{d}}$ it is optimal to sell the stock, and if $p > \frac{1+i-\boldsymbol{d}}{\boldsymbol{u}-\boldsymbol{d}}$ it is optimal to buy the stock.

Remark 4.2.10. In the standard continuous-time Black-Scholes-Merton model (see Section 3.1), the optimal portfolio strategy in the case of a power utility is to invest a constant fraction of the wealth in the stock. This fraction is independent of the time and given by the so-called *Merton ratio*

$$\frac{1}{1-\gamma} \frac{\mu - \rho}{\sigma^2}. \tag{4.10}$$

As before, μ is the drift of the stock, σ the volatility and ρ is the riskfree interest rate, i.e. we have $1 + i = e^{\rho \Delta t}$. Since we can approximate the Black-Scholes-Merton model by the binomial model (see Section 3.3) we would expect that the optimal fraction computed in (4.9) is close to the expression (4.10). Indeed, if we define (cf. Section 3.6)

$$\boldsymbol{u} := \exp\big(\sigma\sqrt{\Delta t}\big), \quad \boldsymbol{d} := \exp\big(-\sigma\sqrt{\Delta t}\big)$$

$$p := \frac{1}{2} + \frac{1}{2}\frac{\mu - \frac{1}{2}\sigma^2}{\sigma}\sqrt{\Delta t}, \quad 1 + i = e^{\rho \Delta t}$$

we obtain

$$\lim_{\Delta t \downarrow 0} \alpha^*(\Delta t) = \frac{1}{1-\gamma}\frac{\mu \quad \rho}{\sigma^2}.$$

However the proof is quite lengthy (using Taylor series expansions with respect to Δt). ◊

HARA-Utility

The HARA-utility has been defined in Section 3.4 and can be seen as a shifted power utility. Thus it is not surprising that maximizing a HARA-utility can be reduced to maximizing a power utility. Since this case is of some importance and since some subtle differences arise we state the respective theorem separately. Suppose for simplicity that the utility function in problem (4.3) is given by

$$U(x) = (x+b)^\gamma$$

where $b \geq 0$ and $0 < \gamma < 1$. We have $dom\, U = \{x \in \mathbb{R} \mid x + b \geq 0\}$. Note that if we insert the wealth X_1 at stage 1 according to formula (3.1) we obtain

$$\mathbb{E}\, U(X_1) = (1+i_1)^\gamma\, \mathbb{E}\Big(x + \frac{b}{1+i_1} + a \cdot R_1\Big)^\gamma.$$

In order to obtain a well-defined optimization problem and in view of the no arbitrage assumption we must assume that $x(1+i_1) \geq -b$. By induction this yields for the multiperiod problem with HARA-utility that the initial wealth x has to satisfy $xS_N^0 \geq -b$. We define $E := dom\, U$ as the state space of the Markov Decision Model. In what follows it will be convenient to denote

$$E_n := \{x \in \mathbb{R} \mid xS_N^0/S_n^0 + b \geq 0\}.$$

The set E_n consists of all possible wealths at stage n such that the condition $X_N \in E$ can be satisfied with probability one. Thus, admissible investments at stage n are

$$\begin{aligned} D_n(x) &:= \Big\{a \in \mathbb{R}^d \mid S_N^0/S_n^0\big(x + a \cdot R_{n+1}\big) \in E \;\; \mathbb{P}\text{-a.s.}\Big\} \\ &= \Big\{a \in \mathbb{R}^d \mid (1+i_{n+1})\big(x + a \cdot R_{n+1}\big) \in E_{n+1} \;\; \mathbb{P}\text{-a.s.}\Big\}, \quad x \in E_n. \end{aligned}$$

We obtain the following result:

Theorem 4.2.11. *Let $U(x) = (x+b)^\gamma$ be the HARA-utility with $b \geq 0$ and $0 < \gamma < 1$. Then it holds:*

a) The value functions are given by

$$V_n(x) = d_n \left(x\, S_N^0/S_n^0 + b\right)^\gamma, \quad x \in E_n$$

with

$$d_N = 1 \quad \text{and} \quad d_n = \prod_{k=n}^{N-1} v_k,$$

and v_n is the value of problem (4.7).

b) The optimal amounts which are invested in the stocks are given by

$$f_n^*(x) = \alpha_n^* \left(x + \frac{bS_n^0}{S_N^0}\right), \quad x \in E_n$$

where α_n^* is the optimal solution of problem (4.7). The optimal portfolio strategy is given by $(f_0^*, f_1^*, \ldots, f_{N-1}^*)$.

Proof. We proceed in the same way as in the proof of Theorem 4.2.6. To apply Theorem 2.3.8 we choose

$$\mathbb{M}_n := \{v : E_n \to \mathbb{R}_+ \mid v(x) = c(x\, S_N^0/S_n^0 + b)^\gamma \;\text{ for } c > 0\},$$

$$\Delta_n := \{f : E_n \to \mathbb{R}^d \mid f(x) = c_0(x + c_1) \;\text{ for } c_0 \in \mathbb{R}^d \text{ and } c_1 > 0\} \cap F_n.$$

We now have to check the three conditions:

(i) Obviously $g_N = U \in I\!M_N$.
(ii) Let $v \in I\!M_{n+1}$. Then

$$\begin{aligned}\mathcal{T}_n v(x) &= c\Big(\frac{S_N^0}{S_n^0}\Big)^\gamma \sup_{a \in D_n(x)} \mathbb{E}\Big(x + \frac{bS_n^0}{S_N^0} + a \cdot R_{n+1}\Big)^\gamma \\ &= c\Big(\frac{S_N^0}{S_n^0}\Big)^\gamma \Big(x + \frac{bS_n^0}{S_N^0}\Big)^\gamma \sup_{\alpha \in A_n} \mathbb{E}\Big(1 + \alpha \cdot R_{n+1}\Big)^\gamma\end{aligned}$$

where we have used the transformation

$$\alpha\Big(x + \frac{bS_n^0}{S_N^0}\Big) := a$$

and

$$A_n = \{\alpha \in \mathbb{R}^d \mid 1 + \alpha \cdot R_{n+1} \geq 0 \ \mathbb{P}\text{-a.s.}\}.$$

If $x = -bS_n^0/S_N^0$, then only $a = 0$ is admissible. Suppose now that $x > -bS_n^0/S_N^0$. The variable α is the fraction of a new wealth

$$\tilde{x} = x + \frac{bS_n^0}{S_N^0}$$

which has to be invested. The optimization problem is now the same as in the power utility case and $\mathcal{T}_n v \in I\!M_n$.
(iii) For $v \in I\!M_{n+1}$ there exists a maximizer $f_n^* \in \Delta_n$ of v. This follows from Theorem 4.1.1 and the considerations in part (ii).

The induction step is along the lines of the proof of Theorem 4.2.6. □

Remark 4.2.12. If the HARA utility function is given by $U(x) = b_0(x+b)^\gamma$ with $\gamma < 0$ and $b_0 < 0$ we obtain an analogous result as in the power utility case with negative exponent (cf. Remark 4.2.7). ◊

Logarithmic Utility

Here we assume that the utility function in problem (4.3) is of the form $U(x) = \log x$ with *dom* $U = (0, \infty)$. In this case it is again convenient to consider the fraction of wealth which is invested into the assets. Thus, we define the set

$$A_n = \{\alpha \in \mathbb{R}^d \mid 1 + \alpha \cdot R_{n+1} > 0 \ \mathbb{P}\text{-a.s.}\}.$$

Note that we need a strict inequality this time. Let us introduce the following generic one-period optimization problem:

$$v_n := \sup_{\alpha \in A_n} \mathbb{E} \log\big(1 + \alpha \cdot R_{n+1}\big). \tag{4.11}$$

According to Theorem 4.1.1 this problem can be solved and we denote the optimal solution by $\alpha_n^* \in \mathbb{R}^d$.

Theorem 4.2.13. *Let U be the logarithmic utility. Then it holds:*

a) The value functions are given by

$$V_n(x) = \log x + d_n, \quad x > 0$$

where

$$d_N = 0 \quad and \quad d_n = \sum_{k=n}^{N-1} \big(\log(1 + i_{k+1}) + v_k\big)$$

and v_n is the value of problem (4.11).

b) The optimal amounts which are invested in the stocks are given by

$$f_n^*(x) = \alpha_n^* x, \quad x > 0$$

where α_n^* is the optimal solution of (4.11). The optimal portfolio strategy is given by $(f_0^*, f_1^*, \dots, f_{N-1}^*)$.

The proof can be done in the same way as for the power utility and we skip it here. In case we have one stock and the price process follows the *binomial model* as described in Section 3.1, the optimization problem in (4.11) reduces to

$$\sup_{\alpha_0 < \alpha < \alpha_1} p \log\Big(1 + i + \alpha(\boldsymbol{u} - 1 - i)\Big) + (1 - p) \log\Big(1 + i + \alpha(\boldsymbol{d} - 1 - i)\Big)$$

where again

$$\alpha_0 := \frac{1+i}{1+i-\boldsymbol{u}} < 0 \quad \text{and} \quad \alpha_1 := \frac{1+i}{1+i-\boldsymbol{d}} > 0.$$

Thus, the optimal fraction which is invested in the stock is of the form

$$\alpha^* := (1+i)\Big(\frac{p}{1+i-\boldsymbol{d}} - \frac{1-p}{\boldsymbol{u}-1-i}\Big). \tag{4.12}$$

Note that $\alpha^* \in (\alpha_0, \alpha_1)$. As in the proof of Lemma 4.2.9 it can be shown that $\alpha^* = \alpha^*(p)$ is increasing in p and $\alpha^*(p) = 0$ if $p = \frac{1+i-\boldsymbol{d}}{\boldsymbol{u}-\boldsymbol{d}}$.

Remark 4.2.14. The logarithmic utility can be see as a limiting case of the power utility with $\gamma \downarrow 0$. Thus, the following observation is not surprising: If $\alpha_{\text{pw}}^*(\gamma)$ is the optimal fraction in the binomial model with power utility, then

$$\lim_{\gamma \to 0} \alpha_{\text{pw}}^*(\gamma) = \alpha_{\log}^*$$

where $\alpha^*_{\text{pw}}(\gamma)$ and $\alpha^*_{\log}$ are given in (4.9) and (4.12) respectively. This follows after some simple but tedious algebra. ◊

Exponential Utility

Finally we investigate the case where the utility function in problem (4.3) is of the form $U(x) = -\frac{1}{\gamma}e^{-\gamma x}$ with $\gamma > 0$ and *dom* $U = \mathbb{R}$. In this case the utility function is bounded from above and we do not need the Integrability Assumption (ii) of (FM). It is sufficient that no arbitrage opportunities are available. Since the domain of the exponential utility is $\mathbb{R}$ we have no restrictions on the investment decisions and obtain $D_n(x) = \mathbb{R}^d$. Again we introduce a generic one-period optimization problem which has the following form:

$$v_n := \inf_{a \in \mathbb{R}^d} \mathbb{E} \exp\Big(-\gamma\, S^0_N / S^0_n \, a \cdot R_{n+1} \Big). \tag{4.13}$$

According to Remark 4.1.2 this problem can be solved and we denote the optimal solution by $a^*_n \in \mathbb{R}^d$. The solution is of the form $a^*_n = \frac{1}{\gamma}\frac{S^0_n}{S^0_N}\tilde{a}_n$ where $\tilde{a}_n$ is the minimum point of

$$a \mapsto \mathbb{E} \exp\big(-a \cdot R_{n+1} \big), \quad a \in \mathbb{R}^d. \tag{4.14}$$

If the random vectors $R_1, R_2, \ldots$ are identically distributed, then $\tilde{a}_n$ is independent of n.

Theorem 4.2.15. *Let U be the exponential utility. Then it holds:*

a) The value functions are given by

$$V_n(x) = d_n \exp\left(-\gamma\, S^0_N / S^0_n \, x\right), \quad x \in \mathbb{R}$$

where

$$d_N = -\frac{1}{\gamma} \quad \text{and} \quad d_n = -\frac{1}{\gamma} \prod_{k=n}^{N-1} v_k$$

and v_n is the value of problem (4.13).

b) The optimal amounts which are invested in the stocks are given by

$$f^*_n(x) = a^*_n, \quad x \in \mathbb{R}$$

where a^*_n is the optimal solution of problem (4.13), and the policy $(f^*_0, \ldots, f^*_{N-1})$ is optimal.

Note that the optimal amounts which are invested in the risky assets depend only on the distribution of the relative risks R_n and not on the current wealth.

Proof. We proceed in the same way as in the proof of Theorem 4.2.6 by applying Theorem 2.3.8. This time we choose

$$\mathbb{M}_n := \{v : \mathbb{R} \to \mathbb{R} \mid v(x) = -b_0 \exp(-b_1 x) \text{ for } b_0, b_1 > 0\},$$

$$\Delta_n := \{f \in F_n \mid f(x) \equiv c_0 \text{ for } c_0 \in \mathbb{R}^d\}.$$

Checking the three conditions is simple:

(i) Obviously $g_N = U \in \mathbb{M}_N$.
(ii) Let $v \in \mathbb{M}_{n+1}$. Then we obtain the following problem

$$\mathcal{T}_n v(x) = -b_0 \exp\Big(- b_1(1 + i_{n+1})x\Big) \inf_{a \in \mathbb{R}^d} \mathbb{E} \exp\Big(- b_1(1 + i_{n+1})a \cdot R_{n+1}\Big).$$

This optimization problem has a solution according to Remark 4.1.2, part a) and thus $\mathcal{T}_n v \in \mathbb{M}_n$.
(iii) For $v \in \mathbb{M}_{n+1}$ there exists a maximizer $f_n(x) \in \Delta_n$ of v. This follows from Remark 4.1.2 and the considerations in part (ii).

Working out the induction step is not hard, one simply has to keep track of the factors d_n. □

Remark 4.2.16. Let us denote by $M_n(x) := \mathbb{E} \exp(x \cdot R_{n+1})$ the moment generating function $M_n : \mathbb{R}^d \to \mathbb{R}$ of R_{n+1} and let us assume that $M_n(x) < \infty$ for all $x \in \mathbb{R}^d$. Then problem (4.14) is the same as

$$\inf_{a \in \mathbb{R}^d} M_n(-a). \tag{4.15}$$

Since the moment generating function is convex and since the derivative of M_n exists (see e.g. Billingsley (1995), section 21) we can conclude that $\tilde{a}_n$ is optimal if and only if the first order condition

$$\nabla_x M_n(-\tilde{a}_n) = 0$$

holds. ◊

Remark 4.2.17. For the binomial model it is easy to show that the solution of (4.14) has the form

$$\tilde{a}_n = (1 + i)\Big(\frac{\log\left(\frac{1-q}{1-p}\right) - \log\left(\frac{q}{p}\right)}{\boldsymbol{u} - \boldsymbol{d}}\Big)$$

where $q = \frac{1+i-\boldsymbol{d}}{\boldsymbol{u}-\boldsymbol{d}}$. Here $\tilde{a}_n$ is independent of n. Obviously $\tilde{a} > 0 \Leftrightarrow p > q$ and $\tilde{a} < 0 \Leftrightarrow p < q$ and a_n^* is increasing in n (decreasing in n) if $p > q$ $(p < q)$.

Remark 4.2.18. When we deal with the exponential utility, the problem of maximizing the expected terminal wealth is also well-defined in case the wealth can get negative (which is not allowed in the previous cases). Thus, we can approximate the random relative risk in period $[n-1, n)$ by a d-dimensional normal distribution, i.e. $R_n \sim \mathcal{N}(\mu_n, \Sigma_n)$ and R_n takes values below -1. In particular we obtain for $a \in \mathbb{R}^d$ that

$$a \cdot R_{n+1} \sim \mathcal{N}\Big(a \cdot \mu_{n+1},\, a^\top \Sigma_{n+1} a\Big).$$

Recall that for $Z \sim \mathcal{N}(\mu, \sigma^2)$ we obtain $\mathbb{E}\, e^{\theta Z} = e^{\mu\theta + \frac{1}{2}\sigma^2\theta^2}$. In our case this yields

$$\mathbb{E} \exp\left(-a \cdot R_{n+1}\right) = \exp\Big(- a \cdot \mu_{n+1} + \frac{1}{2} a^\top \Sigma_{n+1} a\Big).$$

The minimum of this expression is obviously attained at

$$\tilde{a}_n := \Sigma_{n+1}^{-1} \mu_{n+1}$$

and

$$a_n^* = \frac{1}{\gamma} \frac{S_n^0}{S_N^0} \Sigma_{n+1}^{-1} \mu_{n+1}.$$

This expression is similar to the Merton ratio, cf. also Remark 4.2.10. ◇

4.3 Consumption and Investment Problems

We consider now the following extension of the consumption problem of Example 2.1.4. Our investor has an initial wealth $x > 0$ and at the beginning of each of N periods she can decide how much of the wealth she consumes and how much she invests into the financial market given as in Section 4.2. In particular $\mathcal{F}_n := \mathcal{F}_n^S$. The amount c_n which is consumed at time n is evaluated by a utility function $U_c(c_n)$. The remaining wealth is invested in the risky assets and the riskless bond, and the terminal wealth X_N yields another utility $U_p(X_N)$. How should the agent consume and invest in order to maximize the sum of her expected utilities?
As in Section 4.2 we impose the Assumption (FM) on the financial market. Moreover, we assume that the utility functions U_c and U_p satisfy $\mathit{dom}\, U_c = \mathit{dom}\, U_p := [0, \infty)$. Analogously to (3.1) the wealth process (X_n) evolves as follows

$$X_{n+1} = (1 + i_{n+1})(X_n - c_n + \phi_n \cdot R_{n+1})$$

where $(c, \phi) = (c_n, \phi_n)$ is a consumption-investment strategy, i.e. (ϕ_n) and (c_n) are $\mathcal{F}_n$-adapted and $0 \le c_n \le X_n$.

The consumption-investment problem is then given by

$$\begin{cases} \mathbb{E}_x \left[\sum_{n=0}^{N-1} U_c(c_n) + U_p(X_N^{c,\phi}) \right] \to \max \\ (c,\phi) \text{ is a consumption-investment strategy with} \\ \qquad X_N^{c,\phi} \in dom\, U_p \ \mathbb{P}\text{-a.s.} \end{cases} \tag{4.16}$$

Problem (4.16) can be solved by the following Markov Decision Model (using the same arguments as in Section 4.2 for the terminal wealth problem):

- $E := [0, \infty)$ where $x \in E$ denotes the wealth,
- $A := \mathbb{R}_+ \times \mathbb{R}^d$ where $a \in \mathbb{R}^d$ is the amount of money invested in the risky assets and $c \in \mathbb{R}_+$ the amount which is consumed,
- $D_n(x)$ is given by

$$D_n(x) := \Big\{ (c,a) \in A \mid 0 \le c \le x \text{ and} \\ (1 + i_{n+1})(x - c + a \cdot R_{n+1}) \in E \ \mathbb{P}\text{-a.s.} \Big\},$$

- $\mathcal{Z} := [-1, \infty)^d$ where $z \in \mathcal{Z}$ denotes the relative risk,
- $T_n(x,c,a,z) := (1 + i_{n+1})(x - c + a \cdot z)$,
- $Q_n^Z(\cdot|x,c,a) :=$ distribution of R_{n+1} (independent of (x,c,a)),
- $r_n(x,c,a) := U_c(c)$,
- $g_N(x) := U_p(x)$.

The given consumption-investment problem (4.16) can now be solved by using the results of Section 2.3. The value functions are defined by

$$V_n(x) = \sup_\pi \mathbb{E}_{nx}^\pi \left[\sum_{k=n}^{N-1} U_c\big(c_n(X_n)\big) + U_p(X_N) \right]$$

where the supremum is taken over all policies $\pi = (f_0, \dots, f_{N-1})$ with $f_n(x) = \big(c_n(x), a_n(x)\big)$ and $V_0(x)$ is the value of the given problem (4.16). The Markov Decision Model has an upper bounding function $b(x) := 1 + x$ (cf. Proposition 4.2.1). In order to apply Theorem 2.3.8 we have to look at the following *one-period optimization problem* (cf. Section 4.1). For $x \in dom\, U_p$ consider

$$D(x) := \{(c,a) \in A \mid 0 \le c \le x \text{ and } (1+i)(x - c + a \cdot R) \in dom\, U_p \, \mathbb{P}\text{-a.s.}\},$$

$$u(x,c,a) := U_c(c) + \mathbb{E}\, U_p\big((1+i)(x - c + a \cdot R)\big)$$

and let

$$v(x) := \sup_{(c,a) \in D(x)} u(x,c,a).$$

Analogous to the pure investment problem we obtain:

Theorem 4.3.1. *Let U_c and U_p be utility functions with dom U_c = dom $U_p = [0, \infty)$. Then it holds:*

a) There are no arbitrage opportunities if and only if there exists a measurable function $f^* : \text{dom } U_p \to A$ such that

$$u\big(x, f^*(x)\big) = v(x), \quad x \in \text{dom } U_p.$$

b) $v(x)$ is strictly increasing, strictly concave and continuous on dom U_p.

The proof follows along the same lines as the proof of Theorem 4.1.1 and makes use of the iterated supremum

$$\sup_{(c,a)\in D(x)} u(x,c,a) = \sup_{0\le c\le x} \sup_{a\in A(c)} u(x,c,a)$$

with $A(c) := \{a \in \mathbb{R}^d \mid (1+i)(x-c+a\cdot R) \in \text{dom } U_p \, \mathbb{P}\text{-a.s.}\}$.

Remark 4.3.2. If dom U_c = dom $U_p = \mathbb{R}$ and U_c and U_p are bounded from above, then the statements in Theorem 4.3.1 are also true. ◊

Now we are able to state the solution of the consumption-investment problem.

Theorem 4.3.3. *For the multiperiod consumption-investment problem it holds:*

a) The value functions V_n are strictly increasing, strictly concave and continuous.
b) The value functions can be computed recursively by the Bellman equation

$$V_N(x) = U_p(x),$$
$$V_n(x) = \sup_{(c,a)\in D_n(x)} \Big\{U_c(c) + \mathbb{E}\, V_{n+1}\Big((1+i_{n+1})(x-c+a\cdot R_{n+1})\Big)\Big\}.$$

c) There exist maximizers f_n^* of V_{n+1} and the strategy $(f_0^*, f_1^*, \ldots, f_{N-1}^*)$ is optimal for the N-stage consumption-investment problem.

Proof. We show that the Structure Assumption (SA_N) is satisfied with

$$\mathbb{M}_n := \{v \in \mathbb{B}_b^+ \mid v \text{ is strictly increasing, strictly concave and continuous}\}$$

and $\Delta_n := F_n$. The statements then follow from Theorem 2.3.8.

(i) $g_N = U_p \in \mathbb{M}_N$ holds, since U_p is a utility function.

(ii) Now let $v \in I\!M_{n+1}$. Then

$$\mathcal{T}_n v(x) = \sup_{(c,a)\in D_n(x)} \Big\{U_c(c) + \mathbb{E}\, v\Big((1+i_{n+1})(x-c+a\cdot R_{n+1})\Big)\Big\},\ x \in E$$

and by Theorem 4.3.1 we obtain $\mathcal{T}_n v \in I\!M_n$.

(iii) The existence of maximizers follows from Theorem 4.3.1. □

Remark 4.3.4. The results of Theorem 4.3.3 also hold if the utility functions U_c and U_p satisfy $dom\, U_c = dom\, U_p = \mathbb{R}$ and U_c and U_p are bounded from above (e.g. exponential utilities).

Theorem 4.2.4 can be extended to the consumption-investment problem.

Theorem 4.3.5. *Let $\mathbb{E}\, R_n = 0$ for all $n = 1,\dots,N$. Then the optimal consumption-investment strategy $(f_0^*, f_1^*, \dots, f_{N-1}^*)$ is given by*

$$f_n^*(x) = (c_n^*(x), a_n^*(x)) \quad \textit{with } a_n^*(x) \equiv 0,$$

i.e. the strategy 'invest all the money in the bond' is the optimal investment strategy.

Proof. As in the proof of Theorem 4.2.4 we consider for $v \in I\!M_{n+1}$ the optimization problem

$$\begin{aligned}\mathcal{T}_n v(x) &= \sup_{(c,a)\in D_n(x)} \Big\{U_c(c) + \mathbb{E}\, v\big(\mathcal{T}_n(x,c,a,R_{n+1})\big)\Big\}\\ &\le \sup_{0\le c\le x} \Big\{U_c(c) + v\big((1+i_{n+1})(x-c)\big)\Big\}\end{aligned}$$

by using Jensen's inequality and $\mathbb{E}\, R_{n+1} = 0$. Thus $a_n^*(x) \equiv 0$. □

Power Utility

Let us assume that the utility functions are now given by

$$U_c(x) = U_p(x) = \frac{1}{\gamma}x^\gamma, \quad x \in [0,\infty)$$

with $0 < \gamma < 1$. Here we obtain the following general results for the consumption-investment problem.

Theorem 4.3.6. *a) The value functions are given by*

$$V_n(x) = d_n x^\gamma, \quad x \ge 0$$

where (d_n) satisfy the recursion

$$d_n^\delta = \gamma^{-\delta} + \big((1+i_{n+1})^\gamma v_n\big)^\delta d_{n+1}^\delta, \quad d_N = \frac{1}{\gamma}$$

with $\delta = (1-\gamma)^{-1}$ and where v_n is the value of problem (4.7).
b) The optimal consumption $c_n^*(x)$ is given by

$$c_n^*(x) = x(\gamma d_n)^{-\delta}, \quad x \ge 0$$

and the optimal amounts which are invested in the stocks are given by

$$a_n^*(x) = x\frac{(\gamma d_n)^\delta - 1}{(\gamma d_n)^\delta}\alpha_n^*, \quad x \ge 0$$

where α_n^* is the optimal solution of problem (4.7). The optimal consumption-investment strategy $(f_0^*, \dots, f_{N-1}^*)$ is defined by $f_n^* := (c_n^*, a_n^*)$, $n = 0, 1, \dots, N-1$.

Note that the optimal consumption and investment fractions are independent of the wealth. Moreover, since $a_n^*(x) = \alpha_n^*\big(x - c_n^*(x)\big)$, the optimal fractions of the remaining wealth after consumption which are invested in the assets are the same as in the problem (4.7) (without consumption). Since $d_n \ge d_{n+1}$ the optimal consumptions satisfy $c_n^*(x) \le c_{n+1}^*(x)$, in particular the optimal fractions of consumption are increasing in n.

Proof. Again we have to find sets $\mathbb{M}_n$ and Δ_n which satisfy the Structure Assumption (SA$_N$) of Theorem 2.3.8. We try

$$\mathbb{M}_n := \{v : E \to \mathbb{R}_+ \mid v(x) = bx^\gamma \text{ for } b > 0\}$$

$$\Delta_n := \{f \in F_n \mid f(x) = (\zeta x, \alpha x) \text{ for } \alpha \in \mathbb{R}^d, \zeta \in \mathbb{R}_+\}.$$

We check now

(i) $g_N = U_p$ is obviously in $\mathbb{M}_N$.
(ii) Let $v \in \mathbb{M}_{n+1}$. Then we obtain

$$\begin{aligned}\mathcal{T}_n v(x) &= \sup_{(c,a)\in D_n(x)} \left\{\frac{1}{\gamma}c^\gamma + b\,\mathbb{E}\left(\mathcal{T}_n\big(x,c,a,R_{n+1}\big)\right)^\gamma\right\} \\ &= \sup_{(c,a)\in D_n(x)} \left\{\frac{1}{\gamma}c^\gamma + b(1+i_{n+1})^\gamma\,\mathbb{E}\left(x - c + a\cdot R_{n+1}\right)^\gamma\right\}.\end{aligned}$$

If $x = 0$ then only $(c, a) = (0, 0)$ is admissible. Hence suppose $x > 0$. We use the transformation $c = \zeta x$ and $a = \alpha(x - c) = \alpha(1 - \zeta)x$ to obtain

$$\begin{aligned}
\mathcal{T}_n v(x) &= x^\gamma \sup_{0 \le \zeta \le 1} \left\{ \frac{1}{\gamma}\zeta^\gamma + b(1 + i_{n+1})^\gamma (1 - \zeta)^\gamma \sup_{\alpha \in A_n} \mathbb{E}\big(1 + \alpha \cdot R_{n+1}\big)^\gamma \right\} \\
&= x^\gamma \sup_{0 \le \zeta \le 1} \left\{ \frac{1}{\gamma}\zeta^\gamma + b(1 + i_{n+1})^\gamma v_n (1 - \zeta)^\gamma \right\}
\end{aligned}$$

where v_n is the value of problem (4.7). Thus, $\mathcal{T}_n v(x) = \tilde{b}x^\gamma \in \mathbb{M}_n$ with

$$\tilde{b} = \sup_{0 \le \zeta \le 1} \left\{ \frac{1}{\gamma}\zeta^\gamma + b(1 + i_{n+1})^\gamma v_n (1 - \zeta)^\gamma \right\}.$$

Note that $\tilde{b} > 0$.

(iii) For $v \in \mathbb{M}_{n+1}$ we have to prove the existence of a maximizer in the set Δ_n. Since the optimization problem in (ii) separates, it suffices to solve the consumption problem

$$\sup_{0 \le \zeta \le 1} \left\{ \frac{1}{\gamma}\zeta^\gamma + b(1 + i_{n+1})^\gamma v_n (1 - \zeta)^\gamma \right\}.$$

For the solution of this problem we recall the following general result: The optimization problem

$$\sup_{0 \le a \le 1} g(a) \tag{4.17}$$

where $g(a) := ba^\gamma + d(1-a)^\gamma$ with $b > 0, d > 0$, has the optimal solution:

$$\begin{aligned}
a^* &= \frac{b^\delta}{b^\delta + d^\delta} \\
g(a^*) &= (b^\delta + d^\delta)^{1-\gamma} = (b^\delta + d^\delta)^{\frac{1}{\delta}},
\end{aligned}$$

with $\delta = (1 - \gamma)^{-1}$. Note that a^* is unique and $0 < a^* < 1$. From (4.17) we conclude that a maximizer of $v \in \mathbb{M}_{n+1}$ exists in the set Δ_n.

The statements now follow from Theorem 2.3.8 by induction. From the preceding considerations we obtain the recursion

$$d_n = \sup_{0 \le \zeta \le 1} \left\{ \frac{1}{\gamma}\zeta^\gamma + d_{n+1}(1 + i_{n+1})^\gamma v_n (1 - \zeta)^\gamma \right\}$$

and by inserting the maximum point

$$d_n^\delta = \gamma^{-\delta} + \big((1 + i_{n+1})^\gamma v_n\big)^\delta d_{n+1}^\delta.$$

Finally we obtain from the optimization problem (4.17) that

$$c_n^*(x) = x(\gamma d_n)^{-\delta} \quad \text{and} \quad a_n^*(x) = \alpha_n^*(x - c_n^*(x)) = \alpha_n^*\big(1 - (\gamma d_n)^{-\delta}\big)x$$

which concludes the proof. □

Logarithmic Utility

Here we assume that both utility functions are of the form

$$U_c(x) = U_p(x) = \log x, \quad x > 0.$$

The proof of the following result is similar to the proof for the power utility.

Theorem 4.3.7. *a) The value functions are given by*

$$V_n(x) = (N - n + 1)\log x + d_n, \quad x > 0$$

with $d_n \in \mathbb{R}$.
b) The optimal consumption $c_n^*(x)$ is given by

$$c_n^*(x) := \frac{x}{N - n + 1}, \quad x > 0$$

and the optimal amounts which are invested in the stocks are given by

$$a_n^*(x) := x\frac{N - n}{N - n + 1}\alpha_n^*, \quad x > 0$$

where α_n^* is the optimal solution of (4.11). The optimal consumption-investment strategy $(f_0^*, \ldots, f_{N-1}^*)$ is defined by $f_n^* := (c_n^*, a_n^*)$ for $n = 0, 1, \ldots, N - 1$.

Proof. For the Structure Assumption (SA_N) we choose

$$\mathbb{M}_n := \{v : E \to \mathbb{R} \mid v(x) = (N - n + 1)\log x + b \ \text{ for } b \in \mathbb{R}\},$$

$$\Delta_n := \{f \in F_n \mid f(x) = (\zeta x, \alpha x) \text{ for } \alpha \in \mathbb{R}^d, \zeta \in \mathbb{R}_+\}.$$

We have to check the following conditions:

(i) Obviously $g_N(x) = \log x \in \mathbb{M}_N$.
(ii) Let $v \in \mathbb{M}_{n+1}$. Then we obtain

$$\begin{aligned}
\mathcal{T}_n v(x) &= \sup_{(c,a)\in D_n(x)} \big\{\log c + (N - n)\,\mathbb{E}\log\big(T_n(x, c, a, R_{n+1})\big) + b\big\} \\
&= \sup_{(c,a)\in D_n(x)} \Big\{ \log c + (N - n)\log(1 + i_{n+1}) + b \\
&\qquad\qquad + (N - n)\,\mathbb{E}\log\big(x - c + a \cdot R_{n+1}\big)\Big\}.
\end{aligned}$$

Again we use the transformation $c = \zeta x$ and $a = \alpha(x - c) = \alpha(1 - \zeta)x$ to obtain

$$\mathcal{T}_n v(x) = \sup_{0 \le \zeta \le 1} \Big\{ \log(\zeta x) + (N - n) \log\big((1 - \zeta)x\big) + (N - n)v_n \\ +(N - n) \log(1 + i_{n+1}) + b \Big\}$$

where v_n is the value of problem (4.11).
Thus, $\mathcal{T}_n v(x) = (N - n + 1) \log x + \tilde{b} \in I\!M_n$ with

$$\tilde{b} = (N-n)\Big(\log(N-n)+v_n+\log(1+i_{n+1})\Big)-(N-n+1)\log(N-n+1)+b.$$

(iii) The existence of a maximizer in the set Δ_n follows from the considerations in (ii). By induction we finally obtain

$$c_n^*(x) = \frac{x}{N - n + 1}$$
$$a_n^*(x) = \alpha_n^*\big(x - c_n^*(x)\big) = x\frac{N - n}{N - n + 1}\alpha_n^*$$

and the statements are shown. □

4.4 Optimization Problems with Regime Switching

A popular way of allowing random variations of the return distributions over time is to include an environment process in the form of a Markov chain which determines the return distributions. Sometimes these models are called *regime switching models* or *Markov-modulated models*. The idea is that a Markov chain can model the changing economic conditions which determine the distribution of the relative risk process. This underlying Markov chain can be interpreted as an environment process which collects relevant factors for the stock price dynamics like technical progress, political situations, law or natural catastrophes. Statistical investigations have shown a rather good fit of these kind of models.
In what follows we denote the external Markov chain by (Y_n) and call it the environment process. We assume that (Y_n) is observable by the agent and has finite state space E_Y. We consider a financial market with one riskless bond (with interest rate $i_n = i$) and d risky assets with relative risk process $(R_n) = (R_n^1, \dots, R_n^d)$. Here we assume that the distribution of R_{n+1} depends on Y_n. More precisely, we assume that (R_n, Y_n) is a stationary Markov process and that the following conditional independence holds:

$$\begin{aligned}
&\mathbb{P}\left(R_{n+1} \in B, Y_{n+1} = k | Y_n = j, R_n\right) \\
&= \mathbb{P}\left(R_{n+1} \in B, Y_{n+1} = k | Y_n = j\right) \\
&= \mathbb{P}\left(R_{n+1} \in B | Y_n = j\right) \cdot \mathbb{P}\left(Y_{n+1} = k | Y_n = j\right) =: Q_j(B) p_{jk}
\end{aligned} \tag{4.18}$$

for $B \in \mathcal{B}(\mathbb{R}^d), j, k \in E_Y$. The quantities p_{jk} are the transition probabilities of the Markov chain (Y_n) and Q_j is the distribution of R_{n+1} given $Y_n = j$ (independent of n). In the following, let $R(j)$ be a random variable with distribution Q_j, i.e. $\mathbb{P}(R(j) \in B) = Q_j(B) = \mathbb{P}(R_{n+1} \in B | Y_n = j)$. Given (Y_n), the random variables $R_1, R_2, \ldots$ are independent and given Y_n, the random variables R_{n+1} and Y_{n+1} are independent.
We investigate the consumption-investment problem as introduced in the last section. The filtration $(\mathcal{F}_n)$ to which portfolio strategies have to be adapted is here given by

$$\mathcal{F}_n := \sigma(S_0, \ldots, S_n, Y_0, \ldots, Y_n).$$

Note that $\mathcal{F}_n = \sigma(R_1, \ldots, R_n, Y_0, \ldots, Y_n)$. Throughout this section we impose the following assumption on the financial market (cf. Section 4.2).

Assumption (FM):

(i) *There are no arbitrage opportunities in the market, i.e. for all $j \in E_Y$ and all $\phi \in \mathbb{R}^d$ it holds:*

$$\phi \cdot R(j) \geq 0 \ \mathbb{P}\text{-a.s.} \quad \Rightarrow \quad \phi \cdot R(j) = 0 \ \mathbb{P}\text{-a.s.}$$

(ii) $\mathbb{E}\, \|R(j)\| < \infty$ *for all $j \in E_Y$.*

As in Remark 3.1.6 it is possible to show that the first assumption is indeed equivalent to the absence of arbitrage if all states of the Markov chain are reached with positive probability.
We assume again that the utility functions U_c and U_p satisfy *dom* $U_c =$ *dom* $U_p = [0, \infty)$. The wealth process (X_n) evolves as follows

$$X_{n+1} = (1+i)\Big(X_n - c_n + \phi_n \cdot R_{n+1}\Big)$$

where $(c, \phi) = (c_n, \phi_n)$ is a consumption-investment strategy, i.e. ϕ_n and c_n are $(\mathcal{F}_n)$-adapted and $0 \leq c_n \leq X_n$. The consumption-investment problem with regime switching is then defined for $(X_0, Y_0) = (x, j)$ by

$$\begin{cases} \mathbb{E}_{xj}\left[\sum_{n=0}^{N-1} \beta^n U_c(c_n) + \beta^N U_p(X_N^{c,\phi})\right] \to \max \\ (c, \phi) \text{ is a consumption-investment strategy with} \\ \qquad X_N^{c,\phi} \in dom\, U_p \ \mathbb{P}\text{-a.s.} \end{cases} \tag{4.19}$$

Here $\beta \in (0, 1]$ is a discount factor. Because of the Markovian structure of problem (4.19) it can be shown as in Section 4.2 that the optimal consumption-investment strategy belongs to the set of all Markov strategies (c_n, ϕ_n) where ϕ_n and c_n depend only on X_n and Y_n (see e.g. Theorem 2.2.3). Thus, problem (4.19) can be solved by the following stationary Markov Decision Model (cf. Section 4.3):

- $E := [0, \infty) \times E_Y$ where $(x, j) \in E$ denotes the wealth and environment state respectively,
- $A := \mathbb{R}_+ \times \mathbb{R}^d$ where $a \in \mathbb{R}^d$ is the amount of money invested in the risky assets and $c \in \mathbb{R}_+$ the amount which is consumed,
- $D(x, j)$ is given by

$$D(x,j) := \{(c,a) \in A \mid 0 \le c \le x \text{ and } (1+i)(x - c + a \cdot R(j)) \ge 0 \, \mathbb{P}\text{-a.s.}\},$$

- $\mathcal{Z} := [-1, \infty)^d \times E_Y$ where $(z, k) \in \mathcal{Z}$ denotes the relative risk and the new environment state,
- $T\big((x,j),(c,a),(z,k)\big) := \Big((1+i)\big(x - c + a \cdot z\big), k\Big)$,
- $Q^Z(B \times \{k\} | (x,j),(a,c)) := Q_j(B) p_{jk}$, for $j, k \in E_Y$ and $B \in \mathcal{B}(\mathbb{R}^d)$,
- $r\big((x,j),(c,a)\big) := U_c(c)$,
- $g(x,j) := U_p(x)$,
- $\beta \in (0,1]$.

Note that the disturbances are given by $Z_{n+1} = (R_{n+1}, Y_{n+1})$. The consumption-investment problem can now be solved by this stationary Markov Decision Model. The value functions are defined by

$$J_n(x,j) := \sup_\pi \mathbb{E}^\pi_{xj} \left[\sum_{k=0}^{n-1} \beta^k U_c\big(c_k(X_k, Y_k)\big) + \beta^n U_p(X_n) \right], \quad (x,j) \in E$$

where the supremum is taken over all policies $\pi = (f_0, \dots, f_{N-1})$ with

$$f_n(X_n, Y_n) := \big(c_n(X_n, Y_n), a_n(X_n, Y_n)\big).$$

The general solution of the consumption-investment problem with regime switching is contained in the next theorem.

Theorem 4.4.1. *a) The value functions $x \mapsto J_n(x,j)$ are strictly increasing, strictly concave and continuous.*

b) The value functions can be computed recursively by the Bellman equation

$$J_0(x,j) = U_p(x),$$
$$J_{n+1}(x,j) = \sup_{(a,c)} \Big\{ U_c(c) + \beta \sum_{k \in E_Y} p_{jk} \int J_n\Big((1+i)(x - c + a \cdot z), k\Big) Q_j(dz) \Big\}.$$

c) There exist maximizers f_n^* of J_{n-1} and $(f_N^*, \dots, f_1^*)$ is optimal for the N-stage consumption-investment problem.

Proof. It is easily shown by using part (ii) of Assumption (FM) that the stationary Markov Decision Model has an upper bounding function $b(x, j) := 1+x$ (cf. Proposition 4.2.1). Then the Structure Assumption (SA_N) is satisfied for the sets

$$\begin{aligned} I\!M := \big\{ & v \in I\!B_b^+ \mid x \mapsto v(x,j) \text{ is strictly increasing,} \\ & \text{strictly concave and continuous for all } j \in E_Y \big\} \end{aligned}$$

and $\Delta := F$. The proof follows along the same lines as the proof of Theorem 4.3.3. In particular the existence of a maximizer of $v \in I\!M$ follows from Assumption (FM) (cf. Theorem 4.3.1) since for $v \in I\!M$ the function

$$x \mapsto \beta \sum_{k \in E_Y} p_{jk} v(x, k)$$

is strictly increasing, concave and continuous for all j. The optimization problem at stage n then reduces to the problem in Theorem 4.3.1. Finally we obtain all statements from Theorem 2.3.8. □

Power Utility

Let us now assume that the consumption and the terminal wealth are evaluated by the power utility, i.e.

$$U_c(x) = U_p(x) = \frac{1}{\gamma} x^\gamma, \quad x \in [0, \infty)$$

with $0 < \gamma < 1$. In order to formulate the main result we consider the generic one-period optimization problem

$$v(j) := \sup_{\alpha \in A(j)} \mathbb{E}\Big(1 + \alpha \cdot R(j)\Big)^\gamma, \quad j \in E_Y \tag{4.20}$$

where $A(j) := \{\alpha \in \mathbb{R}^d \mid 1 + \alpha \cdot R(j) \geq 0 \ \mathbb{P}\text{-a.s.}\}$. There exists an optimal solution $\alpha^*(j)$ of problem (4.20) in view of (FM).

Theorem 4.4.2. *a) The value functions are given by*

$$J_n(x, j) = d_n(j) x^\gamma, \quad x \geq 0, \ j \in E_Y$$

where the $d_n(j) > 0$ satisfy the recursion

$$\begin{aligned} d_0(j) &= \gamma^{-1} \\ d_{n+1}^\delta(j) &= \gamma^{-\delta} + \big(\beta(1+i)^\gamma v(j)\big)^\delta \Big(\sum_{k \in E_Y} p_{jk} d_n(k) \Big)^\delta \end{aligned}$$

with $\delta := (1-\gamma)^{-1}$ *and* $v(j)$ *is the value of problem (4.20).*

b) The optimal consumption $c_n^*(x,j)$ *is given by*

$$c_n^*(x,j) = x\big(\gamma d_n(j)\big)^{-\delta}$$

and the optimal amounts which are invested in the stocks are given by

$$a_n^*(x,j) = \big(x - c_n^*(x,j)\big)\alpha^*(j)$$

where $\alpha^*(j)$ *is the optimal solution of (4.20). The optimal consumption-investment strategy* $(f_N^*, \dots, f_1^*)$ *is then defined by*

$$f_n^*(x,j) := \big(c_n^*(x,j), a_n^*(x,j)\big), \quad (x,j) \in E.$$

The proof follows along the same lines as the proof of Theorem 4.3.6. Note that the Markov Decision Model is stationary in this section.

Remark 4.4.3. If we consider the same optimization problem without consumption, we obtain

$$J_n(x,j) = d_n(j)x^\gamma, \quad (x,j) \in E$$

and the sequence $d_n(j) > 0$ is given recursively by $d_0(j) \equiv \gamma^{-1}$ and

$$d_{n+1}(j) = \beta(1+i)^\gamma v(j) \sum_{k \in E_Y} p_{jk} d_n(k), \quad j \in E_Y.$$

This result can be derived analogously to Theorem 4.4.2. ◊

In this model it is now rather interesting to do some sensitivity analysis with respect to the input parameters and in particular with respect to the environment process (Y_n). From now on we assume that we have only one stock and that the support of $R(j)$ is independent from j. Then $A(j) = \tilde{A}$ and the optimization problem (4.20) reduces to

$$v(j) := \sup_{\alpha \in \tilde{A}} \mathbb{E}\Big(1 + \alpha \cdot R(j)\Big)^\gamma, \quad j \in E_Y. \tag{4.21}$$

Again the optimal solution of (4.21) is denoted by $\alpha^*(j)$. In the following Lemma we use the increasing concave order for the return distributions (cf. Appendix B.3):

Theorem 4.4.4. *If* $Q_j \le_{icv} Q_k$*, then* $\alpha^*(j) \le \alpha^*(k)$*, i.e. the fraction which is invested in the stock in environment state* k *is larger than the invested fraction in environment state* j*.*

Proof. We can use Proposition 2.4.16 to prove the statement. In the proof of the previous theorem we have seen that the consumption and investment decision separates. Here we are only interested in the investment decision. Thus we have to show that

$$h(\alpha) := \int (1+\alpha y)^\gamma Q_k(dy) - \int (1+\alpha y)^\gamma Q_j(dy)$$

is increasing in $\alpha \in \tilde{A}$. This can be done by computing the derivative

$$h'(\alpha) = \gamma \int (1+\alpha y)^{\gamma-1} y Q_k(dy) - \gamma \int (1+\alpha y)^{\gamma-1} y Q_j(dy).$$

Then $h'(\alpha) \geq 0$ since $Q_j \leq_{icv} Q_k$ and

$$y \mapsto \frac{y}{(1+\alpha y)^{1-\gamma}}$$

is increasing and concave for $\alpha \in \tilde{A}$. □

For the next lemma we need the notion of stochastic monotonicity of the Markov chain (Y_n) which is explained in Appendix B.3. Moreover to simplify the notation we assume that $E_Y = \{1, \ldots, m\}$.

Theorem 4.4.5. *Let (Y_n) be stochastically monotone and suppose that $Q_1 \leq_{icv} Q_2 \leq_{icv} \cdots \leq_{icv} Q_m$.*

a) Then $J_n(x,j)$ is increasing in j and $c_n^*(x,j)$ is decreasing in j.
b) If $\alpha^*(j) \geq 0$ for $j \in E_Y$ then $a_n^*(x,j)$ is increasing in j.

Proof. a) According to Theorem 4.4.2 it suffices to show that $d_n(j)$ is increasing in j. For $n = 0$ this is clear. Now suppose the statement holds for n. Due to the recursion for $d_{n+1}(j)$ we have to show that

$$j \mapsto \int \left(1+\alpha^*(j)y\right)^\gamma Q_j(dy) \sum_{k=1}^m p_{jk} d_n(k) \tag{4.22}$$

is increasing in j. By the induction hypothesis and the assumption that (Y_n) is stochastically monotone we have that

$$j \mapsto \sum_{k=1}^m p_{jk} d_n(k)$$

is increasing in j. Moreover, the first factor in (4.22) is increasing in j, since $\alpha^*(j)$ is increasing by Theorem 4.4.4 and $y \mapsto \left(1+\alpha^*(j)y\right)^\gamma$ is increasing and concave. Note that both factors are non-negative.

b) By Theorem 4.4.2 we have

$$a_n^*(x,j) = \big(x - c_n^*(x,j)\big)\alpha^*(j).$$

Thus the statement follows from a). □

4.5 Portfolio Selection with Transaction Costs

We consider now the utility maximization problem of Section 4.2 under proportional transaction costs. For the sake of simplicity we restrict to one bond and one risky asset. If an additional amount of a (positive or negative) is invested in the stock, then proportional transaction costs of $c|a|$ are incurred which are paid from the bond position. We assume that $0 \le c < 1$. In order to compute the transaction costs, not only is the total wealth interesting, but also the allocation between stock and bond matters. Thus, in contrast to the portfolio optimization problems so far, the state space of the Markov Decision Model is two-dimensional and consists of the amounts held in the bond and in the stock. We assume that $x = (x_0, x_1) \in E := \mathbb{R}_+^2$ where x_0 and x_1 are the amounts held in the bond and stock respectively. Note that short-sellings are not allowed. The action space is $A := \mathbb{R}_+^2$ and $(a_0, a_1) \in A$ denotes the amount invested in the bond and stock respectively *after transaction.* Since transaction costs have to be subtracted we obtain the following set of admissible bond and stock holdings after transaction:

$$D(x_0, x_1) = \{(a_0, a_1) \in A \mid a_0 + a_1 \le x_0 + x_1 - c|a_1 - x_1|\}. \tag{4.23}$$

The independent disturbances (Z_n) are given by the relative price changes $(\tilde{R}_n)$ of the stock. Recall that $\tilde{R}_{n+1} = S_{n+1}/S_n$. Thus, the transition function at time n is given by

$$T_n\big(x, (a_0, a_1), z_{n+1}\big) := \Big(a_0(1 + i_{n+1}), a_1 z_{n+1}\Big). \tag{4.24}$$

The one-stage reward is $r_n \equiv 0$ and $g_N(x_0, x_1) := U(x_0 + x_1)$ where U is the utility function of the investor with $dom\, U = [0, \infty)$. The data of the Markov Decision Model is summarized as follows:

- $E := \mathbb{R}_+^2$ where $x = (x_0, x_1) \in E$ denotes the amount invested in bond and stock,
- $A := \mathbb{R}_+^2$ where $(a_0, a_1) \in A$ denotes the amount invested in bond and stock after transaction,
- $D(x)$ is given in (4.23),
- $\mathcal{Z} := \mathbb{R}_+$ where $z \in \mathcal{Z}$ denotes the relative price change of the stock,
- T_n is given by (4.24),
- $Q_n^Z(\cdot|x, a_0, a_1) :=$ the distribution of $\tilde{R}_{n+1}$ (independent of (x, a_0, a_1)),

- $r_n \equiv 0$,
- $g_N(x) := U(x_0 + x_1)$, $x = (x_0, x_1) \in E$.

The General Model with Transaction Costs

The terminal wealth problem with transaction costs is now

$$\sup_{\pi} \mathbb{E}_x^{\pi}\, U(X_N^0 + X_N^1) \tag{4.25}$$

where X_N^0 and X_N^1 are the terminal amounts in the bond and stock, respectively. Note that transaction costs are considered through the set of admissible actions. We make the following assumptions throughout this section.

Assumption (FM):

(i) *The utility function U is homogeneous of degree γ, i.e. $U(\lambda x) = \lambda^{\gamma} U(x)$ for all $\lambda > 0$ and $x \geq 0$.*
(ii) $\mathbb{E}\,\|\tilde{R}_n\| < \infty$ *for all $n = 1, \ldots, N$.*

Obviously condition (FM) (i) is fulfilled for the power utility (see also Remark 4.5.3). Note that we allow for arbitrage opportunities in the financial market. However since short-sellings are excluded, the set of admissible actions is compact and we will see that an optimal portfolio strategy exists. First we obtain:

Proposition 4.5.1. *The function $b(x) := 1 + x_0 + x_1, x \in E$ is an upper bounding function for the Markov Decision Model.*

Proof. We have to check conditions (i)–(iii) of Definition 2.4.1. Part (i) is obvious. Part (ii) follows since any concave function can be bounded from above by an affine-linear function and (iii) holds since

$$\mathbb{E}\, b\big(a_0(1 + i_{n+1}), a_1 \tilde{R}_{n+1}\big) = 1 + a_0(1 + i_{n+1}) + a_1\, \mathbb{E}\, \tilde{R}_{n+1} \leq d_0 b(x)$$

for all $(a_0, a_1) \in D(x)$ and for some $d_0 > 0$. □

In what follows we will call a function $v : E \to \mathbb{R}$ *homogeneous of degree* γ if it satisfies

$$v(\lambda x_0, \lambda x_1) = \lambda^{\gamma} v(x_0, x_1)$$

for all $\lambda > 0$ and $(x_0, x_1) \in E$. Let us consider the set

$$\mathbb{M} := \Big\{ v \in \mathbb{B}_b^+ \mid v \text{ is increasing, concave and homogeneous of degree } \gamma \Big\}$$

where increasing means that v is increasing in each component. For $v \in \mathbb{M}$ we have

$$\mathcal{T}_n v(x) = \sup_{(a_0,a_1)\in D(x)} \mathbb{E}\, v\big(a_0(1+i_{n+1}), a_1\tilde{R}_{n+1}\big).$$

Since v is increasing, the supremum is attained at the upper boundary of $D(x)$ (see Figure 4.2) which is given by the function (note that transaction costs are paid from the bond position)

$$h(x,a_1) := \begin{cases} x_0 + (1-c)(x_1 - a_1), & \text{if } 0 \le a_1 \le x_1 \\ x_0 + (1+c)(x_1 - a_1), & \text{if } x_1 < a_1 \le x_1 + \frac{x_0}{1+c}. \end{cases}$$

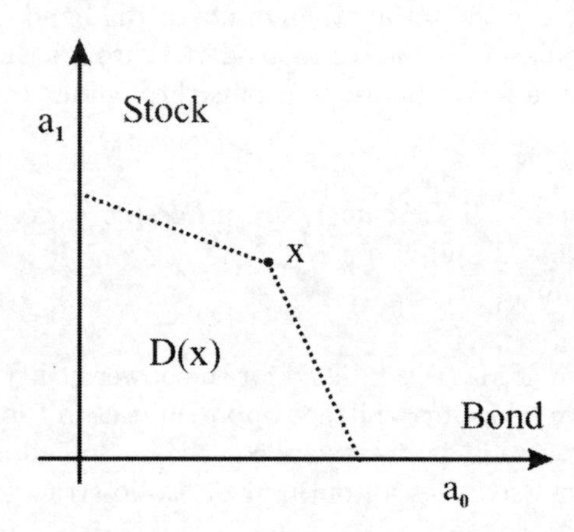

Fig. 4.2 Admissible stock and bond holdings after transaction.

Thus, we can also write

$$D(x_0,x_1) = \left\{(a_0,a_1) \in A \;\middle|\; a_0 \le h(x,a_1),\ 0 \le a_1 \le x_1 + \frac{x_0}{1+c}\right\}$$

and the operator $\mathcal{T}_n$ is given by

$$\mathcal{T}_n v(x) = \sup_{0\le a_1 \le x_1 + \frac{x_0}{1+c}} \mathbb{E}\, v\big(h(x,a_1)(1+i_{n+1}), a_1\tilde{R}_{n+1}\big), \quad x = (x_0,x_1) \in E.$$

When we have a maximum point a_1^* of this problem, then

$$f_n^1(x) = a_1^* \quad \text{and} \quad f_n^0(x) = h(x,a_1^*)$$

give the optimal amounts invested in the stock and the bond after transaction, respectively. In particular this means that it is enough to determine the amount invested in the stock after transaction and this is what we will do in the sequel. We denote this quantity by a instead of a_1.

Proposition 4.5.2. *The Structure Assumption* (SA_N) *is satisfied with the sets* $\mathbb{M}_n := \mathbb{M}$ *and* $\Delta_n := \Delta \cap F_n$ *where* Δ *consists of all functions* $f : E \to \mathbb{R}_+$ *such that there exist constants* $0 \le q_- \le q_+ \le \infty$ *and measurable functions* $f_+, f_- : E \to \mathbb{R}_+$ *with*

$$f(x_0, x_1) = \begin{cases} f_+(x_0, x_1), & \text{if } \frac{x_1}{x_0} > q_+ \\ x_1, & \text{if } q_- \le \frac{x_1}{x_0} \le q_+ \\ f_-(x_0, x_1), & \text{if } \frac{x_1}{x_0} < q_- \end{cases} \tag{4.26}$$

and $f_+(x_0, x_1) < x_1, f_-(x_0, x_1) > x_1$.

Note that $f(x_0, x_1) \in \Delta_n$ describes the amount invested in the stock if the state $(x_0, x_1) \in E$ is given. We will later show that the optimal policy is characterized by sell-, buy- and hold regions which are defined by cones.

Proof. Part (i) of (SA_N) is satisfied since $g_N(x) = U(x_0 + x_1) \in \mathbb{M}$. In what follows suppose $v \in \mathbb{M}$. For fixed n define

$$L(x, a) := \mathbb{E}\, v\big(h(x, a)(1 + i_{n+1}), a\tilde{R}_{n+1}\big), \quad 0 \le a \le x_1 + \frac{x_0}{1+c}.$$

Note that $a \mapsto L(x, a)$ is concave. We have to show that

$$\mathcal{T}_n v(x) = \sup_{0 \le a \le x_1 + \frac{x_0}{1+c}} L(x, a) \in \mathbb{M}$$

and that a maximizer exists. Let us denote the (largest) maximizer of the right-hand side by $f^*(x)$ (if it exists). The proof of the existence of $f^*(x)$ and its form is now structured by the following steps.

(i) For $\lambda > 0$ and $x \in E$ we have due to the homogeneity of v and the piecewise linearity of h:

$$\mathcal{T}_n v(\lambda x) = \sup_{0 \le a \le \lambda x_1 + \frac{\lambda x_0}{1+c}} L(\lambda x, a) = \lambda^\gamma \sup_{0 \le a' \le x_1 + \frac{x_0}{1+c}} L(x, a')$$

where $\lambda a' = a$. This implies

$$f^*(\lambda x) = \lambda f^*(x). \tag{4.27}$$

(ii) Next we consider the special state $x = (0, 1)$, i.e. one unit in the stock and nothing in the bond. Here we obtain

$$\mathcal{T}_n v(0, 1) = \sup_{0 \le a \le 1} \mathbb{E}\, v\big((1 - c)(1 - a)(1 + i_{n+1}), a\tilde{R}_{n+1}\big).$$

Since we have to maximize a concave function on the compact interval $[0, 1]$, a maximum point exists and we denote by a_+ the largest maximum point of this problem. Thus we have

$$f^*(0,1) = a_+.$$

Moreover, we define

$$q_+ := \begin{cases} \frac{a_+}{(1-c)(1-a_+)} & \text{if } a_+ < 1 \\ \infty & \text{if } a_+ = 1. \end{cases} \tag{4.28}$$

We can also write

$$q_+ = \frac{a_+}{h\big((0,1), a_+\big)}$$

which shows that q_+ is the optimal stock to bond ratio after transaction in state $x = (0,1)$.

(iii) We claim now that for any $x \in E$ with $\frac{x_1}{x_0} > q_+$ it holds that

$$f^*(x) = \Big(x_1 + \frac{x_0}{1-c}\Big)a_+ = \frac{x_0 + (1-c)x_1}{1 + (1-c)q_+} q_+ < x_1$$

and $\big(h(x, f^*(x)), f^*(x)\big) \in L_1 := \{x \in \mathbb{R}^2_+ \mid x_1 = q_+ x_0\}$.

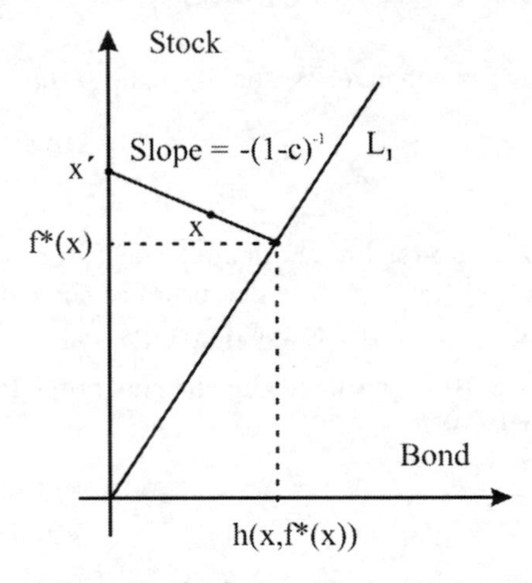

Fig. 4.3 Construction of the optimal solution.

The proof is as follows: Consider the point $x' := \big(0, x_1 + \frac{x_0}{1-c}\big)$. Due to property (4.27) we obtain

$$f^*(x') = x_1' f^*(0,1) = x_1' a_+ = \big(x_1 + \frac{x_0}{1-c}\big)a_+.$$

We will show that $f^*(x)$ and $f^*(x')$ are equal. For this instance, note first that $\frac{x_1}{x_0} > q_+$ implies $f^*(x') < x_1$ and that we have

$$h(x', a) = x_0 + (1-c)(x_1 - a)$$

on the interval $0 \le a \le x_1 + \frac{x_0}{1-c}$. On the other hand we have for $x \in E$:

$$h(x,a) = \begin{cases} x_0 + (1-c)(x_1 - a), & \text{if } 0 \le a \le x_1 \\ x_0 + (1+c)(x_1 - a), & \text{if } x_1 < a \le x_1 + \frac{x_0}{1+c}. \end{cases}$$

Thus, $h(x',a) = h(x,a)$ on $0 \le a \le x_1$ and $h(x',a) \ge h(x,a)$ on $x_1 < a \le x_1 + \frac{x_0}{1+c}$. Since $L(x',a)$ attains its maximum point on the interval $0 \le a \le x_1$ and due to the fact that v is increasing we obtain

$$f^*(x) = f^*(x') = \big(x_1 + \frac{x_0}{1-c}\big)a_+$$

which implies the first part of the statement.
The fact that $\big(h(x,f^*(x)), f^*(x)\big) \in L_1$ can be verified by directly inserting the expression for $f^*(x)$.
Note that $\big(h(x,f^*(x)), f^*(x)\big)$ is in the intersection of L_1 and the line through x and x' (see Figure 4.3).

(iv) As in step (ii) we consider the special state $x = (1,0)$, i.e. one unit in the bond and nothing in the stock. Here we obtain

$$\mathcal{T}_n v(1,0) = \sup_{0 \le a \le \frac{1}{1+c}} \mathbb{E}\, v\big((1-(1+c)a)(1+i_{n+1}), a\tilde{R}_{n+1}\big).$$

We denote by a_- the largest maximum point of this problem. Thus we have $f^*(1,0) = a_-$. Moreover, we define

$$q_- := \begin{cases} \frac{a_-}{1-(1+c)a_-}, & \text{if } a_- < \frac{1}{1+c} \\ \infty, & \text{if } a_- = \frac{1}{1+c}. \end{cases} \tag{4.29}$$

We can also write

$$q_- = \frac{a_-}{h\big((1,0),a_-\big)}$$

which shows that q_- is the optimal stock to bond ratio after transaction.

(v) As in step (iii) we can show that for any $x \in E$ with $\frac{x_1}{x_0} < q_-$ it holds that

$$f^*(x) = \Big(x_0 + (1+c)x_1\Big)a_- = \frac{x_0 + (1+c)x_1}{1+(1+c)q_-} q_- > x_1$$

and $\big(h(x,f^*(x)), f^*(x)\big) \in L_2 := \{x \in \mathbb{R}^2_+ \mid x_1 = q_- x_0\}$.

(vi) We show that $q_- \le q_+$. Suppose this is not the case, then there exists an $x \in E$ such that

$$q_+ < \frac{x_1}{x_0} < q_-.$$

But following (iii) and (v) this would imply that $f^*(x) < x_1 < f^*(x)$ which is a contradiction. Thus, we must have $q_- \le q_+$.

(vi) Finally we have to investigate the case $x \in E$ and $q_- < \frac{x_1}{x_0} < q_+$. Using similar arguments as before we obtain here that $f^*(x) = x_1$, i.e. no transaction is done.

All parts together prove that a maximizer $f^*(x)$ exists and is of the form (4.26). It remains to show that $\mathcal{T}_n v \in I\!M$. Theorem 2.4.14 implies that $\mathcal{T}_n v$ is increasing, since $D(x) \subset D(x')$ for $x \le x'$, $g_N(x) = U(x_0 + x_1)$ is increasing by assumption and $x \mapsto L(x,a)$ is increasing for all a. The concavity of $\mathcal{T}_n v$ can be shown with Theorem 2.4.19. For the homogeneity, note that the considerations in (i)–(vi) imply that

$$\mathcal{T}_n v(x_0, x_1) = \begin{cases} \mathbb{E}\, v\Big(f^*(x) q_+^{-1}(1 + i_{n+1}), f^*(x)\tilde{R}_{n+1}\Big), & \text{if } \frac{x_1}{x_0} > q_+ \\ \mathbb{E}\, v\Big(x_0(1 + i_{n+1}), x_1 \tilde{R}_{n+1}\Big), & \text{if } q_- \le \frac{x_1}{x_0} \le q_+ \\ \mathbb{E}\, v\Big(f^*(x) q_-^{-1}(1 + i_{n+1}), f^*(x)\tilde{R}_{n+1}\Big), & \text{if } \frac{x_1}{x_0} < q_- \end{cases}$$

with $f^*(x_0, x_1) = (x_1 + \frac{x_0}{1-c})a_+$ in the first case and in the second case $f^*(x_0, x_1) = (x_0 + (1+c)x_1)a_-$. Since v is homogeneous of degree γ, we obviously obtain that $\mathcal{T}_n v$ is homogeneous of degree γ. □

Remark 4.5.3. There is another version of Proposition 4.5.2. Assume that the given utility function $U : (0, \infty) \to \mathbb{R}$ is differentiable and the derivative of U is homogeneous of degree γ. The power and logarithmic utility functions satisfy this assumption. Then Proposition 4.5.2 remains valid with the same Δ_n and

$$I\!M := \{v \in I\!B_b^+ \mid v \text{ is increasing, concave, differentiable and the gradient of } v \text{ is homogeneous of degree } \gamma\}.$$

◊

The next theorem summarizes the result for the terminal wealth problem with transaction costs. In order to state it we define for $v \in I\!M$

$$q_+(v) := \text{argmax}_{q \ge 0}\, \mathbb{E}\, v\Big(\frac{1 + i_{n+1}}{1 + q(1-c)}, \frac{q\tilde{R}_{n+1}}{1 + q(1-c)}\Big) \tag{4.30}$$

$$q_-(v) := \text{argmax}_{q \ge 0}\, \mathbb{E}\, v\Big(\frac{1 + i_{n+1}}{1 + q(1+c)}, \frac{q\tilde{R}_{n+1}}{1 + q(1+c)}\Big). \tag{4.31}$$

With the help of these quantities the optimal investment strategy can be characterized by three regions: a 'buy' region, a 'sell' region and a region of 'no transaction'.

Theorem 4.5.4. *For the terminal wealth problem with transaction costs it holds:*

a) *The value functions V_n are concave, increasing and homogeneous of degree γ and for $x = (x_0, x_1) \in E$ given by*

$$V_N(x) = U(x_0 + x_1)$$
$$V_n(x) = \sup_{0 \le a \le x_1 + \frac{x_0}{1+c}} \mathbb{E}\, V_{n+1}\big(h(x,a)(1+i_{n+1}), a\tilde{R}_{n+1}\big).$$

b) *The optimal amount invested in the stock at time n, is given by*

$$f_n^*(x) = \begin{cases} \frac{x_0+(1-c)x_1}{1+(1-c)q_+(V_{n+1})} q_+(V_{n+1}), & \text{if } \frac{x_1}{x_0} > q_+(V_{n+1}) \\ x_1, & \text{if } q_-(V_{n+1}) \le \frac{x_1}{x_0} \le q_+(V_{n+1}) \\ \frac{x_0+(1+c)x_1}{1+(1+c)q_-(V_{n+1})} q_-(V_{n+1}), & \text{if } \frac{x_1}{x_0} < q_-(V_{n+1}) \end{cases} \tag{4.32}$$

and the optimal amount invested in the bond at time n is equal to $h(x, f_n^*(x))$, $x \in E$.

Proof. The proof follows directly from Proposition 4.5.2 and Theorem 2.3.8. Note that for $v \in \mathbb{M}$ the optimization problem

$$\sup_{q \ge 0} \mathbb{E}\, v\Big(\frac{1+i_{n+1}}{1+q(1-c)}, \frac{q\tilde{R}_{n+1}}{1+q(1-c)}\Big)$$

is equivalent to

$$\sup_{0 \le a \le 1} \mathbb{E}\, v\big((1-c)(1-a)(1+i_{n+1}), a\tilde{R}_{n+1}\big)$$

when we use the transformation

$$q = \frac{a}{(1-c)(1-a)} \Leftrightarrow \frac{1}{1+q(1-c)} = 1-a \Leftrightarrow \frac{q}{1+q(1-c)} = \frac{a}{1-c}.$$

An analogous statement holds for q_-. □

The optimal investment policy from Theorem 4.5.4 has the following interesting properties:

(i) It holds that

$$\frac{f_n^*(x)}{h(x, f_n^*(x))} = \begin{cases} q_+(V_{n+1}), & \text{if } \frac{x_1}{x_0} > q_+(V_{n+1}) \\ \frac{x_1}{x_0}, & \text{if } q_-(V_{n+1}) \le \frac{x_1}{x_0} \le q_+(V_{n+1}) \\ q_-(V_{n+1}), & \text{if } \frac{x_1}{x_0} < q_-(V_{n+1}) \end{cases} \tag{4.33}$$

i.e. the optimal stock to bond ratio after transaction is $q_+(V_{n+1})$ if $\frac{x_1}{x_0} > q_+(V_{n+1})$ and $q_-(V_{n+1})$ if $\frac{x_1}{x_0} < q_-(V_{n+1})$. If the stock to bond ratio is between the two levels, then no transaction is taken. This is illustrated in Figure 4.4. In other words, if the current stock to bond holding

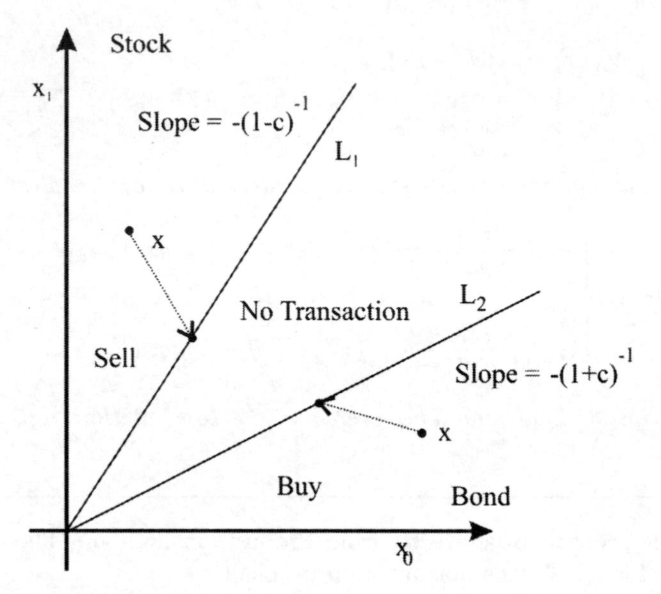

Fig. 4.4 Optimal trading regions.

$\frac{x_1}{x_0}$ is less than $q_-(V_{n+1})$, then purchase stocks until the stock to bond holding equals $q_-(V_{n+1})$. If the current stock to bond holding is greater than $q_+(V_{n+1})$, then sell stocks until the stock to bond holding equals $q_+(V_{n+1})$. If the current stock to bond holding lies between the two limits, do not trade.

(ii) In case we have no transaction costs, i.e. $c = 0$, we obtain $q_+(v) = q_-(v) = \frac{\alpha_n^*}{1-\alpha_n^*}$ where α_n^* is given by (4.7) in case $0 < \alpha_n^* < 1$. Note that in Section 4.2 we do not have excluded short-sellings.

Binomial Model with Transaction Costs

Finally let us compute the optimal portfolio strategy in a one-period binomial model explicitly (i.e. $N = 1$). Here we use the power utility function $U(x) = \frac{1}{\gamma}x^\gamma$ with $0 < \gamma < 1$ and we skip the time index on the bond and stock return rates to ease notation. Note that $\tilde{R}$ takes only the two values $\boldsymbol{u}$ and $\boldsymbol{d}$ with probability p and $(1-p)$. Moreover, we assume that

$$\boldsymbol{d} < (1+i)(1-c) =: c_-, \qquad c_+ := (1+i)(1+c) < \boldsymbol{u}.$$

We have to derive the maximum points $q_+ := q_+(v)$ and $q_- := q_-(v)$ in (4.30) and (4.31), respectively for $v(x_0, x_1) := \frac{1}{\gamma}(x_0 + x_1)^\gamma, (x_0, x_1) \in E$. Thus, we have to determine

$$q_+ := \operatorname{argmax}_{q \ge 0} \; p\Big(\frac{(1+i) + q\boldsymbol{u}}{1 + q(1-c)}\Big)^\gamma + (1-p)\Big(\frac{(1+i) + q\boldsymbol{d}}{1 + q(1-c)}\Big)^\gamma$$
$$q_- := \operatorname{argmax}_{q \ge 0} \; p\Big(\frac{(1+i) + q\boldsymbol{u}}{1 + q(1+c)}\Big)^\gamma + (1-p)\Big(\frac{(1+i) + q\boldsymbol{d}}{1 + q(1+c)}\Big)^\gamma.$$

We obtain by setting the derivative with respect to q equal to zero that

$$q_- = \Big(\frac{(1+i)(1 - M_-)}{\boldsymbol{u}M_- - \boldsymbol{d}}\Big)^+$$

with

$$M_- := \left(\frac{(1-p)(c_+ - \boldsymbol{d})}{p(\boldsymbol{u} - c_+)}\right)^\delta$$

and $\delta = (1-\gamma)^{-1}$. Analogously we obtain

$$q_+ = \Big(\frac{(1+i)(1 - M_+)}{\boldsymbol{u}M_+ - \boldsymbol{d}}\Big)^+$$

with

$$M_+ := \left(\frac{(1-p)(c_- - \boldsymbol{d})}{p(\boldsymbol{u} - c_-)}\right)^\delta.$$

It is not difficult to see that $q_- < q_+$ and that q_+ is increasing in c and q_- is decreasing in c, i.e. if the transaction costs get large, the region of no transaction also gets larger. Depending on p the sell-, buy- and no transaction regions have different sizes. In order to discuss this issue, we introduce the following definitions:

$$p_-^1 := \frac{c_- - \boldsymbol{d}}{\boldsymbol{u} - \boldsymbol{d}}$$
$$p_+^1 := \frac{c_+ - \boldsymbol{d}}{\boldsymbol{u} - \boldsymbol{d}}$$

$$p_-^2 := \frac{\boldsymbol{u}^{1-\gamma}(c_- - \boldsymbol{d})}{\boldsymbol{u}^{1-\gamma}(c_- - \boldsymbol{d}) + \boldsymbol{d}^{1-\gamma}(\boldsymbol{u} - c_-)}$$
$$p_+^2 := \frac{\boldsymbol{u}^{1-\gamma}(c_+ - \boldsymbol{d})}{\boldsymbol{u}^{1-\gamma}(c_+ - \boldsymbol{d}) + \boldsymbol{d}^{1-\gamma}(\boldsymbol{u} - c_+)}.$$

Note that we always have $p_-^1 \le p_+^1$ and $p_-^2 < p_+^2$. Moreover, we assume that

$$(c_+ - \boldsymbol{d})(\boldsymbol{u} - c_-)\boldsymbol{d}^{1-\gamma} \le \boldsymbol{u}^{1-\gamma}(c_- - \boldsymbol{d})(\boldsymbol{u} - c_+)$$

which is certainly satisfied for small c. This implies $p_+^1 \le p_-^2$. Altogether we have

$$p_-^1 \le p_+^1 \le p_-^2 < p_+^2.$$

We distinguish the following cases

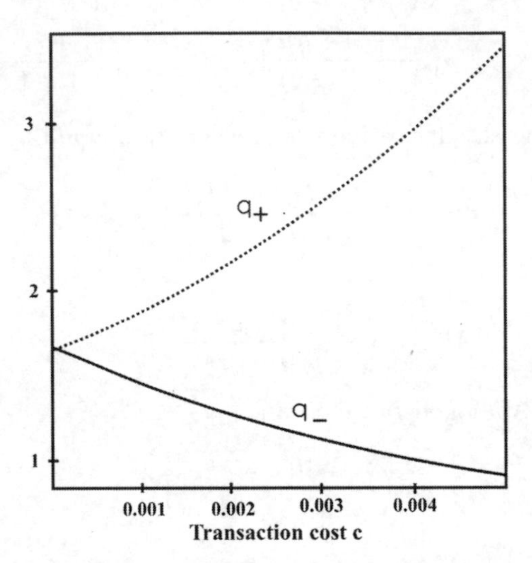

Fig. 4.5 Limits q_+ and q_- as a function of the transaction costs c.

(i) $p \le p_-^1$: In this case $q_+ = 0$ and there is only a sell region.
(ii) $p_-^1 < p \le p_+^1$: Here $q_+ \in (0, \infty)$ and $q_- = 0$, i.e. there are only a sell and a no transaction region.
(iii) $p_+^1 < p < p_-^2$: Here $q_+, q_- \in (0, \infty)$ and there are all three regions.
(iv) $p_-^2 \le p < p_+^2$: Here $q_+ = \infty$ and $q_- \in (0, \infty)$, i.e. there are only a buy and a no transaction region.
(v) $p \ge p_+^2$: Here $q_- = \infty$ and there is only a buy region.

Note that in case $c_+ \ge \boldsymbol{u}$ we always have $p_+^1 \ge 1$ and thus the optimal strategy is not to buy stocks at all. The intuition is as follows: An increase of 1 Euro in the stock yields a $(1+c)$ Euro reduction in the bond. The marginal investment in the stock yields either $\boldsymbol{u}$ Euro or $\boldsymbol{d}$ Euro which is in each case not better than the return c_+ we get from a $(1+c)$ Euro investment in the bond. If $\boldsymbol{d} \ge c_-$, then $p_-^2 \le 0$ and a similar argument yields that it is optimal not to buy bonds at all.
Figure 4.5 shows the limits q_+ and q_- as a function of the transaction costs c. As data we have chosen $p = 0.55, \boldsymbol{u} = 1.2, i = 0, \boldsymbol{d} = 0.8$ and $\gamma = 0.2$. Note that in case $c = 0$, i.e. we have no transaction costs, $q_+ = q_- = \frac{\alpha^*}{1-\alpha^*} = 1.658$ (cf. (4.9)).

4.6 Dynamic Mean-Variance Problems

An alternative approach towards finding an optimal investment strategy was introduced by Markowitz in 1952 and indeed a little bit earlier by de Finetti. In contrast to utility functions the idea is now to measure the risk by the portfolio variance and incorporate this measure as follows: Among all portfolios which yield at least a certain expected return (benchmark), choose the one with smallest portfolio variance. The single-period problem was solved in the 1950s. It still has great importance in real-life applications and is widely applied in risk management departments of banks. The problem of multiperiod portfolio-selection was proposed in the late 1960s and early 1970s and has been solved recently. The difficulty here is that the original formulation of the problem involves a side constraint. However, this problem can be transformed into one without constraint by the Lagrange multiplier technique. Then we solve this stochastic Lagrange problem by a suitable Markov Decision Model. We use the same non-stationary financial market as in Section 4.2 with independent relative risk variables. Our investor has initial wealth $x_0 > 0$. This wealth can be invested into d risky assets and one riskless bond. How should the agent invest over N periods in order to find a portfolio with minimal variance which yields at least an expected return of μ?
For the mathematical formulation of the problem we restrict without loss of generality to Markov portfolio strategies and consider a non-stationary Markov Decision Model with the following data (where r_n and g_N will be specified later):

- $E := \mathbb{R}$ where $x \in E$ denotes the wealth,
- $A := \mathbb{R}^d$ where $a \in A$ is the amount of money invested in the risky assets,
- $D_n(x) := A$,
- $\mathcal{Z} := [-1, \infty)^d$ where $z \in \mathcal{Z}$ denotes the relative risk,
- $T_n(x, a, z) := (1 + i_{n+1})(x + a \cdot z)$,
- $Q_n^Z(\cdot|x, a) :=$ distribution of R_{n+1} (independent of (x, a)).

The mean-variance problem is given by

$$(MV) \quad \begin{cases} \operatorname{Var}_{x_0}^{\pi}[X_N] \to \min \\ \mathbb{E}_{x_0}^{\pi}[X_N] \geq \mu \\ \pi \in F^N. \end{cases}$$

We will also introduce the problem where the expected return has to be *equal* to μ:

$$(MV_=) \quad \begin{cases} \operatorname{Var}_{x_0}^{\pi}[X_N] \to \min \\ \mathbb{E}_{x_0}^{\pi}[X_N] = \mu \\ \pi \in F^N. \end{cases}$$

In order to obtain a well-defined problem we make the following assumptions throughout this section.

Assumption (FM):

(i) *$\mathbb{E}\,\|R_n\| < \infty$ and $\mathbb{E}\,R_n \neq 0$ for all $n = 1, \dots, N$.*
(ii) *The covariance matrix of the relative risk process*

$$\left(\mathrm{Cov}(R_n^j, R_n^k)\right)_{1 \le j,k \le d}$$

is positive definite for all $n = 1, \dots, N$.
(iii) *$x_0 S_N^0 < \mu$.*

If $\mathbb{E}\,R_n = 0$ for all $n = 1, \dots, N$ then (X_n / S_n^0) is a martingale and there exists no strategy for (MV) with $\mathbb{E}_{x_0}^{\pi}[X_N] \ge \mu$ in view of assumption $x_0 S_N^0 < \mu$. Since $\mathrm{Var}[a \cdot R_n] = a\mathrm{Cov}(R_n^i, R_n^j)a^\top$ assumption (ii) means that the financial market is *non-redundant*, i.e. there does not exist an asset which can be replicated by the others. Assumption (iii) excludes a trivial case: If (iii) is not satisfied, then an expected return of at least μ can be achieved by investing in the riskless asset only. This strategy of course has a variance of zero and is thus optimal. Note that we do not exclude arbitrage opportunities here. The next lemma reveals that problems (MV) and ($MV_=$) are indeed equivalent.

Lemma 4.6.1. *A strategy π^* is optimal for (MV) if and only if π^* is optimal for ($MV_=$).*

Proof. Suppose that π^* is optimal for (MV). We will then show that necessarily $\mathbb{E}_{x_0}^{\pi^*}[X_N] = \mu$ by contradiction. Suppose that $\mathbb{E}_{x_0}^{\pi^*}[X_N] > \mu$. Define $\hat{\pi} := \alpha \pi^* = (\alpha f_0^*, \dots, \alpha f_{N-1}^*)$ with

$$\alpha := \frac{\mu - x_0 S_N^0}{\mathbb{E}_{x_0}^{\pi^*}[X_N] - x_0 S_N^0}.$$

Obviously $\hat{\pi} \in F^N$ and $0 < \alpha < 1$. Since the wealth process of $\hat{\pi}$ satisfies

$$\begin{aligned}
X_{n+1} &= (1 + i_{n+1})\Big(X_n + \alpha f_n^*(X_n) \cdot R_{n+1}\Big) \\
&= (1 + i_{n+1})\Big((1 + i_n)\big(X_{n-1} + \alpha f_{n-1}^*(X_{n-1}) \cdot R_n\big) + \alpha f_n^*(X_n) \cdot R_{n+1}\Big) \\
&= S_{n+1}^0 x_0 + \alpha \sum_{j=0}^{n} \prod_{k=j+1}^{n+1} (1 + i_k) f_j^*(X_j) \cdot R_{j+1}
\end{aligned}$$

we obtain

$$\mathbb{E}_{x_0}^{\hat{\pi}}[X_N] = S_N^0 x_0 + \alpha \mathbb{E}_{x_0}^{\hat{\pi}} \left[\sum_{j=0}^{N-1} \prod_{k=j+1}^{N} (1 + i_k) f_j^*(X_j) \cdot R_{j+1} \right] = \mu$$

and

$$\mathrm{Var}_{x_0}^{\hat{\pi}}[X_N] = \alpha^2 \mathrm{Var}_{x_0}^{\pi^*}[X_N] < \mathrm{Var}_{x_0}^{\pi^*}[X_N]$$

which is a contradiction to the optimality of π^*. Thus we necessarily have $\mathbb{E}_{x_0}^{\pi^*}[X_N] = \mu$. This observation now directly implies the statement of the Lemma. □

Problem (MV) can be solved via the well-known Lagrange multiplier technique. Let $L_{x_0}(\pi, \lambda)$ be the Lagrange-function, i.e.

$$L_{x_0}(\pi, \lambda) := \mathrm{Var}_{x_0}^{\pi}[X_N] + 2\lambda\big(\mu - \mathbb{E}_{x_0}^{\pi}[X_N]\big) \quad \text{for } \pi \in F^N, \lambda \geq 0.$$

As usual, (π^*, λ^*) is called a *saddle-point* of the Lagrange-function $L_{x_0}(\pi, \lambda)$ if

$$\sup_{\lambda \geq 0} L_{x_0}(\pi^*, \lambda) = L_{x_0}(\pi^*, \lambda^*) = \inf_{\pi \in F^N} L_{x_0}(\pi, \lambda^*).$$

Lemma 4.6.2. *Let (π^*, λ^*) be a saddle-point of $L_{x_0}(\pi, \lambda)$. Then the value of (MV) is given by*

$$\inf_{\pi \in F^N} \sup_{\lambda \geq 0} L_{x_0}(\pi, \lambda) = \sup_{\lambda \geq 0} \inf_{\pi \in F^N} L_{x_0}(\pi, \lambda) = L_{x_0}(\pi^*, \lambda^*)$$

and π^* is optimal for (MV).

Proof. Obviously the value of (MV) is equal to $\inf_{\pi \in F^N} \sup_{\lambda \geq 0} L_{x_0}(\pi, \lambda)$ and

$$\inf_{\pi \in F^N} \sup_{\lambda \geq 0} L_{x_0}(\pi, \lambda) \geq \sup_{\lambda \geq 0} \inf_{\pi \in F^N} L_{x_0}(\pi, \lambda).$$

For the reverse inequality we obtain

$$\begin{aligned} \inf_{\pi \in F^N} \sup_{\lambda \geq 0} L_{x_0}(\pi, \lambda) &\leq \sup_{\lambda \geq 0} L_{x_0}(\pi^*, \lambda) = L_{x_0}(\pi^*, \lambda^*) \\ &= \inf_{\pi \in F^N} L_{x_0}(\pi, \lambda^*) \leq \sup_{\lambda \geq 0} \inf_{\pi \in F^N} L_{x_0}(\pi, \lambda), \end{aligned}$$

and the first statement follows. Further from the definition of a saddle-point we obtain for all $\lambda \geq 0$

$$\lambda^*\big(\mu - \mathbb{E}_{x_0}^{\pi^*}[X_N]\big) \geq \lambda\big(\mu - \mathbb{E}_{x_0}^{\pi^*}[X_N]\big),$$

and hence $\mathbb{E}_{x_0}^{\pi^*}[X_N] \geq \mu$. Then we conclude $L_{x_0}(\pi^*, \lambda^*) = Var_{x_0}^{\pi^*}[X_N]$ and π^* is optimal for (MV). □

From Lemma 4.6.2 we see that it is sufficient to look for a saddle point (π^*, λ^*) of $L_{x_0}(\pi, \lambda)$. It is not difficult to see that the pair (π^*, λ^*) is a saddle-point if $\lambda^* > 0$ and $\pi^* = \pi^*(\lambda^*)$ satisfy

$$\pi^* \text{ is optimal for } P(\lambda^*) \text{ and } \mathbb{E}_{x_0}^{\pi^*}[X_N] = \mu.$$

Here, $P(\lambda)$ denotes the so-called *Lagrange-problem* for the parameter $\lambda > 0$

$$P(\lambda) \begin{cases} L_{x_0}(\pi, \lambda) \to \min \\ \\ \pi \in F^N. \end{cases}$$

Note that the problem $P(\lambda)$ is not a standard Markov Decision Problem and is hard to solve directly due to the variance in the objective function (which is not separable). We embed the problem $P(\lambda)$ into a tractable auxiliary problem $QP(b)$ that turns out to be a stochastic LQ-problem (cf. Section 2.6.3). For $b \in \mathbb{R}$ define

$$QP(b) \begin{cases} \mathbb{E}_{x_0}^{\pi}\left[(X_N - b)^2\right] \to \min \\ \\ \pi \in F^N. \end{cases}$$

The following result shows the relationship between the problems $P(\lambda)$ and $QP(b)$.

Lemma 4.6.3. *If π^* is optimal for $P(\lambda)$, then π^* is optimal for $QP(b)$ with $b := \mathbb{E}_{x_0}^{\pi^*}[X_N] + \lambda$.*

Proof. Suppose π^* is not optimal for $QP(b)$ with $b := \mathbb{E}_{x_0}^{\pi^*}[X_N] + \lambda$. Then there exists $\pi \in F^N$ such that

$$\mathbb{E}_{x_0}^{\pi}[X_N^2] - 2b\,\mathbb{E}_{x_0}^{\pi}[X_N] < \mathbb{E}_{x_0}^{\pi^*}[X_N^2] - 2b\,\mathbb{E}_{x_0}^{\pi^*}[X_N].$$

Define the function $U : \mathbb{R}^2 \to \mathbb{R}$ by

$$U(x, y) := y - x^2 + 2\lambda(\mu - x).$$

Then U is concave and $U(x, y) = L_{x_0}(\pi, \lambda)$ for $x := \mathbb{E}_{x_0}^{\pi}[X_N]$ and $y := \mathbb{E}_{x_0}^{\pi}[X_N^2]$. Moreover, we set $x^* := \mathbb{E}_{x_0}^{\pi^*}[X_N]$ and $y^* := \mathbb{E}_{x_0}^{\pi^*}[X_N^2]$. The concavity of U implies (since $U_x = -2(\lambda + x)$ and $U_y = 1$)

$$\begin{aligned} U(x, y) &\le U(x^*, y^*) - 2(\lambda + x^*)(x - x^*) + y - y^* \\ &= U(x^*, y^*) - 2b(x - x^*) + y - y^* < U(x^*, y^*), \end{aligned}$$

where the last inequality is due to our assumption $y - 2bx < y^* - 2bx^*$. Hence π^* is not optimal for $P(\lambda)$, leading to a contradiction. □

The implication of Lemma 4.6.3 is that any optimal solution of $P(\lambda)$ (as long as it exists) can be found by solving problem $QP(b)$.
Problem $QP(b)$ now obviously is a Markov Decision Problem with the same data $E, A, D_n(x), \mathcal{Z}, T_n, Q_n^Z$ as before and reward functions

- $r_n(x,a) := 0$,
- $g_N(x) := -(x-b)^2$.

Assumption (A_N) is satisfied, since r_n and g_N are non-positive. Here we treat the problem as a cost minimizing problem. For the remaining part we define for $n = 0, 1, \ldots, N$ the numbers d_n recursively:

$$\begin{aligned} C_n &:= \mathbb{E}[R_n R_n^\top], \qquad \ell_n := (\mathbb{E}[R_n])^\top C_n^{-1}\, \mathbb{E}[R_n], \\ d_N &:= 1, \qquad d_n := d_{n+1}(1-\ell_{n+1}). \end{aligned} \tag{4.34}$$

Before we state the next theorem it is crucial to make the following observation.

Lemma 4.6.4. *For all $n = 0, 1, \ldots, N-1$ it holds that $0 < d_n < 1$.*

Proof. Let us denote $v_n := \mathbb{E}[R_n]$ and $\Sigma_n := \big(\mathrm{Cov}(R_n^j, R_n^k)\big)$. Then it follows from the definition of the covariance matrix that

$$C_n = \Sigma_n + v_n v_n^\top.$$

Since Σ_n is positive definite by Assumption (FM) and $v_n v_n^\top$ is positive semidefinite, C_n is itself positive definite. This implies that C_n^{-1} is positive-definite and since $v_n \neq 0$ by assumption we have $\ell_n > 0$. By the Sherman-Morrison formula we obtain

$$C_n^{-1} = \Sigma_n^{-1} - \big(1 + v_n^\top \Sigma_n^{-1} v_n\big)^{-1} \Sigma_n^{-1} v_n v_n^\top \Sigma_n^{-1}.$$

Thus, it follows that

$$\ell_n = v_n^\top C_n^{-1} v_n = \big(1 + v_n^\top \Sigma_n^{-1} v_n\big)^{-1} v_n^\top \Sigma_n^{-1} v_n < 1.$$

Altogether we have $0 < 1 - \ell_n < 1$ and the statement follows easily by induction. □

Theorem 4.6.5. *For the solution of the Markov Decision Problem $QP(b)$ it holds:*

a) The value functions are given by

$$V_n(x) = \Big(\frac{x S_N^0}{S_n^0} - b\Big)^2 d_n, \quad x \in E$$

where (d_n) is defined in (4.34). Then $V_0(x_0)$ is the value of problem $QP(b)$.

b) The optimal policy $\pi^* = (f_0^*, \ldots, f_{N-1}^*)$ *is given by*

$$f_n^*(x) = \left(\frac{bS_n^0}{S_N^0} - x\right) C_{n+1}^{-1}\, \mathbb{E}\, R_{n+1}, \quad x \in E.$$

c) The first and the second moment of X_N *under* π^* *are given by*

$$\mathbb{E}_{x_0}^{\pi^*}[X_N] = x_0 S_N^0 d_0 + b(1-d_0)$$
$$\mathbb{E}_{x_0}^{\pi^*}[X_N^2] = (x_0 S_N^0)^2 d_0 + b^2(1-d_0).$$

Proof. The problem $QP(b)$ is of linear-quadratic form and close to Section 2.6.3. However, the cost function involves a linear term. We show that the Structure Assumption (SA_N) of Theorem 2.3.8 is satisfied for the cost-to-go functions with

$$\mathbb{M}_n := \{v : \mathbb{R} \to \mathbb{R} \mid v(x) = (c_1 x - c_2)^2 \text{ for } c_1, c_2 \in \mathbb{R}\}$$
$$\Delta_n := \{f : \mathbb{R} \to \mathbb{R}^d \mid f(x) = (c_3 - x)c_4 \text{ for } c_3 \in \mathbb{R},\ c_4 \in \mathbb{R}^d\}.$$

Let us check the conditions:

(i) $(x-b)^2 \in \mathbb{M}_N$.
(ii) Let $v \in \mathbb{M}_{n+1}$. We assume that $v(x) = (c_1 x - b)^2$ and obtain with the abbreviation $c_5 := c_1(1+i_{n+1})$:

$$\mathcal{T}_n v(x) = \inf_{a \in \mathbb{R}^d} \mathbb{E}\left[v\big((1+i_{n+1})(x + a \cdot R_{n+1})\big)\right]$$
$$= \inf_{a \in \mathbb{R}^d} \mathbb{E}\left[\Big(c_1(1+i_{n+1})(x + a \cdot R_{n+1}) - b\Big)^2\right]$$
$$= \inf_{a \in \mathbb{R}^d} \Big\{c_5^2 x^2 + 2c_5^2 x a\, \mathbb{E}[R_{n+1}] + c_5^2 a^\top C_{n+1} a - 2bc_5(x + a\, \mathbb{E}[R_{n+1}]) + b^2\Big\}.$$

The minimum point has to satisfy

$$c_5 C_{n+1} a - b\, \mathbb{E}[R_{n+1}] + c_5 x\, \mathbb{E}[R_{n+1}] = 0.$$

Hence the minimizer is given by

$$f_n^*(x) = \frac{(b - c_5 x)}{c_5} C_{n+1}^{-1}\, \mathbb{E}[R_{n+1}].$$

Inserting the minimizer into the equation for $\mathcal{T}_n v$ gives

$$\mathcal{T}_n v(x) = c_5^2 x^2 - (b - c_5 x)^2\, \mathbb{E}[R_{n+1}]^\top C_{n+1}^{-1}\, \mathbb{E}[R_{n+1}] - 2bc_5 x + b^2$$
$$= (c_5 x - b)^2 \Big(1 - \mathbb{E}[R_{n+1}]^\top C_{n+1}^{-1}\, \mathbb{E}[R_{n+1}]\Big) \in \mathbb{M}_n.$$

(iii) The existence and structure of the minimizer has already been shown in part (ii).

Thus, we can apply Theorem 2.3.8 and obtain

$$V_N(x) = (x-b)^2$$

and by induction using the results from (ii)

$$V_n(x) = \mathcal{T}_n V_{n+1}(x) = d_{n+1}(1-\ell_{n+1})\Big(\frac{S_N^0 x}{S_n^0} - b\Big)^2$$

with $0 < d_{n+1} < 1$ according to Lemma 4.6.4. The minimizers have been computed in part (ii) and give the optimal policy.
Inserting the optimal policy π^* into the wealth equation yields the statements in part c). □

Now we are able to compute $\lambda^* > 0$ and $\pi^* = \pi^*(\lambda^*)$ such that $\mathbb{E}_{x_0}^{\pi^*}[X_N] = \mu$ and π^* is optimal for $P(\lambda^*)$. From Lemma 4.6.3 and Theorem 4.6.5 we obtain

$$b^* := \mathbb{E}_{x_0}^{\pi^*}[X_N] + \lambda^* = \mu + \lambda^*$$

$$\begin{aligned}\mathbb{E}_{x_0}^{\pi^*}[X_N] &= x_0 S_N^0 d_0 + b^*(1-d_0)\\ &= x_0 S_N^0 d_0 + (\mu+\lambda^*)(1-d_0) \stackrel{!}{=} \mu\end{aligned}$$

and therefore

$$\lambda^* = (\mu - x_0 S_N^0)\frac{d_0}{1-d_0}$$

since $d_0 \in (0,1)$. Using $b^* = \mu + \lambda^*$ we get the following solution of the mean variance problem which is the main result of this section.

Theorem 4.6.6. *For the mean-variance problem (MV) it holds:*

a) The value of (MV) is given by

$$\mathrm{Var}_{x_0}^{\pi^*}[X_N] = \frac{d_0}{1-d_0}\Big(\mathbb{E}_{x_0}^{\pi^*}[X_N] - x_0 S_N^0\Big)^2$$

where d_0 is given in (4.34). Note that $\mathbb{E}_{x_0}^{\pi^*}[X_N] = \mu$.

b) The optimal portfolio strategy $\pi^* = (f_0^*, \dots, f_{N-1}^*)$ is given by

$$f_n^*(x) = \left(\Big(\frac{\mu - d_0 x_0 S_N^0}{1-d_0}\Big)\frac{S_n^0}{S_N^0} - x\right) C_{n+1}^{-1}\,\mathbb{E}[R_{n+1}], \quad x \in E.$$

Since $x_0 S_N^0 < \mu$, a higher expected return corresponds to a higher variance of the portfolio. Now if we assume that we have initially x_0 Euro available for investment, then Theorem 4.6.6 yields the following linear dependence of the expected return of the portfolio on the standard deviation:

$$\mu(\sigma) := \mu = x_0 S_N^0 + \sigma\sqrt{\frac{1-d_0}{d_0}}$$

where $\sigma^2 := Var_{x_0}^{\pi^*}[X_N]$. This curve is called the *efficient frontier* or *capital market line* and gives the solution of the portfolio problem as a function of σ. The slope is called the *price of risk* or *Sharpe ratio.* Moreover, the so-called Two-Fund Theorem which is well known in the classical Markowitz model, can be recovered in the multistage problem.

Corollary 4.6.7 (Two-Fund Theorem). *At any time, the optimal policy invests a certain amount of money in the bond and the remaining money is invested in a mutual fund.*

Proof. In the representation of the optimal policy $\pi^* = (f_0^*, \dots, f_{N-1}^*)$ in Theorem 4.6.6, the first factor is a real number and $C_{n+1}^{-1}\,\mathbb{E}[R_{n+1}]$ is a vector which determines the mutual fund. □

Remark 4.6.8 (Hedging of Claims). When the initial wealth of an investor is insufficient to hedge a contingent claim perfectly, the investor can try to minimize the distance of her terminal wealth to the payoff H of the contingent claim. Hence an optimization problem of the following type arises:

$$\min_{\pi \in F^N} \mathbb{E}_x^\pi \left[\ell(H - X_N)\right]$$

where the function ℓ measures the loss. This is again a Markov Decision Problem if $H = h(S_N)$ is a European claim. In this case besides the wealth, also the current stock price has to be part of the state of the Markov Decision Model. If H is an arbitrary claim then it would be necessary to keep track of the complete history. There are different reasonable choices for the loss function ℓ. Popular choices are $\ell(x) = x^2$ or $\ell(x) = x^+$. The first function leads to variance optimal hedging and the second one to shortfall minimization. ◇

4.7 Dynamic Mean-Risk Problems

In the preceding section the role of the variance is to measure the risk of the portfolio strategy. However, the variance has many drawbacks, one for example is that it measures deviations which are below and above the mean. Recently a new axiomatic theory of how to measure risk has emerged which is to a considerable extent motivated by practical considerations. In this section we will solve, at least *in the binomial model*, the problem of the previous section when the variance is replaced by the Average-Value-at-Risk which is for $X \in L^1(\Omega, \mathcal{F}, \mathbb{P})$ given by

$$AVaR_\gamma(X) = \frac{1}{1-\gamma} \int_\gamma^1 VaR_\alpha(X) d\alpha$$

where $\gamma \in (0,1)$ and $VaR_\alpha(X)$ is the Value-at-Risk at level α. For a short review of risk measures the reader is referred to Appendix C.2.
Let us now consider the *binomial model* or *Cox-Ross-Rubinstein model* of Section 3.1 with one stock and one bond (with interest rate $i = 0$). As in the previous section we look at the following Markov Decision Model with disturbances $Z_n = R_n$ (where r and g will be defined later):

- $E := \mathbb{R}$ where $x \in E$ denotes the wealth,
- $A := \mathbb{R}$ where $a \in A$ is the amount of money invested in the risky asset,
- $D(x) := A$,
- $\mathcal{Z} := \{\boldsymbol{d} - 1, \boldsymbol{u} - 1\}$ where $z \in \mathcal{Z}$ denotes the relative risk,
- $T(x, a, z) := x + az$,
- $Q^Z(\{\boldsymbol{u} - 1\} | x, a) := p \in (0,1)$.

For $\pi \in F^N$ we denote by $AVaR_\gamma^{x_0,\pi}(X_N)$ the Average-Value-at-Risk of the terminal wealth X_N under portfolio strategy π and initial wealth x_0. Note that the $AVaR$ can also be characterized as the solution of a convex optimization problem (see Example C.2.2). Since we assume a binomial model we obtain $\mathbb{E}_{x_0}^\pi[X_N] \in \mathbb{R}$ and hence $AVaR_{x_0,\gamma}^\pi(X_N) \in \mathbb{R}$ for all $\gamma \in (0,1)$. We want to solve

$$(MR) \begin{cases} AVaR_{x_0,\gamma}^\pi(X_N) \to \min \\ \\ \mathbb{E}_{x_0}^\pi[X_N] \geq \mu \\ \pi \in F^N. \end{cases}$$

We are in particular interested in large values for γ, e.g. $\gamma = 0.995$. Denote the value of the problem by $V_{MR}(x_0)$. To ease the exposition we will make the following assumptions throughout this section.

Assumption (FM):

(i) *There exist no arbitrage opportunities, i.e.* $\boldsymbol{d} < 1 < \boldsymbol{u}$.
(ii) $p > q$ *where*

$$q := \frac{1 - \boldsymbol{d}}{\boldsymbol{u} - \boldsymbol{d}} \in (0,1).$$

(iii) $0 < x_0 < \mu$.

In contrast to the mean-variance problem we need a no-arbitrage condition here. Otherwise the value of (MR) would be equal to $-\infty$. The probability

$q \in (0,1)$ is the risk neutral probability for an up movement under which the stock price process becomes a martingale. The case $p < q$ can be treated similarly (cf. Theorem 4.7.4). If $p = q$ then (X_n) is a martingale and there exists no admissible strategy of (MR) since $x_0 < \mu$. In this case the value (MR) is equal to $+\infty$. Moreover, as in the mean-variance problem condition (iii) is reasonable because otherwise the constraint can be fulfilled by just investing in the bond.
We solve problem (MR) by the well-known Lagrange approach (see Section 4.6): For this purpose let us denote the Lagrange function by

$$L_{x_0}(\pi,\lambda) := AVaR^{\pi}_{x_0,\gamma}(X_N) + \lambda\Big(\mu - \mathbb{E}^{\pi}_{x_0}[X_N]\Big), \quad \lambda \ge 0.$$

Obviously we have $V_{MR}(x_0) = \inf_{\pi\in F^N}\sup_{\lambda\ge 0} L_{x_0}(\pi,\lambda)$. We will first consider the dual problem $V_D(x_0) = \sup_{\lambda\ge 0}\inf_{\pi\in F^N} L_{x_0}(\pi,\lambda)$. In order to solve it we can write the objective function as follows where we make use of the representation of the Average-Value-at-Risk as the solution of an optimization problem (see Example C.2.2):

$$\sup_{\lambda\ge 0}\inf_{\pi\in F^N}\Big\{\inf_{b\in\mathbb{R}}\Big\{b + \frac{1}{1-\gamma}\mathbb{E}^{\pi}_{x_0}\left[(X_N+b)^-\right]\Big\} + \lambda\Big(\mu - \mathbb{E}^{\pi}_{x_0}[X_N]\Big)\Big\}$$

$$= \sup_{\lambda\ge 0}\inf_{b\in\mathbb{R}}\Big\{b + \inf_{\pi}\Big\{\frac{\mathbb{E}^{\pi}_{x_0}\left[(X_N+b)^-\right]}{1-\gamma} + \lambda\,\mathbb{E}^{\pi}_{x_0}\left[(-X_N-b)\right]\Big\} + \lambda b + \lambda\mu\Big\}$$

$$= \sup_{\lambda\ge 0}\inf_{b\in\mathbb{R}}\Big\{b(1+\lambda) + \lambda\mu + \inf_{\pi\in F^N}\Big\{\Big(\frac{1}{1-\gamma}+\lambda\Big)\mathbb{E}^{\pi}_{x_0}\left[(X_N+b)^-\right] - \lambda\,\mathbb{E}^{\pi}_{x_0}\left[(X_N+b)^+\right]\Big\}\Big\}.$$

Next we aim to solve the inner stochastic optimization problem

$$P(\lambda,b)\quad\begin{cases}\Big(\frac{1}{1-\gamma}+\lambda\Big)\mathbb{E}^{\pi}_{x_0}\left[(X_N+b)^-\right] - \lambda\,\mathbb{E}^{\pi}_{x_0}\left[(X_N+b)^+\right] \to \min\\ \pi\in F^N\end{cases}$$

where $\lambda \ge 0$ and $b \in \mathbb{R}$. The problem $P(\lambda,b)$ is a Markov Decision Problem with the same data as before and reward functions

- $r(x,a) := 0,$
- $g(x) := \lambda(x+b)^+ - \Big(\frac{1}{1-\gamma}+\lambda\Big)(x+b)^-.$

In what follows we treat this problem as a cost minimization problem. For $n = 0,1,\dots,N$ let

$$c_n := \Big(\lambda + \frac{1}{1-\gamma}\Big)\Big(\frac{1-p}{1-q}\Big)^{N-n}, \quad \text{and} \quad d_n := \lambda\Big(\frac{p}{q}\Big)^{N-n}.$$

Note that $g(x) = d_N(x+b)^+ - c_N(x+b)^-$ and that $c_0 \geq d_0$ if γ is large. Moreover it holds: $c_0 \geq d_0$ is equivalent to $c_n \geq d_n$ for all $n = 0, 1, \dots, N$.

Theorem 4.7.1. *For the solution of the Markov Decision Problem $P(\lambda, b)$ it holds:*

a) If $c_0 \geq d_0$, then the value functions are given by

$$V_n(x) = c_n(x+b)^- - d_n(x+b)^+, \quad x \in E$$

and $V_0(x_0)$ is the value of problem $P(\lambda, b)$. If $c_0 < d_0$, then the value of $P(\lambda, b)$ is equal to $-\infty$.

b) If $c_0 \geq d_0$, then the optimal policy $\pi^* = (f_0^*, \dots, f_{N-1}^*)$ is stationary and given by

$$f_n^*(x) = f^*(x) := \max\left\{\frac{x+b}{1-\boldsymbol{u}}, \frac{x+b}{1-\boldsymbol{d}}\right\}, \quad x \in E.$$

Proof. Suppose $c_0 \geq d_0$. We show that the Structure Assumption (SA$_N$) is fulfilled with

$$\begin{aligned}
I\!M_n &:= \{v : E \to \mathbb{R} \mid v(x) = c(x+b)^- - d(x+b)^+, \\
&\qquad\qquad\qquad\qquad \text{for } 0 \leq d \leq d_n \leq c_n \leq c\} \\
\Delta_n &:= \{f : \mathbb{R} \to \mathbb{R}^d \mid f(x) = \max\{c_1(x+b), c_2(x+b)\} \\
&\qquad\qquad\qquad\qquad \text{for } c_1 < 0 < c_2\}.
\end{aligned}$$

In this example our strong Integrability Assumption (A$_N$) is not satisfied, however Theorem 2.5.4 still holds true (see Remark 2.3.14). Let us check the conditions:

(i) $c_N(x+b)^- - d_N(x+b)^+ \in I\!M_N$ is obvious.
(ii) Let $v \in I\!M_{n+1}$ and suppose $v(x) = c(x+b)^- - d(x+b)^+$. We obtain:

$$\begin{aligned}
\mathcal{T}v(x) &= \inf_{a\in\mathbb{R}} \Big\{p\Big[c\big(x+b+a(\boldsymbol{u}-1)\big)^- - d\big(x+b+a(\boldsymbol{u}-1)\big)^+\Big] \\
&\qquad +(1-p)\Big[c\big(x+b+a(\boldsymbol{d}-1)\big)^- - d\big(x+b+a(\boldsymbol{d}-1)\big)^+\Big]\Big\} \\
&=: \inf_{a\in\mathbb{R}} h(a).
\end{aligned}$$

Let us first assume that $x+b \geq 0$. The function h is piecewise linear with slope changing at $a_1 := \frac{x+b}{1-\boldsymbol{u}} < 0$ and $a_2 := \frac{x+b}{1-\boldsymbol{d}} > 0$. The slope on the right hand side of a_2 is non-negative if and only if $\frac{1-p}{1-q}c \geq \frac{p}{q}d$ which is the case by our assumption since

$$\frac{p}{q}d \leq \frac{p}{q}d_{n+1} = d_n \leq c_n = \frac{1-p}{1-q}c_{n+1} \leq \frac{1-p}{1-q}c.$$

Note that the slope on the left hand side of a_1 is negative. Finally we investigate the slope on the interval $[a_1, a_2]$. It can be shown that this slope is negative if $d > 0$ and zero if $d = 0$. Hence the minimum is attained at $a^* = a_2$. Moreover, we obtain

$$\mathcal{T}v(x) = -\frac{p}{q}d(x+b).$$

In case $x + b < 0$ we obtain $a_2 < a_1$. Since $\frac{1-p}{1-q}c \geq \frac{p}{q}d$, we get $a^* = a_1$ as a minimum point. Summarizing we obtain

$$\mathcal{T}v(x) = \frac{1-p}{1-q}c(x+b)^- - \frac{p}{q}d(x+b)^+ \in I\!M_n.$$

(iii) The existence and structure of the minimizer has already been shown in part (ii).

The statement for $c_0 \geq d_0$ then follows by induction. In case $c_0 < d_0$ define $n^* := \inf\{n \in \mathbb{N} \mid c_n \geq d_n\}$ and consider the strategy $\pi = (f_0, f_1, \ldots, f_{N-1})$ with

$$f_n(x) = \begin{cases} 0, & n = 0, \ldots, n^* - 2 \\ a, & n = n^* - 1 \\ f^*(x), & n \geq n^* \end{cases}$$

where $f^*(x) := \max\left\{\frac{x+b}{1-\boldsymbol{u}}, \frac{x+b}{1-\boldsymbol{d}}\right\}$ and $a \in A = \mathbb{R}$. By the reward iteration we get

$$\begin{aligned} V_{0\pi}(x_0) &= \mathcal{T}_{0f_0} \ldots \mathcal{T}_{N-1f_{N-1}} g_N(x_0) \\ &= p\Big[c_{n^*}\big(x_0 + b + a(\boldsymbol{u}-1)\big)^- - d_{n^*}\big(x_0 + b + a(\boldsymbol{u}-1)\big)^+\Big] \\ &\quad +(1-p)\Big[c_{n^*}\big(x_0 + b + a(\boldsymbol{d}-1)\big)^- - d_{n^*}\big(x_0 + b + a(\boldsymbol{d}-1)\big)^+\Big]. \end{aligned}$$

Following the discussion in part (ii) we finally obtain that $\lim_{a\to\infty} V_{0\pi}(x_0) = -\infty$ and the value of problem $P(\lambda, b)$ is equal to $-\infty$. □

In what follows we write $V_0^b(x_0) = V_0(x_0)$ and $f^b(x) = f^*(x)$ to stress the dependence on b. Next we have to solve

$$P(\lambda) \begin{cases} L_{x_0}(\pi, \lambda) \to \min \\ \\ \pi \in F^N \end{cases}$$

where $\lambda \geq 0$. For this purpose define

$$\gamma_1 := 1 - \left(\frac{1-p}{1-q}\right)^N,$$

$$\gamma_2 := p^N \frac{(1-q)^N - (1-p)^N}{\big(p(1-q)\big)^N - \big(q(1-p)\big)^N},$$

$$\lambda^* = \lambda^*(\gamma) := \min\left\{\frac{q^N}{p^N - q^N}; \frac{(1-p)^N(1-\gamma)^{-1} - (1-q)^N}{(1-q)^N - (1-p)^N}\right\}.$$

Note that $\gamma_1 \le \gamma_2$. With these definitions we can formulate the solution of the optimization problem $P(\lambda)$ as follows.

Proposition 4.7.2. *a) The value of problem $P(\lambda)$ is given by*

$$\inf_{\pi \in F^N} L_{x_0}(\pi, \lambda) = \begin{cases} (\mu - x_0)\lambda - x_0 & \text{if } \lambda \in [0, \lambda^*],\ \gamma \ge \gamma_1 \\ -\infty & \text{else.} \end{cases}$$

b) The optimal policy π^* for $P(\lambda)$ is stationary, i.e. $\pi^* = (f^b, \dots, f^b) \in F^N$ with

$$b \in \begin{cases} \{-x_0\} & \text{if } \lambda \in [0, \lambda^*) \\ [-x_0, \infty) & \text{if } \lambda = \lambda^*,\ \gamma \ge \gamma_2 \\ (-\infty, -x_0] & \text{if } \lambda = \lambda^*,\ \gamma_1 \le \gamma \le \gamma_2. \end{cases}$$

Proof. Recall that

$$\inf_{\pi \in F^N} L_{x_0}(\pi, \lambda) = \inf_{b \in \mathbb{R}} \left\{ b(1+\lambda) + \lambda\mu + V_0^b(x_0) \right\}.$$

Therefore define

$$h(\lambda, b) := b(1+\lambda) + \lambda\mu + V_0^b(x_0),$$

i.e. we have to solve $\inf_{b \in \mathbb{R}} h(\lambda, b)$. From Theorem 4.7.1 we know that $V_0^b(x_0) = -\infty$ if and only if

$$c_0 < d_0 \iff \lambda > \frac{\big((1-p)q\big)^N}{\big(p(1-q)\big)^N - \big((1-p)q\big)^N}(1-\gamma)^{-1} =: \lambda_1^*(\gamma).$$

Hence for $\lambda \in [0, \lambda_1^*(\gamma)]$ we obtain

$$h(\lambda, b) = b(1+\lambda) + \lambda\mu + c_0(x_0+b)^- - d_0(x_0+b)^+$$

which is piecewise linear in b and the slope changes at $-x_0$. The slope for $b < -x_0$ is given by $1 + \lambda + c_0$ and for $b > -x_0$ by $1 + \lambda - d_0$. In order to obtain a minimum point of $h(\lambda, b)$, the slope on the left-hand side of $-x_0$ has to be non-positive and the slope on the right-hand side of $-x_0$ has to be non-negative. This is the case if and only if $\lambda \le \lambda^*$. If $\lambda < \lambda^*$ the minimum point is unique. Thus, the value of our problem can only be finite if $\lambda \in \big[0, \min\{\lambda^*, \lambda_1^*(\gamma)\}\big]$. If $\gamma < \gamma_1$ we obtain $\lambda^* < 0$ and no λ exists for which the value is finite. Now if $\gamma \ge \gamma_1$ it can be shown that $\lambda^* < \lambda_1^*(\gamma)$. In this case the value of the problem is given by

$$\inf_{b\in\mathbb{R}} h(\lambda, b) = -x_0(1+\lambda) + \lambda\mu = (\mu - x_0)\lambda - x_0.$$

The optimal policy follows from the preceding discussions. □

Now we are ready to state the solution of the initially posed mean-risk problem (MR).

Theorem 4.7.3. *For the mean-risk problem (MR) it holds:*

a) The value of problem (MR) is given by

$$V_{MR}(x_0) = \begin{cases} (\mu - x_0)\lambda^* - x_0 & \text{if } \gamma \ge \gamma_1 \\ -\infty & \text{else.} \end{cases}$$

b) The optimal policy π^* is stationary, i.e. $\pi^* = (f^*, \dots, f^*) \in F^N$ with

$$f^*(x) = \begin{cases} \frac{1}{1-\boldsymbol{d}}\left((\mu - x_0)\frac{q^N}{p^N - q^N} - x_0 + x\right) & \text{if } \gamma \ge \gamma_2 \\ \frac{1}{1-\boldsymbol{u}}\left((x_0 - \mu)\frac{(1-q)^N}{(1-q)^N - (1-p)^N} - x_0 + x\right) & \text{if } \gamma_1 \le \gamma \le \gamma_2. \end{cases}$$

Proof. If $\gamma < \gamma_1$, then by Proposition 4.7.2 we obtain $\inf_{\pi \in F^N} L_{x_0}(\pi, \lambda) = -\infty$ for all λ and the statement follows. Let now $\gamma \ge \gamma_1$. We derive from Proposition 4.7.2 that

$$\sup_{\lambda \ge 0} \inf_{\pi \in F^N} L_{x_0}(\pi, \lambda) = \sup_{0 \le \lambda \le \lambda^*} \Big\{(\mu - x_0)\lambda - x_0\Big\} = (\mu - x_0)\lambda^* - x_0.$$

Next we search for an optimal policy π^* in Proposition 4.7.2 such that $\pi^* = \pi^*(\lambda^*)$ and λ^* satisfy:

$$\mathbb{E}_{x_0}^{\pi^*}[X_N] \ge \mu \quad \text{and} \quad \lambda^*\Big(\mathbb{E}_{x_0}^{\pi^*}[X_N] - \mu\Big) = 0.$$

If $\gamma > \gamma_1$ we have $\lambda^* > 0$ and we have to find π^* such that $\mathbb{E}_{x_0}^{\pi^*}[X_N] = \mu$. It can be shown that a policy $\pi^* = (f^b, \dots, f^b) \in F^N$ as defined in Proposition 4.7.2 yields for $b \ge -x_0$:

$$\mathbb{E}_{x_0}^{f^b}[X_N] = x_0 + (x_0 + b)\frac{p^N - q^N}{q^N}. \tag{4.35}$$

If $\gamma \geq \gamma_2$ we have to find $b \in [-x_0, \infty)$ such that the expression in (4.35) is equal to μ. Indeed, if we solve the equation we obtain

$$b^* = (\mu - x_0)\frac{q^N}{p^N - q^N} - x_0 > -x_0$$

which defines the optimal policy. Similar arguments lead to the optimal policy in the case $\gamma \in (\gamma_1, \gamma_2]$.
If $\gamma = \gamma_1$ we have $\lambda^* = 0$. In this case we only have to find an optimal π^* that is admissible for (MR). This is fulfilled for every f^b with

$$b \leq (x_0 - \mu)\frac{(1-q)^N}{(1-q)^N - (1-p)^N} - x_0.$$

Altogether we obtain now

$$\inf_\pi \sup_{\lambda \geq 0} L_{x_0}(\pi, \lambda) \geq \sup_{\lambda \geq 0} \inf_\pi L_{x_0}(\pi, \lambda) = L_{x_0}(\pi^*, \lambda^*) \geq \inf_\pi \sup_{\lambda \geq 0} L_{x_0}(\pi, \lambda),$$

i.e. (π^*, λ^*) is a saddle-point of $L_{x_0}(\pi, \lambda)$. Hence by Lemma 4.6.2, π^* is optimal for (MR). □

Though we have restricted our analysis to the case $p > q$ so far, for the sake of completeness we will present the solution of problem (MR) also in the case $p < q$. Here other definitions for γ_1, γ_2 and $\lambda^* = \lambda^*(\gamma)$ are necessary:

$$\gamma_1 := \frac{q^N - p^N}{q^N},$$

$$\gamma_2 := (1-p)^N \frac{q^N - p^N}{\big((1-p)q\big)^N - \big((1-q)p\big)^N},$$

$$\lambda^* := \min\left\{\frac{(1-q)^N}{(1-p)^N - (1-q)^N}; \frac{p^N(1-\gamma)^{-1} - q^N}{q^N - p^N}\right\}.$$

Theorem 4.7.4. *Suppose $p < q$. For the mean-risk problem (MR) it holds:*

a) The value of problem (MR) is the same as in Theorem 4.7.3 with λ^* and γ_1 as defined above.
b) The optimal policy π^* is stationary, i.e. $\pi^* = (f^*, \dots, f^*) \in F^N$ with

$$f^*(x) = \begin{cases} \frac{1}{1-d}\left((\mu - x_0)\frac{q^N}{p^N - q^N} - x_0 + x\right) & \text{if } \gamma_1 \leq \gamma \leq \gamma_2 \\ \frac{1}{1-u}\left((x_0 - \mu)\frac{(1-q)^N}{(1-q)^N - (1-p)^N} - x_0 + x\right) & \text{if } \gamma \geq \gamma_2. \end{cases}$$

Remark 4.7.5. a) Note that, as in the mean-variance problem, the efficient frontier is linear

$$\mu(\rho) = \frac{\rho + x_0}{\lambda^*} + x_0$$

where $\rho := AVaR^{\pi^*}_{x_0,\gamma}(X_N)$. The slope of this function is given by $\frac{1}{\lambda^*}$. In the case $p > q$ we have

$$\lambda^* = \frac{q^N}{p^N - q^N} \quad \Leftrightarrow \gamma \geq \gamma_2.$$

Otherwise, the slope is given by

$$\frac{(1-p)^N(1-\gamma)^{-1} - (1-q)^N}{(1-q)^N - (1-p)^N}.$$

b) In case $p = q$ the stock is a martingale and hence we cannot improve our wealth on average and the value of (MR) is $+\infty$. If $p > q$ the amount invested in the stock is always positive and in the case $p < q$ it is always negative. Moreover, the larger $|p - q|$, the less is the absolute amount which is invested. The explanation is as follows: compare this situation to the classical red-and-black casino game. There, a timid strategy is optimal if the game is favourable. Since short-sellings are allowed in the financial model, the case $p < q$ is also favourable by short-selling the stock. If $|p-q|$ is large we can invest less since the probability for a gain is large. If $|p-q|$ is small, then we have to take a higher risk and play a bold strategy. ◊

4.8 Index-Tracking

The problem of index-tracking which is formulated below can be seen as an application of mean-variance hedging in an incomplete market. Suppose we have a financial market with one bond and d risky assets as in Section 3.1. Besides the tradeable assets there is a non-tradable asset whose price process $(\hat{S}_n)$ evolves according to

$$\hat{S}_{n+1} = \hat{S}_n \hat{R}_{n+1}.$$

The positive random variable $\hat{R}_{n+1}$ which is the relative price change of the non-traded asset may be correlated with R_{n+1}. It is assumed that the random vectors $(R_1, \hat{R}_1), (R_2, \hat{R}_2), \ldots$ are independent and the joint distribution of $(R_n, \hat{R}_n)$ is given. The aim now is to track the non-traded asset as closely as possible by investing into the financial market. The tracking error is measured in terms of the quadratic distance of the portfolio wealth to the price process $(\hat{S}_n)$, i.e. the optimization problem is then

$$\begin{cases} \mathbb{E}_{x\hat{s}} \left[\sum_{n=0}^{N} \left(X_n^\phi - \hat{S}_n \right)^2 \right] \to \min \\ \phi = (\phi_n) \text{ is a portfolio strategy} \end{cases} \tag{4.36}$$

where ϕ_n is $\mathcal{F}_n = \sigma(R_1, \ldots, R_n, \hat{R}_1, \ldots, \hat{R}_n)$-measurable. This problem can be formulated as a linear-quadratic problem (see Section 2.6.3). It is important to note here however, that the state space of the Markov Decision Model has to include (besides the wealth) the price of the non-traded asset. Thus, the data is given as follows:

- $E := \mathbb{R} \times \mathbb{R}_+$ where $(x, \hat{s}) \in E$ and x is the wealth and $\hat{s}$ the value of the non-traded asset,
- $A := \mathbb{R}^d$ where $a \in A$ is the amount of money invested in the risky assets,
- $D(x, \hat{s}) := A$,
- $\mathcal{Z} := (-1, \infty)^d \times \mathbb{R}_+$ where $z = (z_1, z_2) \in \mathcal{Z}$ and z_1 is the relative risk of the traded assets and z_2 is the relative price change of the non-traded asset.
- The transition function is given by

$$T_n\big((x, \hat{s}), a, (z_1, z_2)\big) := \begin{pmatrix} 1 + i_{n+1} & 0 \\ 0 & z_2 \end{pmatrix} \begin{pmatrix} x \\ \hat{s} \end{pmatrix} + \begin{pmatrix} (1 + i_{n+1}) z_1^\top \\ 0 \end{pmatrix} a,$$

- $Q_n^Z\big(\cdot \,|(x, \hat{s}), a\big) :=$ joint distribution of $(R_{n+1}, \hat{R}_{n+1})$ (independent of $\big((x, \hat{s}), a\big)$),
- $r_n\big((x, \hat{s}), a\big) := -(x - \hat{s})^2$,
- $g_N(x, \hat{s}) := -(x - \hat{s})^2$.

Problem (4.36) can now be solved by the Markov Decision Model. The value functions (cost-to-go functions) are given by

$$V_n(x, \hat{s}) := \inf_\pi \mathbb{E}_{nx\hat{s}}^\pi \left[\sum_{k=n}^{N} (X_k - \hat{S}_k)^2 \right], \quad (x, \hat{s}) \in E$$

and $V_0(x, \hat{s})$ is the minimal value of problem (4.36). When we define

$$Q := \begin{pmatrix} 1 & -1 \\ -1 & 1 \end{pmatrix}$$

the problem is equivalent to minimizing

$$\mathbb{E}_{nx\hat{s}}^\pi \left[\sum_{k=n}^{N} \begin{pmatrix} X_k \\ \hat{S}_k \end{pmatrix}^\top Q \begin{pmatrix} X_k \\ \hat{S}_k \end{pmatrix} \right].$$

The linear-quadratic problem in Section 2.6.3 yields the following result.

Theorem 4.8.1. *a) The value functions of problem (4.36) are given by*

$$V_n(x,\hat{s}) = (x,\hat{s})Q_n \begin{pmatrix} x \\ \hat{s} \end{pmatrix}, \quad (x,\hat{s}) \in E$$

where the matrices Q_n can be computed recursively as in Theorem 2.6.3. In particular Q_n are symmetric and positive semidefinite.

b) The optimal portfolio strategy $\pi^* = (f_0^*, \ldots, f_{N-1}^*)$ is linear and given by

$$f_n^*(x,\hat{s}) = -\Big(\mathbb{E}\big[R_{n+1}R_{n+1}^\top\big]\Big)^{-1} \cdot \mathbb{E}\left[\Big(R_{n+1}, \tfrac{q_{n+1}^{21}}{(1+i_{n+1})q_{n+1}^{11}}\hat{R}_{n+1}R_{n+1}\Big)\right]\begin{pmatrix} x \\ \hat{s} \end{pmatrix}$$

where the elements of Q_{n+1} are denoted by q_{n+1}^{ij}.

Instead of tracking an index, the problem can be modified slightly such that the investor tries to outperform the index. This can be achieved by adding to $(\hat{S}_n)$ an additional (deterministic) amount, say (b_n), and minimizing

$$\mathbb{E}_{x\hat{s}}\left[\sum_{n=0}^{N}(X_n^\phi - \hat{S}_n - b_n)^2\right]$$

over all portfolio strategies ϕ. This problem can be solved in the same way.

4.9 Indifference Pricing

In this section we introduce indifference pricing, a method to price contingent claims in incomplete markets which has attained considerable attention in recent years. More precisely we assume here that the incompleteness of the market comes from an additional asset which cannot be traded. The underlying idea of the indifference pricing approach can be traced back to the zero-utility premium principle in insurance: Suppose an investor with a utility function and initial wealth $x > 0$ is given as well as a contingent claim with payoff $H \geq 0$ at maturity where the payoff depends both on the traded and non-traded asset. What is the fair price at which the investor is willing to sell H at time 0?

The indifference pricing approach says that this is the amount $v_0(H)$ such that the following quantities are equal:

- The maximal expected utility the investor achieves when she only uses her wealth x and invests in the traded asset.
- The maximal expected utility the investor achieves when she sells short H for the price v_0 and uses the wealth $x + v_0$ to invest in the traded asset.

Of course, the price of a contingent claim in this approach depends on the utility function of the investor. In what follows we will be more precise and use an exponential utility function. In order to keep the outline simple, we will start with a one-period model.

One-Period Financial Market

Suppose we have a one-period financial market with one bond, a traded risky asset and a non-traded risky asset. For simplicity we assume that the bond has zero interest, i.e. $i \equiv 0$. We denote the price process of the traded asset by $S = (S_n)$ and the price process of the non-traded asset by $\hat{S} = (\hat{S}_n)$. The initial prices $S_0 = s$ and $\hat{S}_0 = \hat{s}$ are given and at time 1 we assume that $S_1 = S_0 \tilde{R}$ and $\hat{S}_1 = \hat{S}_0 \hat{R}$ where $\tilde{R}$ and $\hat{R}$ are the random relative price changes. We suppose that $\Omega := \{\omega_1, \omega_2, \omega_3, \omega_4\}$ and

$$\begin{aligned} \tilde{R}(\omega_1) &= \tilde{R}(\omega_2) = \boldsymbol{u} & \tilde{R}(\omega_3) &= \tilde{R}(\omega_4) = \boldsymbol{d} \\ \hat{R}(\omega_1) &= \hat{R}(\omega_3) = \hat{\boldsymbol{u}} & \hat{R}(\omega_2) &= \hat{R}(\omega_4) = \hat{\boldsymbol{d}} \end{aligned}$$

and the joint distribution of $(\tilde{R}, \hat{R})$ can be computed from the given probabilities $p_k := \mathbb{P}(\{\omega_k\}) > 0$, $k = 1, \ldots, 4$. Throughout this section we assume

Assumption (FM): $0 < \boldsymbol{d} < 1 < \boldsymbol{u}$ *and* $\hat{\boldsymbol{d}} < \hat{\boldsymbol{u}}$.

If the amount $a \in \mathbb{R}$ is invested into asset S at time 0, the wealth of the investor at time 1 is given by

$$X_1 = x + a(\tilde{R} - 1).$$

We assume that the investor has an exponential utility function, i.e.

$$U(x) = -e^{-\gamma x}, \quad x \in \mathbb{R},\ \gamma > 0.$$

Next we consider a contingent claim whose payoff depends on both assets i.e. the payoff is given by a non-negative random variable H such that $H = h(S_1, \hat{S}_1)$. Now define by

$$V_0^H(x, s, \hat{s}) := \sup_{a \in \mathbb{R}} \mathbb{E}\left[-e^{-\gamma x - \gamma a(\tilde{R}-1) + \gamma H}\right] \tag{4.37}$$

the expected utility the investor achieves when she has initial wealth x and is short in H. If $H \equiv 0$ then we write $V_0^0(x, s, \hat{s})$. Formally, the indifference price is defined as follows.

Definition 4.9.1. The indifference price of the contingent claim H is the amount $v_0 = v_0(H, s, \hat{s})$ such that

$$V_0^0(x, s, \hat{s}) = V_0^H(x + v_0, s, \hat{s})$$

for all $x \in \mathbb{R}$, $s > 0, \hat{s} > 0$.

Note that v_0 does not depend on x. In our one-period market it is pretty simple to compute the indifference price. For convenience we denote by

$$q := \frac{1 - \boldsymbol{d}}{\boldsymbol{u} - \boldsymbol{d}}$$

the risk neutral probability and define $\tilde{h}(S_1, \hat{S}_1) = \tilde{H} := e^{\gamma H}$ and

$$\begin{aligned} h_u(s, \hat{s}) &:= p_1 \cdot \tilde{h}(s\boldsymbol{u}, \hat{s}\hat{\boldsymbol{u}}) + p_2 \cdot \tilde{h}(s\boldsymbol{u}, \hat{s}\hat{\boldsymbol{d}}) \\ h_d(s, \hat{s}) &:= p_3 \cdot \tilde{h}(s\boldsymbol{d}, \hat{s}\hat{\boldsymbol{u}}) + p_4 \cdot \tilde{h}(s\boldsymbol{d}, \hat{s}\hat{\boldsymbol{d}}). \end{aligned}$$

Then we obtain

Theorem 4.9.2. *For the one-period financial market it holds:*

a) The solution of problem (4.37) is given by

$$V_0^H(x, s, \hat{s}) = -e^{-\gamma x} \left(\frac{h_u(s, \hat{s})}{q}\right)^q \left(\frac{h_d(s, \hat{s})}{1 - q}\right)^{1-q}.$$

b) The indifference price of the contingent claim H is given by

$$v_0(H, s, \hat{s}) = \frac{q}{\gamma} \log\left(\frac{h_u(s, \hat{s})}{p_1 + p_2}\right) + \frac{1 - q}{\gamma} \log\left(\frac{h_d(s, \hat{s})}{p_3 + p_4}\right).$$

Proof. The proof of part a) is straightforward. In order to solve the problem (4.37) we have to minimize the function

$$a \mapsto e^{-\gamma a(\boldsymbol{u}-1)} h_u(s, \hat{s}) + e^{-\gamma a(\boldsymbol{d}-1)} h_d(s, \hat{s}).$$

Note that this function is convex, thus it is sufficient to find the zero of the first derivative. The minimum is attained at

$$a^* = \frac{\log(\boldsymbol{u} - 1) - \log(1 - \boldsymbol{d}) + \log(h_u(s, \hat{s})) - \log(h_d(s, \hat{s}))}{\gamma(\boldsymbol{u} - \boldsymbol{d})}.$$

Inserting a^* yields V_0^H. For part b) note that we obtain from the first part that

$$V_0^0(x, s, \hat{s}) = -e^{-\gamma x} \left(\frac{p_1 + p_2}{q}\right)^q \left(\frac{p_3 + p_4}{1 - q}\right)^{1-q}.$$

Using the definition of the indifference price the statement follows immediately. □

Remark 4.9.3 (Minimal Entropy Martingale Measure). Suppose we define a new measure $\mathbb{Q}$ on Ω by setting

$$\mathbb{Q}(\{\omega_k\}) = q_k := q\frac{p_k}{p_1+p_2}, \quad \text{for } k=1,2$$

$$\mathbb{Q}(\{\omega_k\}) = q_k := (1-q)\frac{p_k}{p_3+p_4}, \quad \text{for } k=3,4.$$

It is easy to check that $\mathbb{Q}$ is indeed a probability measure on Ω and that

(i) $\mathbb{E}_{\mathbb{Q}}\, S_1 = S_0$.
(ii) $\mathbb{Q}(\hat{S}_1 \in B|S_1) = \mathbb{P}(\hat{S}_1 \in B|S_1)$ for all $B \subset \{\hat{s}\hat{\boldsymbol{d}}, \hat{s}\hat{\boldsymbol{u}}\}$.

Moreover $\mathbb{Q}$ is the minimal entropy martingale measure in the sense that it solves

$$\min_{\mathbb{Q}} \mathbb{E}_{\mathbb{Q}}\left[\log\frac{d\,\mathbb{Q}}{d\,\mathbb{P}}\right]$$

where the minimum is taken over all $\mathbb{P}$-equivalent measures which satisfy condition (i) (i.e. all equivalent martingale measures). Having this measure it is interesting to note that the indifference price of Theorem 4.9.2 can be written as

$$v_0(H,s,\hat{s}) = \frac{1}{\gamma}\,\mathbb{E}_{\mathbb{Q}}\left[\log \mathbb{E}_{\mathbb{Q}}\left[e^{\gamma H}|S_1\right]\Big|S_0 = s, \hat{S}_0 = \hat{s}\right].$$

Thus, in case the payoff of the contingent claim depends only on the traded asset, i.e. $H = h(S_1)$, we obtain

$$v_0(H,s,\hat{s}) = \mathbb{E}_{\mathbb{Q}}[H|S_0 = s]$$

independent of $\hat{s}$. Moreover, it is easy to see that in this case the measure $\mathbb{Q}$ coincides on the σ-algebra $\Big\{\emptyset, \Omega, \{\omega_1,\omega_2\}, \{\omega_3,\omega_4\}\Big\}$ with the unique risk neutral measure in the binomial model. Thus the indifference pricing formula is consistent with the arbitrage free pricing formula (see Appendix C.1). ◇

Multiperiod Financial Market

In this section we extend the one-period financial market of the previous section to an N-period model by setting

$$S_{n+1} = S_n\tilde{R}_{n+1}, \quad \hat{S}_{n+1} = \hat{S}_n\hat{R}_{n+1}, \quad n = 0,1,\ldots,N-1$$

where we assume that the random vectors $(\tilde{R}_1, \hat{R}_1), (\tilde{R}_2, \hat{R}_2), \dots$ are independent and identically distributed and have the same distribution as in the last section. We have here $\mathcal{F}_n = \sigma(R_1, \dots, R_n, \hat{R}_1, \dots, \hat{R}_n)$.
Now let $H = h(S_N, \hat{S}_N)$ be the payoff of a contingent claim. The evolution of the wealth process (X_n) is here given by

$$X_{n+1} = X_n + \phi_n(\tilde{R}_{n+1} - 1)$$

where $\phi = (\phi_n)$ is a portfolio strategy. As in the one-period market the investor has an exponential utility function $U(x) = -e^{-\gamma x}, \gamma > 0$. Following the theory in Section 4.2 we can formulate the utility maximization problem as a Markov Decision Problem. Since $H = h(S_N, \hat{S}_N)$ we have to include the prices of the traded and non-traded assets in the state space:

- $E := \mathbb{R} \times \mathbb{R}_+^2$ where $(x, s, \hat{s}) \in E$ and x is the wealth and s and $\hat{s}$ are the values of the traded and non-traded assets respectively,
- $A := \mathbb{R}$ where $a \in A$ is the amount of money invested in the traded asset,
- $D(x, s, \hat{s}) := A$,
- $\mathcal{Z} := \mathbb{R}_+ \times \mathbb{R}_+$ where $z = (z_1, z_2) \in \mathcal{Z}$ are the relative price changes of the traded and non-traded assets respectively,
- $T\big((x, s, \hat{s}), a, (z_1, z_2)\big) := \big(x + a(z_1 - 1), sz_1, \hat{s}z_2\big)$,
- $Q^Z\big(\cdot\,|(x, s, \hat{s}), a\big) :=$ joint distribution of $(\tilde{R}_{n+1}, \hat{R}_{n+1})$ given by (p_1, p_2, p_3, p_4) (independent of $\big((x, s, \hat{s}), a\big)$),
- $r_n \equiv 0$,
- $g_N(x, s, \hat{s}) := -e^{-\gamma(x - h(s,\hat{s}))}$.

For $n = 0, 1, \dots, N$ we define the value functions by

$$V_n^H(x, s, \hat{s}) := \sup_\pi \mathbb{E}_{n,x,s,\hat{s}}^\pi \left[-e^{-\gamma(X_N - H)}\right]. \tag{4.38}$$

If $H \equiv 0$, then we write $V_n^0(x, s, \hat{s})$. The indifference price of H at time n which we denote by $v_n = v_n(H, s, \hat{s})$, is then defined as the value which satisfies:

$$V_n^0(x, s, \hat{s}) = V_n^H(x + v_n, s, \hat{s}), \quad x \in \mathbb{R}, s > 0, \hat{s} > 0.$$

This time it is not possible to obtain a closed formula for the indifference price, however, some important structural properties can be shown. In what follows we denote

$$v := \inf_{a \in \mathbb{R}} \mathbb{E}\left[e^{-\gamma a(\tilde{R}_1 - 1)}\right]. \tag{4.39}$$

In view of (FM) this problem has an optimal solution and the value v is positive and finite. We obtain:

Theorem 4.9.4. *For the multiperiod financial market it holds:*

a) The value functions of problem (4.38) are given by

$$V_n^H(x, s, \hat{s}) = -e^{-\gamma x} d_n(s, \hat{s}),$$

where (d_n) satisfy the recursion

$$\begin{aligned} d_N(s, \hat{s}) &:= e^{\gamma h(s,\hat{s})}, \\ d_n(s, \hat{s}) &:= \inf_{a \in \mathbb{R}} \mathbb{E}\left[e^{-\gamma a(\tilde{R}_{n+1}-1)} d_{n+1}(s\tilde{R}_{n+1}, \hat{s}\hat{R}_{n+1})\right]. \end{aligned}$$

In particular, $V_n^0(x, s, \hat{s}) = -e^{-\gamma x} v^{N-n}$.

b) The indifference price of the contingent claim H is given by

$$v_n(H, s, \hat{s}) = \frac{1}{\gamma} \log\left(\frac{d_n(s, \hat{s})}{v^{N-n}}\right).$$

c) The indifference prices satisfy the following consistency condition on Ω:

$$v_n\big(v_{n+1}(H, s\tilde{R}_{n+1}, \hat{s}\hat{R}_{n+1}), s, \hat{s}\big) = v_n(H, s, \hat{s}), \quad n = 0, 1, \ldots, N-1.$$

Proof. a) The expression for V_n^0 follows from Theorem 4.2.15. The statement for V_n^H follows by induction. Note that the value iteration has the form

$$\begin{aligned} V_N^H(x, s, \hat{s}) &= -e^{-\gamma x} e^{\gamma h(s,\hat{s})} = e^{-\gamma x} d_N(s, \hat{s}) \\ V_n^H(x, s, \hat{s}) &= \sup_{a \in \mathbb{R}} \mathbb{E}\left[V_{n+1}^H(x + a(\tilde{R}_{n+1} - 1), s\tilde{R}_{n+1}, \hat{s}\hat{R}_{n+1})\right]. \end{aligned}$$

b) The indifference price follows from part a).

c) Let $v_n := v_n(H, s, \hat{s})$ and $w_n := v_n\big(v_{n+1}(H, s\tilde{R}_{n+1}, \hat{s}\hat{R}_{n+1}), s, \hat{s}\big)$. Then, v_n is the solution of

$$V_n^0(x, s, \hat{s}) = V_n^H(x + v_n, s, \hat{s})$$

which can be written as

$$-e^{-\gamma x} v^{N-n} = -e^{-\gamma(x+v_n)} \inf_{a \in \mathbb{R}} \mathbb{E}\left[e^{-\gamma a(\tilde{R}_{n+1}-1)} d_{n+1}(s\tilde{R}_{n+1}, \hat{s}\hat{R}_{n+1})\right].$$

On the other hand w_n is the price at time n of a contingent claim with payoff

$$v_{n+1}(H, s\tilde{R}_{n+1}, \hat{s}\hat{R}_{n+1}) = \frac{1}{\gamma} \log\left(\frac{d_{n+1}(s\tilde{R}_{n+1}, \hat{s}\hat{R}_{n+1})}{v^{N-n-1}}\right).$$

Using the result of the one-period model, the price w_n is here the solution of

$$-e^{-\gamma x}v = -e^{-\gamma(x+w_n)} \inf_{a\in\mathbb{R}} \mathbb{E}\left[e^{-\gamma a(\tilde{R}_{n+1}-1)} \frac{d_{n+1}(s\tilde{R}_{n+1}, \hat{s}\hat{R}_{n+1})}{v^{N-n-1}}\right].$$

Comparing the last two equations we obtain $v_n = w_n$.

□

The indifference price $v_n(H, s, \hat{s})$ is positive since $d_n(s, \hat{s}) \geq v^{N-n} > 0$. The inequality follows by induction from part a). Due to part c) and to Remark 4.9.3 we can also compute the indifference price by iterating the expectation operator with respect to the minimal entropy martingale measure.

4.10 Approximation of Continuous-Time Models

Many financial models are defined in continuous-time and involve continuous-time stochastic processes like the Wiener process. The corresponding optimization problems are then also defined in continuous time. However, in most cases these problems can be approximated by discrete-time problems which reduce to Markov Decision Problems when the original problem is Markovian too. This observation can be used for numerical purposes. For another discussion into this direction see the example in Section 9.3. Moreover, sometimes the assumption of continuous trading may be questioned since interventions at discrete time points are more realistic. In order to illustrate the relation between continuous- and discrete-time problems, we will briefly explain in an informal way the so-called *approximating Markov chain* approach. For more details we refer the reader to Kushner and Dupuis (2001) or Prigent (2003). To outline the approach we assume that the controlled state process $X = (X_t)$ is given by the stochastic differential equation

$$\begin{aligned} dX_t &= \mu(X_t, u_t)dt + \sigma(X_t, u_t)dW_t, \\ X_0 &= x, \end{aligned}$$

where $W = (W_t)$ is a Wiener process. For simplicity we assume that X is real-valued and that the control $u = (u_t)$ has to satisfy $u_t \in \mathcal{U}$ for all $t \geq 0$ where $\mathcal{U}$ is an arbitrary compact set. The function μ and σ should be such that a solution of the stochastic differential equation exists. Of course, admissible controls have to be adapted with respect to a reasonable filtration (e.g. the filtration generated by the Wiener process). The aim is to maximize the expression

$$V_u(x) = \mathbb{E}_x^u \left[\int_0^T r(t, X_t, u_t) dt + g(X_T) \right], \quad x \in \mathbb{R}$$

over all admissible controls where r and g are suitably defined reward functions. We define the maximal value function by $V(x) = \sup_u V_u(x)$, $x \in \mathbb{R}$.

Now we search for a Markov chain which approximates the state process and for a reward functional such that V_u is approximated. These quantities together define the corresponding Markov Decision Problem. It seems to be reasonable to choose a state space $E := G$, a finite grid $G \subset \mathbb{R}$, an action space $A = \mathcal{U}$ and define $D(x) := A$. The step size is given by h. The approximating Markov chain $(\tilde{X}_n)$ and thus the transition kernel is obtained by the following *local consistency* conditions:

$$\mathbb{E}_{nx}^\pi [\tilde{X}_{n+1} - \tilde{X}_n] = \mathbb{E}_{tx}^u \left[\int_t^{t+h} \mu(X_s, u_s) ds \right] = \mu(x, u_t) h + o(h),$$

$$\begin{aligned} \mathbb{E}_{nx}^\pi \left[(\tilde{X}_{n+1} - \tilde{X}_n)^2 \right] &= \mathbb{E}_{tx}^u \left[\left(\int_t^{t+h} \sigma(X_s, u_s) dW_s \right)^2 \right] + o(h) \\ &= \sigma^2(x, u_t) h + o(h) \end{aligned}$$

where $t = nh$ and $f_n(x) = u_t$. In general these conditions do not determine the transition kernel Q of the Markov Decision Model uniquely, so there is still some freedom to choose the parameters. Note that it is also possible to use the variance of $\tilde{X}_{n+1} - \tilde{X}_n$ instead of the second moment since for $h \to 0$ the limits are the same. The reward function is defined by $r_n(x, a) := r(nh, x, a)h$ and the terminal reward function is g. Thus we obtain with $N = \frac{T}{h}$

$$V_{0\pi}(x) = V_{0\pi}^h(x) = \mathbb{E}_x^\pi \left[\sum_{n=0}^{N-1} r_n(\tilde{X}_n, A_n) + g(\tilde{X}_N) \right]$$

and the maximal value $V_0(x) = V_0^h(x) = \sup_\pi V_{0\pi}(x)$. Now under some additional conditions (like continuity and growth conditions) we obtain

$$\lim_{h \to 0} V_0^h(x) = V(x), \quad x \in E.$$

Often it is also possible to show that the optimal policies of the Markov Decision Problem approximate the optimal control in the continuous-time problem, but this is more demanding and yields further regularity conditions.

4.11 Remarks and References

Pliska (2000) uses the so-called *martingale method* (or *dual method*) to solve portfolio optimization problems. This method utilizes the fact that in complete financial markets every contingent claim can be perfectly hedged. Then the dynamic stochastic optimization problem can be split into two steps: (1) Find the optimal terminal wealth (via a Lagrangian approach) and (2) compute a hedging strategy w.r.t. the optimal terminal wealth. A more economic treatment of intertemporal portfolio selection can be found in Ingersoll (1987) and Huang and Litzenberger (1988).
Korn and Schäl (1999) consider value preserving and growth optimal portfolio strategies. Risk-sensitive portfolio optimization problems are studied in Bielecki and Pliska (1999), Bielecki et al. (1999), Stettner (1999, 2004) and Di Masi and Stettner (1999). Prigent (2007) investigates a lot of practical aspects of portfolio optimization in a static setting as well as in continuous-time.
Sections 4.1, 4.2 and 4.3: An investigation of the one-period optimization problem and its relation to the existence of no arbitrage opportunities can be found in Föllmer and Schied (2004). The equivalence between the existence of optimal portfolios for multiperiod utility maximization problems and absence of arbitrage opportunities follows essentially from the one-period problem since both the optimization problem and the arbitrage condition can be formulated in a local form. Some authors have investigated the relation between the existence of optimal portfolios and existence of equivalent martingale measures, the latter being equivalent to the absence of arbitrage opportunities. A first approach can be found in Hakansson (1971b) who used a 'no-easy-money-condition'. Further papers which treat this question are Schäl (1999, 2000, 2002), Rásonyi and Stettner (2005), and Rogers (1994).
Discrete-time multistage portfolio problems have been considered since the late 1960s by e.g. Mossin (1968), Samuelson (1969) Hakansson (1970, 1971a, 1974), Kreps and Porteus (1979) and Bodily and White (1982). He and Pearson (1991a,b) investigate consumption-investment problems with incomplete markets and short-sale constraints. Edirisinghe (2005) treats the terminal wealth problem with additional liability constraints. In Li and Wang (2008) martingale measures are identified from consumption problems. References for continuous-time portfolio selection problems are Korn (1997) and Pham (2009).

Section 4.4: Regime-switching is a well-known technique to generalize Markov models. A consumption-investment model with regime-switching can be found in Cheung and Yang (2007). In the earlier paper Cheung and Yang (2004) the authors consider the problem without consumption. Çanakoğlu and Özekici (2009) investigate exponential utility maximization whereas in Çanakoğlu and Özekici (2010) they consider the HARA utility problem and also allow the utility function to depend on the underlying Markov chain.

Shortfall risk minimization with regime switching is considered by Awanou (2007). In continuous-time, the classical utility maximization problem has been generalized in this way by Bäuerle and Rieder (2004). An interesting question in this context is how does the Markov-modulation influence the value function and the optimal policy? Often things become more risky with stochastically varying parameters.

Section 4.5: The transaction cost model is based on Kamin (1975), see also Gennotte and Jung (1994), Dumas and Luciano (1991) and Constantinides (1979). They show that the optimal portfolio strategy is to do nothing whenever it is in a certain region and to transact to the nearest boundary if the portfolio is outside this region. A more general model is treated in Abrams and Karmarkar (1980) where not necessarily convex no-transactions regions appear. More recent treatments of problems with transaction costs are contained in Bobryk and Stettner (1999), Sass (2005) and Trivellato (2009).

Section 4.6: Static mean-variance portfolio theory was investigated in Markowitz (1952) (see also the textbooks Markowitz (1987a,b)). Recently it has been rediscovered that the Italian mathematician Bruno de Finetti already anticipated much of Markowitz's mean-variance analysis by over a decade (see de Finetti (1940) and for an English translation see Barone (2006)). A discussion of the historical development can also be found in Rubinstein (2006). Tobin (1958) contributed the two-fund argument. Recently the multiperiod mean-variance problem has been solved explicitly by Li and Ng (2000). For an overview see e.g. Steinbach (2001). A version with regime-switching can be found in Çakmak and Özekici (2006), Costa and Araujo (2008) and Yin and Zhou (2004). Additional intertemporal restrictions are treated in Costa and Nabholz (2007). Zhu et al. (2004) study the mean-variance problem with bounds on the probability of bankruptcy in each period. A more general problem, namely variance-optimal hedging of claims in discrete time for arbitrary stock processes, has been considered in Schäl (1994), Schweizer (1995) and Motoczyński (2000). A mean-variance problem for an insurance company is investigated in Bäuerle (2005). In continuous-time there are different models and methods for dynamic mean-variance optimization problems. We just mention here Zhou and Li (2000) who solved the problem in a complete market driven by a diffusion process with deterministic coefficients. For an overview see Zhou (2003).

Section 4.7: A thorough analysis of the mean-risk problem can be found in Mundt (2007) and Bäuerle and Mundt (2009). There also the problem of intermediate risk constraints is considered in the binomial financial market. There is a growing number of papers which study different problems where the trade-off between expected return and risk is treated. Runggaldier et al. (2002) investigate the shortfall probability when a contingent claim has to be hedged in a binomial model. The authors also consider the problem with partial information. Favero and Vargiolu (2006) extend this analysis to convex and concave loss functions.

Section 4.8: Evolutionary heuristics for solving index tracking problems are treated in Beasley et al. (2003) and variance-optimal strategies in Bergtholdt (1998).

Section 4.9: The idea of indifference pricing is similar to the zero-utility premium principle in actuarial sciences. Most papers on this topic deal with continuous-time markets (a recent survey of theory and applications of indifference pricing is Carmona (2009)). The presentation of this section is based on Musiela and Zariphopoulou (2004). They present the recursive pricing algorithm with the minimal entropy martingale measure.

Section 4.10: A thorough analysis of the approximating Markov chain approach can be found in Kushner and Dupuis (2001) (see also Prigent (2003)). For an application to the Merton problem see Munk (2003). Discrete-time approximations of the dynamic programming equation using analytical methods can be found in Bensoussan and Robin (1982).
There are also a number of papers which investigate the convergence of optimal portfolio strategies in a discrete-time optimization problem to optimal portfolio strategies in a continuous-time problem. He (1991) considers optimal consumption-investment problems and uses the martingale method whereas in Fitzpatrick and Fleming (1991) viscosity solution techniques are used. Duffie and Protter (1992) consider weak convergence of stochastic integrals and apply it to the gain process in financial markets. Rogers (2001) compares a classical continuous-time investor in a Black-Scholes-Merton market with an investor who rebalances her portfolio only at times which are multiples of h. His findings show that the difference is typically low.

Part II
Partially Observable Markov Decision Problems

Chapter 5
Partially Observable Markov Decision Processes

In many applications the decision maker has only partial information about the state process, i.e. part of the state cannot be observed. Examples can be found in engineering, economics, statistics, speech recognition and learning theory among others. An important financial application is given when the drift of a stock price process is unobservable and hard to estimate.
In contrast to the previous sections we cannot observe the complete state directly. Thus it is natural to assume that the admissible policies can only depend on the observed history. Since however the reward and the transition kernel may also depend on the unobservable part of the state it is a priori not clear how to solve the optimization problem. However, by introducing an information process (filter) the problem can be reformulated in terms of a Markov Decision Process with complete information and the theory developed in Chapter 2 can be applied. This approach works here since the *separation principle of estimation and control* holds, i.e. the estimation step is done first and then the optimization. The price one has to pay for this reformulation is an enlarged state space. More precisely, besides the observable part of the state, a probability distribution enters the new state space which defines the conditional distribution of the unobserved state given the history of observations so far.
In Section 5.1 we introduce the *Partially Observable Markov Decision Model.* In what follows the two important special cases of a *Hidden Markov Model* and a *Bayesian Model* will be of particular interest. In a Hidden Markov Model, the unobservable process is a Markov chain with finite state space and in a Bayesian Model the unobservable process is simply a parameter which does not change in time. In Section 5.2 we deal with the probability distribution which enters the state space and derive a recursive formula which is called *filter equation.* In Section 5.3 we explain the reformulation as a classical Markov Decision Model (so-called *filtered model*) with enlarged state space. In Section 5.4 we treat in more detail the Bayesian Model. In particular we introduce the concept of a *sufficient statistic* here which enables us to simplify in some cases the state space. This concept is well known in statistics.

N. Bäuerle and U. Rieder, *Markov Decision Processes with Applications to Finance*, Universitext, DOI 10.1007/978-3-642-18324-9_5,

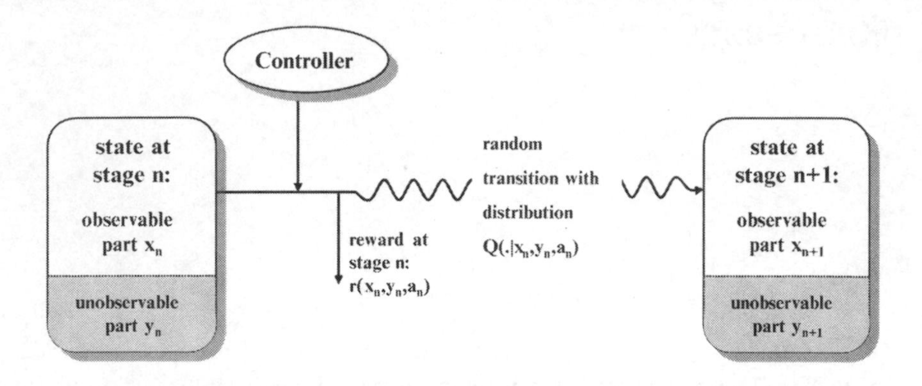

Fig. 5.1 General evolution of a Partially Observable Markov Decision Model.

The key observation is that only a part of the observable history is necessary to determine the conditional distribution of the unknown parameter. We close this chapter with the investigation of two-armed Bernoulli bandits which is a prominent example of Partially Observable Markov Decision Models.

5.1 Partially Observable Markov Decision Processes

We first define the ingredients of a Partially Observable Markov Decision Model where we restrict to stationary data. For the evolution of such models see Figure 5.1.

Definition 5.1.1. A *Partially Observable Markov Decision Model* consists of a set of data $(E_X \times E_Y, A, D, Q, Q_0, r, g, \beta)$ with the following interpretation:

- $E_X \times E_Y$ is the *state space*. We assume that E_X and E_Y are Borel subsets of Polish spaces. Let $(x, y) \in E_X \times E_Y$. Then x is the observable component of the state and y the part which cannot be observed.
- A is the *action space*. We assume that A is a Borel subset of a Polish space.
- $D \subset E_X \times A$ is the set of possible state-action pairs. As usual, $D(x) = \{a \in A | (x, a) \in D\}$ is the set of *feasible actions* depending only on the observable part $x \in E_X$. We assume that D contains the graph of a measurable function from E_X to A.
- Q is a stochastic kernel from $D \times E_Y$ to $E_X \times E_Y$ which determines the distribution of the new state, given the current state and action, i.e. if

$(x, y) \in E_X \times E_Y$ is the current state, $a \in D(x)$ the action, we have that

$$Q(B \times C|x, y, a)$$

is the probability that the new state is in the Borel set $B \times C \subset E_X \times E_Y$.

- Q_0 is the initial distribution of Y_0.
- $r : D \times E_Y \to \mathbb{R}$ is a measurable function. $r(x, y, a)$ gives the *one-stage reward* of the system in state (x, y) if action a is taken.
- $g : E_X \times E_Y \to \mathbb{R}$ is a measurable mapping. $g(x, y)$ gives the *terminal reward* of the system.
- $\beta \in (0, 1]$ is the discount factor.

The planning horizon of the system is $N \in \mathbb{N}$.

As in Chapter 2 we can define a *Partially Observable Markov Decision Process* (X_n, Y_n) on a common probability space (this will be done later in this section). By (X_n) we denote the observable process and by (Y_n) the unobservable process.
In what follows it is convenient to introduce

$$Q^X(B|x, y, a) := Q(B \times E_Y|x, y, a), \quad B \in \mathcal{B}(E_X) \tag{5.1}$$

the marginal transition probability of the observable part. In our applications the following special cases play a key role.

Example 5.1.2 (Hidden Markov Model). In this case (Y_n) constitutes a (stationary) Markov chain on a finite state space which can neither be controlled nor observed but which influences the reward and the dynamics of our system. Moreover, we assume that the transition of this Markov chain is independent from the transition of the observable part. Thus, our general model specializes in the following way: Suppose without loss of generality that $E_Y = \{1, \dots, m\}$ is the state space of the Markov chain. If we denote by

$$p_{ij} := \mathbb{P}(Y_{n+1} = j \mid Y_n = i)$$

the transition probabilities of the Markov chain, the transition kernel for the state process is

$$Q(B \times \{j\}|x, i, a) = p_{ij} Q^X(B|x, i, a)$$

where Q^X still has to be specified. ♦

In order to define policies we have to introduce the *sets of observable histories* which are denoted by

$$H_0 := E_X,$$
$$H_n := H_{n-1} \times A \times E_X.$$

An element $h_n = (x_0, a_0, x_1, \dots, x_n) \in H_n$ is called *observable history up to time n*.

Definition 5.1.3. a) A measurable mapping $f_n : H_n \to A$ with the property $f_n(h_n) \in D(x_n)$ for $h_n \in H_n$ is called a *decision rule* at stage n.
b) A sequence $\pi = (f_0, f_1, \dots, f_{N-1})$ where f_n is a decision rule at stage n for all n, is called *N-stage policy.* We denote by Π_N the set of all N-stage policies.

Let $\pi = (f_0, f_1, \dots, f_{N-1}) \in \Pi_N$ be an arbitrary N-stage policy and let $x \in E_X$ be an initial state. Then the initial conditional distribution Q_0 together with the transition probability Q define a probability measure $\mathbb{P}_x^\pi$ on $(E_X \times E_Y)^{N+1}$ endowed with the product σ-algebra. This follows from the Theorem of Ionescu Tulcea (see Appendix, Proposition B.2.5). More precisely $\mathbb{P}_x^\pi(\cdot) = \int P_{xy}^\pi(\cdot) Q_0(dy)$. For $\omega = (x_0, y_0, \dots, x_N, y_N) \in (E_X \times E_Y)^{N+1}$ we define the random variables X_n and Y_n in a canonical way by their projections

$$X_n(\omega) = x_n, \quad Y_n(\omega) = y_n.$$

If $\pi = (f_0, f_1, \dots, f_{N-1}) \in \Pi_N$ is a given policy, we define recursively

$$\begin{aligned} A_0 &:= f_0(X_0) \\ A_n &:= f_n(X_0, A_0, X_1, \dots, X_n), \end{aligned}$$

the sequence of actions which are chosen successively under policy π. It should always be clear from the context which policy generates (A_n). The optimization problem is now defined as follows. For $\pi \in \Pi_N$ and $X_0 = x$ denote

$$J_{N\pi}(x) := \int \mathbb{E}_{xy}^\pi \left[\sum_{n=0}^{N-1} \beta^n r(X_n, Y_n, A_n) + \beta^N g(X_N, Y_N) \right] Q_0(dy)$$

and

$$J_N(x) := \sup_{\pi \in \Pi_N} J_{N\pi}(x). \tag{5.2}$$

Problem (5.2) is called an N-stage *Partially Observable Markov Decision Problem* and the process (X_n, Y_n) is called a *Partially Observable Markov Decision Process.* Note that the objective in (5.2) has a different form than in Chapter 2. Here we have an additional expectation with respect to Q_0. Furthermore the admissible policies have to be independent of the unobservable states. Hence we cannot use the results of Chapter 2 directly. But in Section 5.3 we will reformulate this optimization problem as a standard Markov Decision Process. In order to obtain a well-defined problem we assume throughout this section:

Assumption: *For all $x \in E_x$ it holds*

$$\sup_\pi \int \mathbb{E}^\pi_{xy}\left[\sum_{n=0}^{N-1} \beta^n r^+(X_n, Y_n, A_n) + \beta^N g^+(X_N, Y_N)\right] Q_0(dy) < \infty.$$

5.2 Filter Equations

In the analysis which follows, the conditional distribution of the unobserved state Y_n, given the information $\sigma\big(X_0, A_0, X_1, \dots, A_{n-1}, X_n\big)$ will be crucial. This conditional distribution can be computed recursively and this recursion is called a *filter equation.* In this section we will derive this equation and give some examples.
To begin with, let us denote by $\mathbb{P}(E_Y)$ the space of all probability measures on E_Y. Note that if E_Y is a Borel space, $\mathbb{P}(E_Y)$ is a Borel space, too. In what follows we *assume* that the transition kernel Q has a density q with respect to some σ-finite measures λ and ν, i.e.

$$Q(d(x', y')|x, y, a) = q(x', y'|x, y, a)\lambda(dx')\nu(dy').$$

The key building block of the filter equation is the so-called *Bayes operator* $\Phi : E_X \times \mathbb{P}(E_Y) \times A \times E_X \to \mathbb{P}(E_Y)$ defined by

$$\Phi(x, \rho, a, x')(C) := \frac{\int_C \Big(\int q(x', y'|x, y, a)\rho(dy)\Big)\nu(dy')}{\int_{E_Y} \Big(\int q(x', y'|x, y, a)\rho(dy)\Big)\nu(dy')}, \quad C \in \mathcal{B}(E_Y).$$

We will see in Theorem 5.2.1 below that if ρ is a (conditional) distribution of Y_n, then $\Phi(x, \rho, a, x')$ is a conditional distribution of Y_{n+1} given ρ, the current observation x, action a and the next observation x'.
Let us next define for $h_{n+1} = (h_n, a_n, x_{n+1}) \in H_{n+1} = H_n \times A \times E_X$ and $C \in \mathcal{B}(E_Y)$ recursively

$$\mu_0 := Q_0 \tag{5.3}$$

$$\mu_{n+1}(C \mid h_n, a_n, x_{n+1}) := \Phi\Big(x_n, \mu_n(\cdot \mid h_n), a_n, x_{n+1}\Big)(C). \tag{5.4}$$

This recursion is called a *filter equation.* The following theorem shows that μ_n is indeed a conditional distribution of Y_n given $(X_0, A_0, X_1, \dots, A_{n-1}, X_n)$.

Theorem 5.2.1. *For all $\pi \in \Pi_N$ we have for $C \in \mathcal{B}(E_Y)$*

$$\mu_n(C \mid X_0, A_0, X_1, \dots, X_n) = \mathbb{P}_x^\pi\left(Y_n \in C \mid X_0, A_0, X_1, \dots, X_n\right).$$

For the proof we need the following lemma.

Lemma 5.2.2. *Let $v : H_n \times E_X \times E_Y \to \mathbb{R}$ be measurable. Then for all $h_n \in H_n$ and $a_n \in D(x_n)$ it holds:*

$$\int \mu_n(dy_n|h_n) \int Q\big(d(x_{n+1}, y_{n+1}) \mid x_n, y_n, a_n\big) v(h_n, x_{n+1}, y_{n+1})$$
$$= \int \mu_n(dy_n|h_n) \int Q^X(dx_{n+1}|x_n, y_n, a_n)$$
$$\int \Phi(x_n, \mu_n, a_n, x_{n+1})(dy_{n+1}) v(h_n, x_{n+1}, y_{n+1})$$

provided the integrals exist.

Proof. The left-hand side of the equation can be written as

$$\int \mu_n(dy_n|h_n) \int \lambda(dx_{n+1}) \int \nu(dy_{n+1}) q(x_{n+1}, y_{n+1}|x_n, y_n, a_n)$$
$$v(h_n, x_{n+1}, y_{n+1}).$$

For the right-hand side we obtain

$$\int \mu_n(dy_n|h_n) \int \lambda(dx_{n+1}) \int \nu(dy) q(x_{n+1}, y|x_n, y_n, a_n)$$
$$\int \Phi(x_n, \mu_n, a_n, x_{n+1})(dy_{n+1}) v(h_n, x_{n+1}, y_{n+1}).$$

Inserting the definition of the Bayes operator and applying Fubini's theorem we see that the right-hand side is equal to the left-hand side. □

Proof. (of Theorem 5.2.1) It follows by induction from Lemma 5.2.2 that

$$\mathbb{E}_x^\pi[v(X_0, A_0, X_1, \dots, X_n, Y_n)] = \mathbb{E}_x^\pi[v'(X_0, A_0, X_1, \dots, X_n)] \tag{5.5}$$

for all $v : H_n \times E_Y \to \mathbb{R}$ and $v'(h_n) := \int v(h_n, y_n)\mu_n(dy_n|h_n)$ provided that the expectations exist. For $n = 0$ both sides reduce to $\int v(x, y) Q_0(dy)$. For a given observable history h_{n-1} it holds:

$$\begin{aligned}
&\mathbb{E}_x^\pi[v(h_{n-1}, A_{n-1}, X_n, Y_n)] \\
&= \int \mu_{n-1}(dy_{n-1}|h_{n-1}) \int Q\big(d(x_n, y_n)|x_{n-1}, y_{n-1}, f_{n-1}(h_{n-1})\big) \\
&\qquad\qquad\qquad\qquad\qquad\qquad v\big(h_{n-1}, f_{n-1}(h_{n-1}), x_n, y_n\big)
\end{aligned}$$

$$\begin{aligned}
&\mathbb{E}_x^\pi\left[v'(h_{n-1}, A_{n-1}, X_n)\right] \\
&= \int \mu_{n-1}(dy_{n-1}|h_{n-1}) \int Q^X\left(dx_n|x_{n-1}, y_{n-1}, f_{n-1}(h_{n-1})\right) \\
&\qquad\qquad\qquad\qquad\qquad\qquad v'\big(h_{n-1}, f_{n-1}(h_{n-1}), x_n\big).
\end{aligned}$$

Lemma 5.2.2 and the definition of μ_n imply that both expectations are equal. The induction step gives finally (5.5). In particular for $v = 1_{B\times C}$ we obtain

$$\begin{aligned}
&\mathbb{P}_x^\pi\Big((X_0, A_0, X_1, \dots, X_n) \in B, Y_n \in C\Big) \\
&= \mathbb{E}_x^\pi\Big[1_B(X_0, A_0, X_1, \dots, X_n)\mu_n\big(C|X_0, A_0, X_1, \dots, X_n\big)\Big],
\end{aligned}$$

i.e. $\mu_n\big(\cdot\,|X_0, A_0, X_1, \dots, X_n\big)$ is a conditional $\mathbb{P}_x^\pi$-distribution of Y_n given $(X_0, A_0, X_1, \dots, X_n)$. □

The next example shows what the Bayes operator looks like in the Hidden Markov Model.

Example 5.2.3 (Hidden Markov Model). Let us recall Example 5.1.2. We have assumed that $E_Y = \{1, \dots, m\}$. Thus,

$$I\!P(E_Y) = \{p \in [0,1]^m \mid \sum_{i=1}^m p_i = 1\}.$$

Moreover, let us suppose that the transition kernel Q^X of the observable part defined in (5.1) has a λ-density q^X. Then the Bayes operator for the Hidden Markov Model is given by:

$$\Phi(x, \rho, a, x')(\{k\}) = \frac{\sum_{i=1}^m \rho(i) p_{ik} q^X(x' \mid x, i, a)}{\sum_{i=1}^m \rho(i) q^X(x' \mid x, i, a)}, \qquad k = 1, \dots, m$$

and Theorem 5.2.1 implies for $k = 1, \dots, m$:

$$\mu_n\big(\{k\} \mid X_0, A_0, X_1, \dots, X_n\big) = \mathbb{P}_x^\pi\left(Y_n = k|X_0, A_0, X_1, \dots, X_n\right). \quad \blacklozenge$$

Often in applications the state transition depends on disturbances (Z_n) which are observable, but the distribution of Z_{n+1} depends on the unobservable state Y_n. The random variables (Z_n) take values in a Borel subset $\mathcal{Z}$ of a Polish space. More precisely, it is assumed that instead of the transition kernel Q a stochastic kernel $Q^{Z,Y}$ is given, where $Q^{Z,Y}(B \times C|x,y,a)$ is the probability that the disturbance variable Z_{n+1} is in B and the unobservable state Y_{n+1} is in C, given the system state is $(X_n, Y_n) = (x,y)$ and action $a \in D(x)$ is taken. In what follows we will denote by

$$Q^Z(B|x,y,a) := Q^{Z,Y}(B \times E_Y|x,y,a), \ B \in \mathcal{B}(\mathcal{Z})$$

the marginal distribution of the disturbance variable Z_{n+1}. The new observable state is then determined by a transition function $T_X : E_X \times A \times \mathcal{Z} \mapsto E_X$ via

$$x_{n+1} = T_X(x_n, a_n, z_{n+1}).$$

The transition kernel Q of the Partially Observable Markov Decision Model is defined for $B \in \mathcal{B}(E_X)$ and $C \in \mathcal{B}(E_Y)$ by

$$Q(B \times C|x,y,a) = Q^{Z,Y}\Big(\{z \in \mathcal{Z}|T_X(x,a,z) \in B\} \times C|x,y,a\Big).$$

Here it is now reasonable to introduce the *sets of observable histories* including the disturbance by

$$\begin{aligned}\tilde{H}_0 &:= E_X,\\ \tilde{H}_n &:= \tilde{H}_{n-1} \times A \times \mathcal{Z} \times E_X.\end{aligned}$$

Elements are denoted by $\tilde{h}_n = (x_0, a_0, z_1, x_1, \dots, z_n, x_n) \in \tilde{H}_n$. When we suppose that $Q^{Z,Y}$ has a density with respect to some σ-finite measures λ and ν, i.e.

$$Q^{Z,Y}(d(z,y')|x,y,a) = q^{Z,Y}(z,y'|x,y,a)\lambda(dz)\nu(dy')$$

then the Bayes operator can be written as

$$\Phi(x,\rho,a,z)(C) := \frac{\int_C \Big(\int q^{Z,Y}(z,y'|x,y,a)\rho(dy)\Big)\nu(dy')}{\int_{E_Y} \Big(\int q^{Z,Y}(z,y'|x,y,a)\rho(dy)\Big)\nu(dy')}, \quad C \in \mathcal{B}(E_Y). \tag{5.6}$$

Note that we use the same notation Φ here (the meaning should always be clear from the context). If we define for $C \in \mathcal{B}(E_Y)$

$$\begin{aligned}\mu_0 &:= Q_0\\ \mu_{n+1}(C \mid \tilde{h}_n, a_n, z_{n+1}, x_{n+1}) &:= \Phi\Big(x_n, \mu_n(\cdot \mid \tilde{h}_n), a_n, z_{n+1}\Big)(C)\end{aligned}$$

then

$$\mu_n(C \mid X_0, \dots, Z_n, X_n) = \mathbb{P}_x^\pi\big(Y_n \in C \mid X_0, \dots, Z_n, X_n\big).$$

Note that $\mu_n(C \mid \tilde{h}_n) = \mu_n(C \mid x_0, a_0, z_1, x_1, \dots, z_n, x_n)$ does not depend on x_n. The following example contains an important special case.

Example 5.2.4 (Bayesian Model). In a Bayesian Model the process (Y_n) is given by $Y_n = \vartheta$ for all n, i.e. $Q(B \times C|x, \theta, a) := Q_\theta^X(B|x, a)\delta_\theta(C)$. This means that the rewards and the transition kernel depend on a parameter θ which is not known despite its initial distribution Q_0, which is called the *prior distribution* of ϑ. In this case it is common to write $E_Y = \Theta$. We suppose that the observable part of the state is updated by a transition function $T_X : E_X \times A \times \mathcal{Z} \mapsto E_X$. The Bayes operator is here given by

$$\Phi(x, \rho, a, z)(C) = \frac{\int_C q^Z(z|x, \theta, a)\rho(d\theta)}{\int_\Theta q^Z(z|x, \theta, a)\rho(d\theta)}, \quad C \in \mathcal{B}(\Theta). \tag{5.7}$$

◆

The following example shows an important application where the sequence (μ_n) can be computed explicitly.

Example 5.2.5 (Kalman-Filter). This example is famous and can be found in many textbooks. We suppose that $E_X = \mathbb{R}^d$, $E_Y = \mathbb{R}^l$ and the state transition is given by

$$\begin{aligned} Y_{n+1} &= BY_n + \varepsilon_{n+1}, \\ X_{n+1} &= CY_{n+1} + \eta_{n+1} \end{aligned}$$

with matrices B and C of appropriate dimension. Moreover, the system cannot be controlled, i.e. $D(x)$ is a singleton. As before, X_n is the observable part of the state and Y_n is the unobservable part of the state at time n. We assume here that $Y_0 \sim Q_0 = \mathcal{N}(0, S)$ where $\mathcal{N}(0, S)$ is the multivariate normal distribution with expectation vector 0 and covariance matrix S and that the disturbance variables $\varepsilon_1, \eta_1, \varepsilon_2, \eta_2, \dots$ are all independent and have a multivariate normal distribution with

$$Q^\varepsilon(\cdot|x, y, a) = \mathcal{N}(0, R), \quad Q^\eta(\cdot|x, y, a) = \mathcal{N}(0, S)$$

where R and S are covariance matrices with RR' being positive definite. These assumptions together with the transition functions determine the stochastic kernel Q of the Partially Observable Markov Decision Model.

It follows from the properties of the normal distribution that $\mu_n(\cdot|h_n)$ is a multivariate normal distribution with mean $m_n(h_n)$ and covariance matrix Σ_n (which is independent of h_n), i.e.

$$\mu_n(\cdot|h_n) = \mathcal{N}\big(m_n(h_n), \Sigma_n\big).$$

The mean and covariance matrix can be computed recursively by the following *Kalman-filter equations* for $n \in \mathbb{N}$:

$$\begin{aligned}
\Sigma_{n|n-1} &:= B\Sigma_{n-1}B' + S, \\
K_n &:= \Sigma_{n|n-1}C'\left(C\Sigma_{n|n-1}C' + R\right)^{-1}, \\
\Sigma_n &:= \Sigma_{n|n-1} - K_n C \Sigma_{n|n-1}, \\
m_n(h_n) &:= Bm_{n-1}(h_{n-1}) + K_n\big(x_n - CBm_{n-1}(h_{n-1})\big)
\end{aligned}$$

where the recursion is started with $\Sigma_0 = S$ and $m_0 = 0$. Using linear transformations, the filter can be extended to the case where Y_n has a deterministic drift D, i.e. $Y_{n+1} = BY_n + D + \varepsilon_{n+1}$. An illustration of the *Kalman-filter* can be seen in Figure 5.2 with $Y_{n+1} = 1.001Y_n + 0.1 + \varepsilon_{n+1}$ and $X_n = 2Y_n + \eta_{n+1}$ and independent and $\mathcal{N}(0,1)$-distributed random variables $\varepsilon_n, \eta_n, \ldots$.

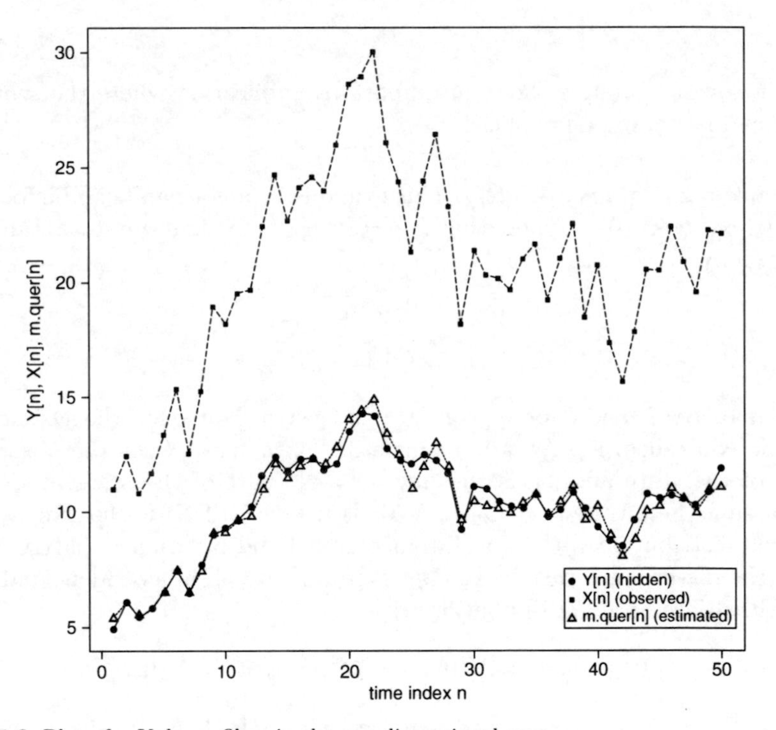

Fig. 5.2 Plot of a Kalman filter in the one-dimensional case.

A realization of the sequences $(X_n), (Y_n)$ and $m_n(h_n)$ is plotted. The upper line of circles is given by the realizations of (X_n) which are observable. The lower line of circles is given by the unobservable (Y_n). The line of triangles consists of the estimates $m_n(h_n)$. ♦

5.3 Reformulation as a Standard Markov Decision Model

We introduce now a second problem which is a fully observable Markov Decision Problem as in Chapter 2 and which turns out to be equivalent to the partially observable problem of the previous section. The idea is to enlarge the state space of the problem by adding the (filtered) probability distribution of Y_n given the history of observations up to time n. This conditional distribution μ_n has been computed recursively by the filter equation (5.4). It will be shown that μ_n contains exactly the relevant information in order to derive optimal policies.
Now suppose a Partially Observable Markov Decision Model is given by the data $(E_X \times E_Y, A, D, Q, Q_0, r, g, \beta)$ as described in Section 5.1. Let us define the following stationary N-stage Markov Decision Model.

Definition 5.3.1. The *filtered Markov Decision Model* consists of a set of data $(E, A, D, Q', r', g', \beta)$ with the following meaning:

- $E := E_X \times I\!P(E_Y)$ is the *state space*. An element is denoted by (x, ρ). x is the observable part of the state of the Partially Observable Markov Decision Model and ρ is the (conditional) distribution of the unobservable state.
- A is the *action space.*
- $D \subset E \times A$ is the set of possible state-action pairs. $D(x, \rho) := D(x)$ for all $(x, \rho) \in E$ is the set of *feasible actions* in state (x, ρ).
- Q' is a stochastic kernel which determines the distribution of the new state as follows: For fixed $(x, \rho) \subset E$, $a \in D(x)$ and Borel subsets $B \subset E_X$ and $C \subset I\!P(E_Y)$ we have

$$Q'(B \times C \mid x, \rho, a) := \int \int_B 1_C\big(\Phi(x, \rho, a, x')\big) Q^X(dx'|x, y, a)\rho(dy).$$

- $r' : D \to \mathbb{R}$ determines the one-stage reward and is given by

$$r'(x, \rho, a) := \int r(x, y, a)\rho(dy).$$

- $g' : E \to \mathbb{R}$ is the *terminal reward* given by

$$g'(x, \rho) := \int g(x, y)\rho(dy).$$

- $\beta \in (0, 1]$ is the discount factor.

The policies for the filtered Markov Decision Model are well-defined by the general theory in Chapter 2. It is important to note that the policies $\pi \in \Pi_N$

which depend on the observable history and which have been introduced in Section 5.1 are feasible for the filtered Markov Decision Model (see Section 2.2). Of course such a policy is in general not Markovian. On the other hand, if $(f_0', \dots, f_{N-1}')$ is a Markov policy for the filtered Markov Decision Model, then $(f_0, \dots, f_{N-1}) \in \Pi_N$ where

$$f_n(h_n) := f_n'\big(x_n, \mu_n(\cdot|h_n)\big).$$

Hence the set of all Markov policies for the filtered Markov Decision Model can be identified as a subset of Π_N. For the optimization we use initially the larger class Π_N of all history-dependent policies. The value functions for the filtered Markov Decision Model are defined as in Chapter 2 and denoted by

$$J_{N\pi}'(x, \rho), \quad \pi \in \Pi_N, \ (x, \rho) \in E$$

and

$$J_N'(x, \rho) = \sup_{\pi \in \Pi_N} J_{N\pi}'(x, \rho), \quad (x, \rho) \in E.$$

Now we are able to show that the filtered Markov Decision Model with value functions J_N' solves the Partially Observable Markov Decision Problem of Section 5.1.

Theorem 5.3.2. *Let $\pi \in \Pi_N$. Then for all $x \in E_X$*

$$J_{N\pi}(x) = J_{N\pi}'(x, Q_0)$$

and hence $J_N(x) = J_N'(x, Q_0)$.

Proof. The filtered Markov Decision Model induces a probability measure $\bar{P}_{xQ_0}^{\pi}$ on E^{N+1} through $\pi \in \Pi_N, x \in E_X$, the initial distribution Q_0 and the transition probability Q'. The expectation with respect to $\bar{P}_{xQ_0}^{\pi}$ is denoted by $\bar{E}_{xQ_0}^{\pi}$. We show that

$$\mathbb{E}_x^{\pi}[v(X_0, A_0, X_1, \dots, X_n, Y_n)] = \bar{\mathbb{E}}_{xQ_0}^{\pi}[v'(X_0, A_0, X_1, \dots, X_n, \mu_n)] \quad (5.8)$$

for all $v : H_n \times E_Y \to \mathbb{R}$ and $v'(h_n, \rho) := \int v(h_n, y)\rho(dy)$ provided that the expectations exist. The proof follows along the same lines as the proof of (5.5). Then we obtain immediately from (5.8) that

$$\mathbb{E}_x^{\pi}\left[r(X_n, Y_n, A_n)\right] = \bar{\mathbb{E}}_{xQ_0}^{\pi}\left[r'(X_n, \mu_n, A_n)\right] \quad (5.9)$$

and

$$\mathbb{E}_x^{\pi}\left[g(X_N, Y_N)\right] = \bar{\mathbb{E}}_{xQ_0}^{\pi}\left[g'(X_N, \mu_N)\right]. \quad (5.10)$$

Now the first statement is true. For the second statement $J_N(x) = J'_N(x, Q_0)$ it is important to note that the value of the filtered Markov Decision Model is not improved if we use the larger class of policies Π_N instead of Markov policies, i.e.

$$\sup_{\pi \in \Pi_N} J'_{N\pi}(x,\rho) = \sup_{\pi' Markov} J'_{N\pi'}(x,\rho), \quad (x,\rho) \in E.$$

(cf. Theorem 2.2.3). Then the proof is complete. □

Of course we can now solve the Partially Observable Markov Decision Problem with the methods developed in Chapter 2. Note that the Integrability Assumption for the Partially Observable Markov Decision Model is equivalent to (A_N) for the filtered Markov Decision Model. Let us define

$$Q^X(B|x,\rho,a) := \int Q^X(B \times E_Y|x,y,a)\rho(dy)$$

for $B \in \mathcal{B}(E_X)$. In particular we obtain the following *Bellman equation.*

Theorem 5.3.3. *Suppose the filtered Markov Decision Model satisfies the Structure Assumption of Theorem 2.3.8.*

a) Then the Bellman equation holds, i.e. for $(x,\rho) \in E_X \times \mathbb{P}(E_Y)$:

$$\begin{aligned} J'_0(x,\rho) &= \int g(x,y)\rho(dy), \\ J'_n(x,\rho) &= \sup_{a \in D(x)} \Big\{ \int r(x,y,a)\rho(dy) \\ &\quad + \beta \int J'_{n-1}(x', \Phi(x,\rho,a,x'))Q^X(dx'|x,\rho,a) \Big\}. \end{aligned}$$

b) Let f'_n *be a maximizer of* J'_{n-1} *for* $n = 1, \dots, N$. *Then the policy* $\pi^* := (f_0^*, \dots, f_{N-1}^*)$ *is optimal for the* N*-stage Partially Observable Markov Decision Problem, where*

$$f_n^*(h_n) := f'_{N-n}(x_n, \mu_n(\cdot|h_n)), \quad h_n \in H_n.$$

5.4 Bayesian Decision Models

In this section we consider the Bayesian Model which was introduced in Example 5.2.4 in more detail, since the formulation of the filtered Markov Decision Problem can be simplified considerably. It is also interesting to note that the optimization problem (5.2) can in this case be interpreted as follows:

Suppose the realization of ϑ is known and equal to θ (i.e. the prior distribution is concentrated in θ). Then we have an ordinary Markov Decision Model with disturbance distribution $Q^Z(\cdot|x,\theta,a)$ and we denote the corresponding value function under a policy $\pi \in \Pi_N$ by $J^\theta_{N\pi}(x)$. Problem (5.2) is then equivalent to

$$J_N(x) = \sup_{\pi\in\Pi_N} \int J^\theta_{N\pi}(x) Q_0(d\theta).$$

Note that in general there does not exist a policy that is optimal for all $\theta \in \Theta$. In the Bayesian model the conditional distribution $\mu_n(\cdot \mid \tilde{h}_n)$ of ϑ given the observable history $\tilde{h}_n = (x_0, a_0, z_1, x_1, \ldots, x_n) \in \tilde{H}_n$ admits the following explicit formula which can be shown by induction.

Lemma 5.4.1. *The posterior distribution μ_n has a Q_0-density, i.e.*

$$\mu_n(C \mid \tilde{h}_n) = \frac{\int_C \prod_{k=0}^{n-1} q^Z(z_{k+1} \mid x_k, \theta, a_k) Q_0(d\theta)}{\int_\Theta \prod_{k=0}^{n-1} q^Z(z_{k+1} \mid x_k, \theta, a_k) Q_0(d\theta)}, \quad C \in \mathcal{B}(\Theta).$$

This representation is very useful when properties of μ_n have to be shown. Moreover, we obtain from Lemma 5.2.2 the following martingale property of the filter process (posterior distributions) (μ_n).

Lemma 5.4.2. *For fixed $C \in \mathcal{B}(\Theta)$ the process*

$$M_n := \mu_n\big(C|X_0, A_0, Z_1, X_1, \ldots, X_n\big)$$

is a martingale, i.e. for all π and $x \in E_X$:

$$\mathbb{E}^\pi_x \Big[M_{n+1} \mid X_0, A_0, Z_1, X_1, \ldots, X_n\Big] = M_n, \quad n \in \mathbb{N}_0.$$

In the next subsection we will present some examples.

Sufficient Statistics

In order to compute $\mu_n(\cdot \mid \tilde{h}_n)$ it is sometimes sufficient to know only a part of $\tilde{h}_n$ or a certain function (characteristics) of it.

Definition 5.4.3. Let a Bayesian Model be given and let I be an arbitrary space endowed with a σ-algebra $\mathcal{I}$. A sequence $t = (t_n)$ of measurable mappings $t_n : \tilde{H}_n \to I$ is called a *sufficient statistic* for (μ_n), if there exists a transition kernel $\hat{\mu}$ from I to Θ such that

$$\mu_n(C \mid \tilde{h}_n) = \hat{\mu}\big(C \mid t_n(\tilde{h}_n)\big)$$

for all $\tilde{h}_n \in \tilde{H}_n$, $C \in \mathcal{B}(\Theta)$ and $n \in \mathbb{N}_0$.

Note that $t_n(\tilde{h}_n)$ is independent of x_n, since $\mu_n(\cdot|\tilde{h}_n)$ is independent of x_n by Lemma 5.4.1. In particular, $t_0(x_0) \equiv t_0$. In the following examples we present typical cases where simple sufficient statistics exist.

Example 5.4.4. (i) Suppose we have the special case that $Q^Z(C|x, \theta, a)$ does only depend on θ, i.e. we have $Q^Z(C|\theta)$. Moreover, we assume that the disturbance variables are exponentially distributed with unknown parameter $\theta \in \Theta = \mathbb{R}_+$, i.e. $Q^Z(\cdot \mid \theta) = Exp(\theta)$. According to Lemma 5.4.1 we obtain

$$\mu_n(C \mid \tilde{h}_n) = \frac{\int_C \theta^n \exp\Big(- \theta \sum_{k=1}^n z_k \Big) Q_0(d\theta)}{\int_\Theta \theta^n \exp\Big(- \theta \sum_{k=1}^n z_k \Big) Q_0(d\theta)}.$$

Obviously μ_n depends on the history $\tilde{h}_n$ only through $\sum_{k=1}^n z_k$ and n. If we define $t_n(\tilde{h}_n) := (\sum_{k=1}^n z_k, n)$, then for $i = (x, n) \in I := \mathbb{R}_+ \times \mathbb{N}$ we obtain that $\mu_n(C|\tilde{h}_n) = \hat{\mu}\big(C|t_n(\tilde{h}_n)\big)$ where

$$\hat{\mu}(C \mid i) = \frac{\int_C \theta^n \exp\big(- \theta x\big) Q_0(d\theta)}{\int_\Theta \theta^n \exp\big(- \theta x\big) Q_0(d\theta')}.$$

When $Q_0 = \Gamma(\alpha, \beta)$, i.e. the initial distribution of ϑ is a Gamma-distribution with parameters α and β, then $\hat{\mu}(\cdot|x, n) = \Gamma(\alpha + n, \beta + x)$.

(ii) As in example (i) we assume that $Q^Z(C|x, \theta, a)$ does only depend on θ. Now suppose the disturbance variables are only zero or one, i.e. $Q^Z(\cdot \mid \theta) = B(1, \theta)$. According to Lemma 5.4.1 we obtain

$$\mu_n(C \mid \tilde{h}_n) = \frac{\int_C \theta^{\sum_{k=1}^n z_k} (1 - \theta)^{n - \sum_{k=1}^n z_k} Q_0(d\theta)}{\int_\Theta \theta^{\sum_{k=1}^n z_k} (1 - \theta)^{n - \sum_{k=1}^n z_k} Q_0(d\theta)}.$$

Obviously μ_n depends on the history $\tilde{h}_n$ only through $\sum_{k=1}^n z_k$ and n. Thus, if we define $t_n(\tilde{h}_n) := (\sum_{k=1}^n z_k, n)$, then for $i = (s, n) \in I := \{(s, n) \in \mathbb{N}_0^2 | s \le n\}$ we obtain that $\mu_n(C|\tilde{h}_n) = \hat{\mu}\big(C|t_n(\tilde{h}_n)\big)$ where

$$\hat{\mu}(C \mid i) := \frac{\int_C \theta^s (1 - \theta)^{n-s} Q_0(d\theta)}{\int_\Theta \theta^s (1 - \theta)^{n-s} Q_0(d\theta)}.$$

Note that in case $Q_0 = U(0, 1)$, i.e. the initial distribution of ϑ is the uniform distribution on the interval $(0, 1)$, then $\hat{\mu}(\cdot|s, n)$ is a *Beta distribution* with parameters $(s + 1, n - s + 1)$. ♦

Definition 5.4.5. Let a Bayesian Model be given and suppose there is a sufficient statistic (t_n) given for (μ_n) with the property

$$t_{n+1}(\tilde{h}_{n+1}) = t_{n+1}(\tilde{h}_n, a_n, z_{n+1}) = \hat{\Phi}\Big(x_n, t_n(\tilde{h}_n), a_n, z_{n+1}\Big)$$

for $\tilde{h}_{n+1} = (\tilde{h}_n, a_n, z_{n+1}, x_{n+1}) \in \tilde{H}_{n+1}$ and for a measurable function $\hat{\Phi} : E_X \times I \times A \times \mathcal{Z} \to I$. Then the sequence (t_n) is called a *sequential sufficient statistic.*

Obviously we have in the Examples 5.4.4 also a sequential sufficient statistic. If there exists a sequential sufficient statistic, then the formulation of the filtered Markov Decision Model can be simplified. Instead of enlarging the state space by the posterior distributions μ_n, it is sufficient to include the information state $i_n = t_n(\tilde{h}_n)$. The resulting information-based Markov Decision Model is defined as follows.

Definition 5.4.6. Suppose a Bayesian Model with a sequential sufficient statistic (t_n) is given. The *information-based Markov Decision Model* is a set of data $(E, A, D, \mathcal{Z}, \hat{T}, \hat{Q}^Z, \hat{r}, \hat{g}, \beta)$ with the following properties:

- $E := E_X \times I$ is the *state space.* E_X is the observable part of the state space and I is the space of information states.
- A is the *action space.*
- $D \subset E \times A$ is the set of possible state-action pairs. $D(x,i) := D(x)$ is the set of *feasible actions* in state (x,i).
- $\mathcal{Z}$ is the *disturbance space.*
- $\hat{T} : E_X \times I \times A \times \mathcal{Z} \to E_X \times I$ is a measurable transition function. More precisely we define

$$\begin{aligned}(x_{n+1}, i_{n+1}) &= \hat{T}(x_n, i_n, a_n, z_{n+1}) \\ &:= \big(T_X(x_n, a_n, z_{n+1}), \hat{\Phi}(x_n, i_n, a_n, z_{n+1})\big).\end{aligned}$$

- $\hat{Q}^Z$ is a stochastic kernel which determines the distribution of the disturbance variable Z_{n+1} as follows: For fixed $(x,i) \in E_X \times I$, $a \in D(x)$ and $B \in \mathcal{B}(\mathcal{Z})$ we define

$$\hat{Q}^Z(B|x,i,a) := \int Q^Z(B|x,\theta,a)\hat{\mu}(d\theta|i).$$

- $\hat{r} : D \times I \to \mathbb{R}$ is the one-stage reward function

$$\hat{r}(x,i,a) := \int r(x,\theta,a)\hat{\mu}(d\theta|i)$$

  whenever the integral exists.

- $\hat{g} : E \times I \to \mathbb{R}$ is the terminal reward function

$$\hat{g}(x,i) := \int g(x,\theta)\hat{\mu}(d\theta|i)$$

provided the integral exists.
- $\beta \in (0,1]$ is the discount factor.

This is a Markov Decision Model as introduced in Chapter 2. The value functions for the information-based Markov Decision Model are denoted by $\hat{J}_{n\pi}(x,i)$ for $\pi \in \tilde{\Pi}_n$ and $\hat{J}_n(x,i)$, where $\tilde{\Pi}_n$ is the set of all policies $\pi = (f_0,\dots,f_{n-1})$ with measurable decision rules $f_k : \tilde{H}_k \to A$, $f_k(\tilde{h}_k) \in D(x_k)$, $k = 0,\dots,n-1$.

Theorem 5.4.7. *Suppose $\pi \in \tilde{\Pi}_N$. Then it holds for $x \in E_X$:*

$$J_{N\pi}(x) = \hat{J}_{N\pi}(x,t_0) \quad \text{and} \quad J_N(x) = \hat{J}_N\big(x,t_0\big).$$

The proof follows along the same lines as the proof of Theorem 5.3.2. Again we can apply Theorem 2.3.8 to obtain:

Theorem 5.4.8. *Suppose the information-based Markov Decision Model satisfies the Structure Assumption of Theorem 2.3.8.*

a) Then the Bellman equation holds, i.e. for $(x,i) \in E_X \times I$:

$$\hat{J}_0(x,i) = \int g(x,\theta)\hat{\mu}(d\theta|i)$$

$$\hat{J}_n(x,i) = \sup_{a\in D(x)} \left\{ \int r(x,\theta,a)\hat{\mu}(d\theta|i) + \beta \int\int \hat{J}_{n-1}\Big(T_X(x,a,z),\hat{\Phi}(x,i,a,z)\Big)Q^Z(dz|x,\theta,a)\hat{\mu}(d\theta|i) \right\}.$$

b) Let $\hat{f}_n$ be a maximizer of $\hat{J}_{n-1}$ for $n = 1,\dots,N$. Then the policy $\pi^* := (f_0^*,\dots,f_{N-1}^*)$ is optimal for the N-stage Bayesian Model, where

$$f_n^*(\tilde{h}_n) := \hat{f}_{N-n}\big(x_n, t_n(\tilde{h}_n)\big), \quad \tilde{h}_n \in \tilde{H}_n.$$

Finally we note that information-based Markov Decision Models are more general than Bayesian Markov Decision Models. This follows since a trivial sufficient statistics for (μ_n) is given by $t_n(\tilde{h}_n) := \mu_n(\cdot|\tilde{h}_n)$. In this case $I := \mathbb{P}(\Theta)$.

Monotonicity Results

In order to apply Theorem 5.4.8, the structure assumption has to be satisfied. In what follows we deal as an example with the 'increasing case'. To ease the notation we make the following simplifying assumptions:

Assumptions:

(i) $\mathcal{Z} \subset \mathbb{R}$, $E_X \subset \mathbb{R}^d$ and $\Theta \subset \mathbb{R}$.
(ii) $Q^Z(\cdot|x,\theta,a)$ is independent of x and has a density $q^Z(\cdot|\theta,a)$ with respect to a σ-finite measure.
(iii) $\hat{\mu}$ has a Q_0-density $\hat{p}(\theta|i)$, i.e. $\hat{\mu}(\theta|i) = \hat{p}(\theta|i)Q_0(d\theta)$.

Assumption (ii) implies that Φ is independent of x and also $\hat{\Phi}$ can be supposed to be independent of x. In order to speak about monotonicity we have to introduce a partial order relation on the state space $E_X \times I$ and on $\mathcal{Z}$. Since $E_X \subset \mathbb{R}^d$ and $\mathcal{Z} \subset \mathbb{R}$ we can use the usual componentwise partial order. On I we define

$$i \le i' \quad :\Leftrightarrow \quad \hat{\mu}(\cdot|i) \le_{lr} \hat{\mu}(\cdot|i')$$

where $\le_{lr}$ is the likelihood ratio order, i.e.

$$\frac{\hat{p}(\theta|i')}{\hat{p}(\theta|i)} \quad \text{is increasing in } \theta$$

(see Appendix B.3). A function $v : I \to \mathbb{R}$ is called increasing if

$$i \le i' \quad \Rightarrow \quad v(i) \le v(i').$$

We need the following preliminary results (for MTP_2 functions see the Appendix B.3).

Lemma 5.4.9. *a) If $q^Z(z|\theta,a)$ is an MTP_2 function in (z,θ) for all a, then*

$$(i,z) \mapsto \hat{\Phi}(i,a,z)$$

is increasing for all $a \in A$.
b) Fix $a \in A$. We have $q^Z(\cdot|\theta,a) \le_{lr} q^Z(\cdot|\theta',a)$ for all $\theta \le \theta'$ if and only if $q^Z(z|\theta,a)$ is an MTP_2 function in (z,θ).

Proof. a) We have to show that $(i,z) \le (i',z')$ implies $\hat{\Phi}(i,a,z) \le \hat{\Phi}(i',a,z')$, i.e.

$$\hat{\mu}\big(\cdot\,|\hat{\Phi}(i,a,z)\big) \le_{lr} \hat{\mu}\big(\cdot\,|\hat{\Phi}(i',a,z')\big).$$

Note that $i_n = t_n(h_n)$ and by definition

$$\begin{aligned}\hat{\mu}\big(\cdot\,|\hat{\Phi}(i_n,a_n,z_n)\big) &= \hat{\mu}\big(\cdot\,|t_{n+1}(h_{n+1})\big) = \mu_{n+1}(\cdot|h_{n+1}) \\ &= \Phi\big(\cdot\,|\mu_n(\cdot|h_n),a_n,z_n\big) = \Phi\big(\cdot\,|\hat{\mu}(\cdot|i_n),a_n,z_n\big).\end{aligned}$$

Hence it is equivalent to show that

$$\Phi(\,\cdot\,|\hat{\mu}(\cdot|i), a, z) \leq_{lr} \Phi(\,\cdot\,|\hat{\mu}(\cdot|i'), a, z'). \tag{5.11}$$

By definition of the Bayes operator the distribution $\Phi(\,\cdot\,|\hat{\mu}(\cdot|i), a, z)$ has the density

$$\frac{q^Z(z|\theta, a)\hat{p}(\theta|i)}{\int q^Z(z|\tilde{\theta}, a)\hat{p}(\tilde{\theta}|i)d\tilde{\theta}}.$$

Using the definition of the likelihood ratio order, (5.11) is then equivalent to

$$\begin{aligned}&q^Z(z|\theta, a)\,\hat{p}(\theta|i)\,q^Z(z'|\theta', a)\,\hat{p}(\theta'|i')\\ \leq\,& q^Z(z|\theta\wedge\theta', a)\,\hat{p}(\theta\wedge\theta'|i)\,q^Z(z'|\theta\vee\theta', a)\,\hat{p}(\theta\vee\theta'|i')\end{aligned}$$

for $\theta, \theta' \in \Theta$. This is equivalent to $q^Z(z|\theta, a)\hat{p}(\theta|i)$ being MTP_2 in (z, θ, i). Now $q^Z(z|\theta, a)$ is MTP_2 in (z, θ) by assumption, hence also in (z, θ, i). On the other hand, $\hat{\mu}(\cdot|i) \leq_{lr} \hat{\mu}(\cdot|i')$ means that for $i \leq i'$ and $\theta, \theta' \in \Theta$ we have

$$\hat{p}(\theta|i)\,\hat{p}(\theta'|i') \;\leq\; \hat{p}(\theta\wedge\theta'|i)\,\hat{p}(\theta\vee\theta'|i').$$

Hence $\hat{p}(\theta|i)$ is MTP_2 in (z, θ, i). Since the product of MTP_2 functions is MTP_2 (see Lemma A.3.4 part b)) the statement follows.

b) The statement follows directly from the definition. See also part a). □

The previous Lemma is now used to prove the following monotonicity result for the information-based Markov Decision Model.

Theorem 5.4.10. *Suppose an information-based Markov Decision Model with an upper bounding function b is given and*

(i) $D(\cdot)$ *is increasing, i.e. $x \leq x'$ implies $D(x) \subset D(x')$,*
(ii) $q^Z(\cdot|\theta, a) \leq_{lr} q^Z(\cdot|\theta', a)$ *for all $\theta \leq \theta'$ and $a \in A$,*
(iii) $(x, z) \mapsto T_X(x, a, z)$ *is increasing for all $a \in D(x)$,*
(iv) $(\theta, x) \mapsto r(\theta, x, a)$ *is increasing for all $a \in D(x)$,*
(v) $(\theta, x) \mapsto g(\theta, x)$ *is increasing,*
(vi) *for all increasing $v \in \mathbb{B}_b^+$ there exists a maximizer $f \in \Delta$ of v.*

Then the sets $\mathbb{M} := \{v \in \mathbb{B}_b^+ \mid v \text{ is increasing}\}$ and Δ satisfy the Structure Assumption (SA_N).

Proof. We use the general Theorem 2.4.14 for increasing Markov Decision Models to show the result. The crucial part is to prove the stochastic monotonicity of the transition kernel since the state space is rather complicated now. Let us check the conditions (i)–(v) of Theorem 2.4.14:

(i) $D(\cdot)$ is increasing by assumption.

(ii) The function

$$(x,i) \mapsto \int\int v\big(T_X(x,a,z), \hat{\Phi}(i,a,z)\big) q^Z(z|\theta,a)dz\hat{\mu}(d\theta|i)$$

is increasing for all increasing $v \in I\!B_b^+$.
The monotonicity in x follows directly, since T_X and v are both increasing by assumption. Next for $i \le i'$ it is clear from Lemma 5.4.9 and our assumption (ii) that

$$v\big(T_X(x,a,z), \hat{\Phi}(i,a,z)\big) \le v\big(T_X(x,a,z), \hat{\Phi}(i',a,z)\big).$$

Now it remains to show that the expression

$$\int\int v\big(T_X(x,a,z), \hat{\Phi}(i',a,z)\big) q^Z(z|\theta,a)dz\hat{\mu}(d\theta|i)$$

is increased if $\hat{\mu}(d\theta|i)$ is replaced by $\hat{\mu}(d\theta|i')$. For $i \le i'$ we have by definition $\hat{\mu}(\cdot|i) \le_{lr} \hat{\mu}(\cdot|i')$ which implies $\hat{\mu}(\cdot|i) \le_{st} \hat{\mu}(\cdot|i')$ (see Theorem B.3.6). Thus, in view of Theorem B.3.3 we have to show that

$$\theta \mapsto \int v\big(T_X(x,a,z), \hat{\Phi}(i',a,z)\big) q^Z(z|\theta,a)dz$$

is increasing. But this follows since $q^Z(\cdot|\theta,a) \le_{lr} q^Z(\cdot|\theta',a)$ for $\theta \le \theta'$ and

$$z \mapsto v\big(T_X(x,a,z), \hat{\Phi}(i,a,z)\big)$$

is increasing by Lemma 5.4.9 and our assumptions.

(iii) The function $(x,i) \mapsto \int r(\theta,x,a)\hat{\mu}(d\theta|i)$ is increasing, since r is increasing in x and θ and $\hat{\mu}(\cdot|i) \le_{st} \hat{\mu}(\cdot|i')$ (see (ii)).

(iv) As in part (iii) it follows that $\hat{g}$ is increasing.

(v) The existence of maximizers is true by assumption. □

5.5 Bandit Problems with Finite Horizon

A nice application of Partially Observable Markov Decision Problems are so-called *bandit problems*. We will restrict here to Bernoulli bandits with two arms. The game is as follows: Imagine we have two slot machines with unknown success probability θ_1 and θ_2. The success probabilities are chosen independently from two prior Beta-distributions. At each stage we have to choose one of the arms. We receive one Euro if the arm wins, else no cash flow appears. The aim is to maximize the expected total reward over N trials. One of the first (and more serious) applications is to medical trials of a new drug. In the beginning the cure rate of the new drug is not known and may

be in competition with well-established drugs with known cure rate (this corresponds to one bandit with known success probability). The problem is not trivial since it is not necessarily optimal to choose the arm with the higher expected success probability. Instead one has to incorporate 'learning effects' which means that sometimes one has to pull one arm just to get some information about its success probability. In case of a finite horizon some nice properties of the optimal policy like the 'stay-on-a-winner' rule can be shown. In case of an infinite horizon (which is treated in Section 7.6.4) it is possible to prove the optimality of a so-called *index-policy*, a result which has been generalized further for multi-armed bandits.
The bandit problem can be formulated as a Partially Observable Markov Decision Model as follows: Note that there is no observable part of the state, such that E_X and the process (X_n) can be skipped. The remaining data is as follows:

- $E_Y := \Theta := [0,1]^2$ where $\theta = (\theta_1, \theta_2) \in \Theta$ denotes the (unknown) success probabilities of arm 1 and 2,
- $A := \{1,2\}$ where $a \in A$ is the number of the arm which is chosen next,
- $D(\cdot) := A$,
- $\mathcal{Z} := \{0,1\}$, i.e. we observe either success ($z = 1$) or failure ($z = 0$),
- $Q^Z(\{1\}|(\theta_1,\theta_2),a) := \theta_a$,
- Q_0 is the initial (prior) distribution of (θ_1, θ_2). We assume that Q_0 is a product of two uniform distributions, i.e. $Q_0 = Be(1,1) \otimes Be(1,1)$,
- $r(\theta, a) := \theta_a$,
- $g \equiv 0$,
- $\beta \in (0,1]$.

Note that the reward should be understood as an expected reward in the sense of Remark 2.1.2. By using this definition we make sure that the expected total reward is really the quantity we maximize.
Obviously we can apply Example 5.4.4 (ii) and obtain as sufficient statistic the number of successes m_a and failures n_a at both arms $a = 1,2$, hence $x = (m_1, n_1, m_2, n_2)$ is an information state and $I = \mathbb{N}_0^2 \times \mathbb{N}_0^2$. The conditional distribution $\hat{\mu}(\cdot|x)$ of $\theta = (\theta_1, \theta_2)$ is then the product of two Beta distributions $Be(m_1+1, n_1+1) \otimes Be(m_2+1, n_2+1)$. The operator $\hat{\Phi} : I \times A \times \mathcal{Z} \to I$ which yields the sequential sufficient statistic is given by

$$\hat{\Phi}(x,a,z) = \begin{cases} x + e_{2a-1} & \text{if } z = 1 \\ x + e_{2a} & \text{if } z = 0 \end{cases}$$

where e_a is the a-th unit vector. The operator $\hat{\Phi}$ simply adds the observation (success/failure) to the right component. Altogether we obtain for the information-based Markov Decision Model:

- $E := I = \mathbb{N}_0^2 \times \mathbb{N}_0^2$ where $x = (m_1, n_1, m_2, n_2) \in E$ denotes the number of successes m_a and failures n_a at arm a,
- $A := \{1,2\}$ where $a \in A$ is the number of the arm which is chosen next,

- $D(\cdot) := A$,
- $\mathcal{Z} := \{0,1\}$ where $z = 1$ and $z = 0$ refer to success and failure respectively,
- $\hat{T}(x,a,z) := \hat{\Phi}(x,a,z)$.
- The distribution of the disturbances is given by

$$\hat{Q}^Z(\{1\}|x,a) = \int Q^Z(\{1\}|\theta,a)\hat{\mu}(d\theta|x) = \frac{m_a+1}{m_a+n_a+2} =: p_a(x),$$

- $\hat{r}(x,a) := \int r(\theta,a)\hat{\mu}(d\theta|x) = \frac{m_a+1}{m_a+n_a+2}$,
- $\hat{g} \equiv 0$,
- $\beta \in (0,1]$.

Note that the rewards are non-negative and bounded. In this section we will consider the problem with finite horizon, in Section 7.6.4 the infinite horizon problem is investigated.
It is convenient to introduce the following abbreviations for $v : E \to \mathbb{R}$:

$$(Q_a v)(x) := p_a(x)v(x+e_{2a-1}) + (1-p_a(x))v(x+e_{2a}), \quad x \in E.$$

Note that since E is countable and A is finite, the Structure Assumption (SA_N) is satisfied and we obtain with Theorem 5.4.8 that the following recursion is valid (instead of $\hat{J}_n$ we will simply write J_n for the value functions of the information-based Markov Decision Model):

$$J_n(x) = \max_{a=1,2} \{p_a(x) + \beta(Q_a J_{n-1})(x)\}, \quad x \in E$$

where $J_0 \equiv 0$. We immediately obtain the following result.

Theorem 5.5.1. *Let* $d_n := p_2 + \beta Q_2 J_{n-1} - p_1 - \beta Q_1 J_{n-1}$ *for* $n \in \mathbb{N}$. *Then it holds:*

a) The policy $\pi^* = (f_N^*, \dots, f_1^*)$ *is optimal, where*

$$f_n^*(x) = \begin{cases} 2 & \text{if } d_n(x) \geq 0 \\ 1 & \text{if } d_n(x) < 0 \end{cases}$$

is the largest maximizer of J_{n-1}.

b) The sequence (d_n) *satisfies the recursion*

$$d_1 = p_2 - p_1,$$
$$d_{n+1} = (1-\beta)d_1 + \beta Q_2 d_n^+ - \beta Q_1 d_n^-.$$

c) Let $\beta = 1$. *Then the 'stay-on-a-winner' property holds for the optimal policy* π^*, *i.e.*

$$f_{n+1}^*(x) = a \quad \Rightarrow \quad f_n^*(x+e_{2a-1}) = a.$$

The interpretation of the last statement is as follows: First note that the optimal N-stage policy is given by $(f_N^*, \dots, f_1^*)$. Hence, if it is optimal to choose arm a in state x (i.e. $f_{n+1}^*(x) = a$), and we observe a success (i.e. the new state is $x + e_{2a-1}$) then it is also optimal to choose arm a in the new state (i.e. $f_n^*(x + e_{2a-1}) = a$).

Proof. a) Note that $d_n(x) = LJ_{n-1}(x, 2) - LJ_{n-1}(x, 1)$ and hence the statement follows directly.

b) Since $J_0 \equiv 0$ we get $d_1 = p_2 - p_1$. For $n \geq 2$ we obtain with the abbreviation $L_a v(x) := Lv(x, a)$

$$\begin{aligned} d_{n+1} &= L_2 J_n - L_1 J_n = L_2(L_1 J_{n-1} + d_n^+) - L_1(L_2 J_{n-1} + d_n^-) \\ &= L_2 L_1 J_{n-1} + \beta Q_2 d_n^+ - L_1 L_2 J_{n-1} - \beta Q_1 d_n^- . \end{aligned}$$

Since $Q_2 p_1 = p_1$, $Q_1 p_2 = p_2$ and $Q_1 Q_2 v = Q_2 Q_1 v$ for bounded $v : E \to \mathbb{R}$, it holds:

$$\begin{aligned} L_2 L_1 J_{n-1} - L_1 L_2 J_{n-1} &= p_2 + \beta Q_2 p_1 - p_1 - \beta Q_1 p_2 \\ &\quad + \beta^2 (Q_2 Q_1 J_{n-1} - Q_1 Q_2 J_{n-1}) \\ &= (1 - \beta)(p_2 - p_1) = (1 - \beta) d_1 \end{aligned}$$

which implies the statement.

c) We show that $f_{n+1}^*(x) = 1$ implies $f_n^*(x + e_1) = 1$. The statement for $a = 2$ can be shown in the same way. Let us consider the following partial order relation on $E = \mathbb{N}_0^2 \times \mathbb{N}_0^2$:

$$x \leq x' :\Longleftrightarrow \begin{cases} m_1 \geq m_1', & m_2 \leq m_2', \\ n_1 \leq n_1', & n_2 \geq n_2' \end{cases}$$

i.e. state x' is larger than state x if the second arm performs better and the first arm performs worse compared with state x. It follows immediately from part b) by induction on n that $d_n(x)$ is increasing in x with respect to this order. Now assume that $f_{n+1}^*(x) = 1$ and $f_n^*(x + e_1) = 2$. The last equation implies $d_n(x + e_1) \geq 0$. Since $x + e_1 \leq x + e_2$ and d_n is increasing this leads to $d_n(x + e_2) \geq 0$. Since $\beta = 1$ we obtain from part b) that

$$\begin{aligned} d_{n+1}(x) &= (Q_2 d_n^+)(x) - (Q_1 d_n^-)(x) \\ &= p_2(x) d_n^+(x + e_3) + (1 - p_2(x)) d_n^+(x + e_4) \\ &\quad - p_1(x) d_n^-(x + e_1) - (1 - p_1(x)) d_n^-(x + e_2) \\ &= p_2(x) d_n^+(x + e_3) + (1 - p_2(x)) d_n^+(x + e_4) \geq 0. \end{aligned}$$

This is a contradiction to the assumption $f_{n+1}^*(x) = 1$ and the statement follows. □

We consider now the *special case* where the *success probability of arm 1 is known*, i.e. the prior distribution at arm 1 is concentrated on the point $\theta_1 = p_1 \in (0,1)$. Then $p_1(x) = p_1$. Here obviously the state space reduces to $E := \mathbb{N}_0^2$, where (m,n) denotes the number m of successes and n of failures at arm 2. The Bellman equation reduces to

$$J_k(m,n) = \max\left\{p_1+\beta J_{k-1}(m,n),\ p(m,n)+\beta(PJ_{k-1})(m,n)\right\}, \quad (m,n) \in E$$

with $J_0 \equiv 0$, and where we use the obvious change of notation:

$$(Pv)(m,n) := p(m,n)v(m+1,n) + (1-p(m,n))v(m,n+1),$$

for bounded $v : \mathbb{N}_0^2 \to \mathbb{R}$ and $p(m,n) := \frac{m+1}{m+n+2}$.

For the next theorem we introduce the following partial order relation on $E = \mathbb{N}_0^2$.

$$(m,n) \le (m',n') :\Longleftrightarrow m \le m',\ n \ge n'. \tag{5.12}$$

Note that this order is equivalent to the likelihood ratio order for the corresponding Beta distributions with parameters $(m+1, n+1)$ and $(m'+1, n'+1)$ (see Example B.3.8). With this definition we are able to show:

Theorem 5.5.2. *Suppose the success probability at arm 1 is known. Then it holds:*

a) The optimal policy $\pi^* = (f_N^*, \dots, f_1^*)$ has the property that $f_k^*(m,n)$ is increasing in (m,n).
b) The stay-on-a-winner *property for arm 2 holds, i.e.*

$$f_{k+1}^*(m,n) = 2 \quad \Rightarrow \quad f_k^*(m+1,n) = 2.$$

c) The stopping rule *for arm 1 holds, i.e.*

$$f_{k+1}^*(m,n) = 1 \quad \Rightarrow \quad f_k^*(m,n) = 1.$$

d) The optimal policy is partly myopic:

$$p(m,n) \ge p_1 \quad \Rightarrow \quad f_k^*(m,n) = 2, \text{ for all } k \in \mathbb{N}.$$

The 'stopping rule' for arm 1 means that once we have decided to play the arm with the known success probability we will continue to do so because we do not gain any new information about the success probability of arm 2. The converse of part d) is not true: There may be cases where $p_1 > p(m,n)$ and it is optimal to choose arm 2.

Proof. a) According to Proposition 2.4.16 it remains to show that LJ_{k-1} is supermodular or equivalently

$$d_k(m,n) = LJ_{k-1}((m,n),2) - LJ_{k-1}((m,n),1)$$

is increasing in (m,n). This can be shown by induction on k. The fact that $d_1(m,n) = p(m,n) - p_1$ is increasing in (m,n) is easy to see. For the induction step, in view of Theorem 5.5.1 b) we have to show that the stochastic kernel P is stochastically monotone. But this means by definition that $(m,n) \le (m',n')$ implies $p(m,n) \le p(m',n')$, which follows directly from the definition of $p(m,n)$.

b) First note that we have by definition the representation

$$d_{k+1}(m,n) = d_1(m,n) + \beta\big((PJ_k)(m,n) - J_k(m,n)\big).$$

On the other hand we know from Theorem 5.5.1 b) that

$$d_{k+1}(m,n) = (1-\beta)d_1(m,n) + \beta(Pd_k^+)(m,n) - \beta d_k^-(m,n).$$

Solving the first equation for d_1 and inserting this into the second equation yields:

$$d_{k+1} = Pd_k^+ - d_k^- - (1-\beta)(PJ_k - J_k).$$

Since it always holds that $PJ_k \ge J_k$ we obtain

$$d_{k+1}(m,n) \le (Pd_k^+)(m,n) - d_k^-(m,n).$$

Now let us assume that $f_{k+1}^*(m,n) = 2$ and $f_k^*(m+1,n) = 1$. The last equation means that $d_k(m+1,n) < 0$. Since obviously

$$(m,n+1) \le (m,n) \le (m+1,n)$$

and $d_k(m,n)$ is increasing in (m,n) we obtain that $Pd_k^+(m,n) = 0$. Thus we get $d_k^-(m,n) > 0$ and this in turn implies $d_{k+1}(m,n) \le -d_k^-(m,n) < 0$. From the last inequality we conclude that $f_{k+1}^*(m,n) = 1$ which is a contradiction to our assumption.

c) It is easy to show by induction that $d_k \le d_{k+1}$. Hence we know that $d_k(m,n) \le d_{k+1}(m,n)$ from which we conclude that $f_{k+1}^*(m,n) = 1$ implies $f_k^*(m,n) = 1$.

d) The condition $p(m,n) \ge p_1$ implies that $d_1(m,n) \ge 0$. We know from part c) that $d_k(m,n) \ge 0$ which is equivalent to $f_k^*(m,n) = 2$. □

5.6 Exercises

Exercise 5.6.1. Prove the explicit representation of the conditional distribution of the unknown parameter in the Bayesian model given in Lemma 5.4.1.

Exercise 5.6.2 (Conjugate Distributions). Let us consider the Bayesian Model. A class of prior probability distributions Q_0 is said to be *conjugate* to a class of likelihood functions $Q^Z(\cdot|\theta)$ if the resulting posterior distributions $\mu_n(\cdot|\tilde{h}_n)$ are in the same family as Q_0. The parameters of this class then determine a sufficient statistic for μ_n.

a) Suppose that $Q^Z(\cdot|\theta) = \text{Poi}(\theta)$, i.e. the disturbance variables are Poisson-distributed with parameter θ and $Q_0 = \Gamma(\alpha, \beta)$, i.e. the prior distribution of θ is a Gamma distribution with parameters α and β. Show that $\mu_n(\cdot|\tilde{h}_n)$ is again Gamma-distributed with parameters $\left(\alpha + \sum_{k=1}^n z_k, \beta + n\right)$.
b) Suppose that $Q^Z(\cdot|\theta) = \mathcal{N}(\theta, \sigma^2)$, i.e. the disturbance variables are normally-distributed with unknown mean θ and known variance σ^2 and $Q_0 = \mathcal{N}(\mu_0, \sigma_0^2)$, i.e. the prior distribution of θ is a Normal distribution with parameters μ_0 and σ_0^2. Show that $\mu_n(\cdot|\tilde{h}_n)$ is again normally-distributed with parameters

$$\Big(\frac{\mu_0}{\sigma_0^2} + \frac{\sum_{k=1}^n z_k}{\sigma^2}\Big)\Big/\Big(\frac{1}{\sigma_0^2} + \frac{n}{\sigma^2}\Big) \quad \text{and} \quad \Big(\frac{1}{\sigma_0^2} + \frac{n}{\sigma^2}\Big)^{-1}.$$

Exercise 5.6.3. Consider the binomial distribution $B(n,p)$ and the Beta distribution $Be(\alpha, \beta)$ (see Appendix B.3). Show:

a) $B(n,p) \leq_{lr} B(m,q)$ if $n \leq m$ and $p \leq q$.
b) $Be(\alpha, \beta) \leq_{lr} Be(\gamma, \delta)$ if $\alpha \leq \gamma$ and $\beta \geq \delta$.

Exercise 5.6.4 (Sequential Ratio Test). Suppose we can observe realizations of random variables (Y_n) which are independent and identically distributed and depend on an unknown parameter θ. We want to test $H_0 : \theta = \theta_0$ against $H_1 : \theta = \theta_1$. There are three actions available:
$a = a_0$, i.e. we decide for H_0
$a = a_1$, i.e. we decide for H_1
$a = a_2$, i.e. we make a further observation.
For each observation we have to pay a cost $c > 0$. Wrong decisions are also punished: We have to pay c_1 if we decide for H_1 and H_0 is correct and we have to pay c_0 if we decide for H_0 and H_1 is correct.

a) Set this up as an information-based Markov Decision Model. Choose $I = [0,1]$ where $x \in I$ is the probability that H_0 is correct.
b) Show that the Bayes-operator is given by

$$\Phi(x,z) = \frac{xq(z|\theta_0)}{xq(z|\theta_0) + (1-x)q(z|\theta_1)}$$

where $q(z|\theta_0)$ is the conditional distribution of Y given H_0 is correct.
c) Show that J_n is continuous and concave in $x \in I$, $J_n(0) = J_n(1) = 0$.
d) Show that there exist x_n^* and $x_n^{**} \in I$ such that

$$f_n^*(x) = \begin{cases} a_0, \, x \ge x_n^{**}, \\ a_1, \, x \le x_n^*, \\ a_2, \, x_n^* < x < x_n^{**} \end{cases}$$

is a minimizer of J_{n-1}.

5.7 Remarks and References

An early reference for sequential Bayesian Decision Models is Martin (1967). For the reformulation approach in Section 5.3 we refer the reader in particular to Rieder (1975a). Textbooks which contain material and historical remarks on this topic are e.g. Hinderer (1970), van Hee (1978), Bertsekas and Shreve (1978), Kumar and Varaiya (1986), Hernández-Lerma (1989), Presman and Sonin (1990) and Runggaldier and Stettner (1994). Sometimes Bayesian Decision Models are treated under the name *Adaptive control model.* More details about models with sufficient statistics can be found in Hinderer (1970) and Bertsekas and Shreve (1978). In DeGroot (2004) various families of conjugate prior distributions are given. Stochastic control of partially observable systems in continuous-time have been investigated among others by Bensoussan (1992). Recently Winter (2008) studied the optimal control of Markovian jump processes with different information structures, see also the paper of Rieder and Winter (2009). For a finite numerical approximation of filters and their use in Partially Observable Markov Decision Problems see Pham et al. (2005) and Corsi et al. (2008).
Structural and comparison results for partially observable Markov Decision Models can be found in Albright (1979), Lovejoy (1987), Rieder (1991), Rieder and Zagst (1994), Müller (1997) and Krishnamurthy and Wahlberg (2009). Zagst (1995) and Brennan (1998) discuss the role of learning in multistage decision problems. Computational results for partially observable Markov Decision Models are given in Monahan (1982b), Lovejoy (1991a,b). Smallwood and Sondik (1973) and Sondik (1978) investigate partially observable Markov Decision Processes with finite and infinite horizon.
There are a number of textbooks on (uncontrolled) general filtering problems. Most of the books treat the problem of filtering from a continuous-time process which we do not investigate here. The references Bain and Crisan (2009), Fristedt et al. (2007) and Elliott et al. (1995) also contain the discrete-time case. The derivation of the Kalman filter in Example 5.2.5 can be found there.

Section 5.5: Structural results for bandit problems with finite horizon (like the stopping rule and the stay-on-a-winner property) are derived in Benzing et al. (1984), Kolonko and Benzing (1985), Kolonko (1986), Benzing and Kolonko (1987) and Rieder and Wagner (1991), see also the books of Berry and Fristedt (1985) and Gittins (1989). Applications to sequential sampling procedures can be found in Rieder and Wentges (1991). In Section 7.6.4 we shall discuss Bandit problems with infinite horizon. There further references on bandit problems are given.

Chapter 6
Partially Observable Markov Decision Problems in Finance

All the models which have been considered in Chapter 4 may also be treated in the case of partial information. Indeed models of this type occur somehow natural in mathematical finance because there are underlying economic factors influencing asset prices which are not specified and cannot be observed. Moreover, for example the drift of a stock is notoriously difficult to estimate. In this chapter we assume that the relative risk return distribution of the stocks is determined up to an unknown parameter which may change. This concept can also be interpreted as one way of dealing with model ambiguity. We choose two of the models from Chapter 4 and extend them to partial observation. The first is the general terminal wealth problem of Section 4.2 and the second is the dynamic mean-variance problem of Section 4.6.

We consider a financial market with one riskless bond (with interest rate $i_n = i$) and d risky assets with relative risk process (R_n). Here we assume that the distribution of R_{n+1} depends on an underlying stationary Markov process (Y_n) which cannot be observed. In Section 4.4 (Regime-switching model) the process (Y_n) is observable. The state space of (Y_n) is E_Y, a Borel subset of a Polish space. We assume that (R_n, Y_n) is a Markov process and moreover

$$\begin{aligned}
&\mathbb{P}\left(R_{n+1} \in B, Y_{n+1} \in C | Y_n = y, R_n\right) \\
&= \mathbb{P}\left(R_{n+1} \in B, Y_{n+1} \in C | Y_n = y\right) \qquad (6.1)\\
&= \mathbb{P}\left(R_{n+1} \in B | Y_n = y\right) \cdot \mathbb{P}\left(Y_{n+1} \in C | Y_n = y\right) =: Q^R(B|y) Q^Y(C|y)
\end{aligned}$$

for $B \in \mathcal{B}(\mathbb{R}^d), C \in \mathcal{B}(E_Y)$. Q^Y is the transition kernel of the 'hidden' Markov process (Y_n) and $Q^R(\cdot|y)$ is the (conditional) distribution of R_{n+1} given $Y_n = y$ (independent of n). In the following let $R(y)$ be a random variable with distribution $Q^R(\cdot|y)$, i.e. $\mathbb{P}(R(y) \in B) = Q^R(B|y) = \mathbb{P}(R_{n+1} \in B | Y_n = y)$. Given (Y_n), the random vectors $R_1, R_2, \ldots$ are independent, and given Y_n, the random variables R_{n+1} and Y_{n+1} are independent.

N. Bäuerle and U. Rieder, *Markov Decision Processes with Applications to Finance*, Universitext, DOI 10.1007/978-3-642-18324-9_6,

6.1 Terminal Wealth Problems

We start with problems of terminal wealth maximization under partial observation. Suppose we have an investor with utility function $U : dom\, U \to \mathbb{R}$ with $dom\, U = [0, \infty)$ or $dom\, U = (0, \infty)$ and initial wealth $x > 0$. Our investor can only observe the stock price and not the driving Markov process (Y_n), i.e. the filtration $(\mathcal{F}_n)$ to which portfolio strategies have to be adapted is given by $\mathcal{F}_n := \mathcal{F}_n^S = \sigma(S_0, S_1, \ldots, S_n) = \sigma(R_1, \ldots, R_n)$. The aim is to maximize the expected utility of her terminal wealth.
The following assumption on the financial market is used throughout this section.

Assumption (FM):

(i) *There are no arbitrage opportunities in the market, i.e. for all $y \in E_Y$ and $\phi \in \mathbb{R}^d$ it holds:*

$$\phi \cdot R(y) \geq 0 \ \mathbb{P}\text{-a.s.} \quad \Rightarrow \quad \phi \cdot R(y) = 0 \ \mathbb{P}\text{-a.s.}$$

(ii) *The support of $R(y)$ is independent of $y \in E_Y$.*
(iii) $\sup_y \mathbb{E}\, \|R(y)\| < \infty$.

The second assumption guarantees that the support of R_{n+1} is independent of Y_n and n. There are a lot of examples where this assumption is satisfied. According to (3.1) the wealth process (X_n) evolves as follows

$$X_{n+1} = (1+i)\Big(X_n + \phi_n \cdot R_{n+1}\Big)$$

where $\phi = (\phi_n)$ is a portfolio strategy such that ϕ is $(\mathcal{F}_n)$-adapted. The partially observable terminal wealth problem is then given by

$$\begin{cases} \mathbb{E}_x\, U(X_N^\phi) \to \max \\ \\ \phi \text{ is a portfolio strategy and } X_N^\phi \in dom\, U \ \mathbb{P}\text{-a.s.}. \end{cases} \tag{6.2}$$

Recall that $\mathcal{F}_n = \mathcal{F}_n^S$ i.e. the admissible portfolio strategies depend only on the observable stock prices (S_n). Problem (6.2) can be solved by the following stationary Partially Observable Markov Decision Model:

- $E_X := dom\, U$ where $x \in E_X$ denotes the wealth,
- E_Y is the state space of (Y_n), where $y \in E_Y$ is the unobservable state,
- $A := \mathbb{R}^d$ where $a \in A$ is the amount of money invested in the risky assets,
- $D(x) := \Big\{a \in \mathbb{R}^d \mid (1+i)\big(x + a \cdot R(y)\big) \in dom\, U \ \mathbb{P}\text{-a.s.}\Big\}$,
- $\mathcal{Z} := [-1, \infty)^d$ where $z \in \mathcal{Z}$ denotes the relative risk,
- $T_X\big(x, a, z\big) := (1+i)\big(x + a \cdot z\big)$,
- $Q^{Z,Y}(B \times C|x, y, a) := Q^R(B|y)Q^Y(C|y)$ for $B \in \mathcal{B}(\mathcal{Z}), C \in \mathcal{B}(E_Y)$,

- Q_0 is the initial (prior) distribution of Y_0,
- $r \equiv 0$,
- $g(x, y) := U(x)$,
- $\beta = 1$.

Note that in view of FM (ii) $D(x)$ does not depend on $y \in E_Y$ and hence $D(x)$ is well-defined. It is assumed (see Section 5.2) that there exist densities of Q^R and Q^Y. The Bayes operator Φ depends only on $\rho \in \mathbb{P}(E_Y)$ and $z \in \mathcal{Z}$. From Section 5.3 we know that we can solve problem (6.2) by a filtered Markov Decision Model. Theorem 5.3.3 implies the following result.

Theorem 6.1.1. *For the multiperiod terminal wealth problem with partial observation it holds:*

a) The value functions $J_n(x, \rho)$ are strictly increasing, strictly concave and continuous in $x \in dom\, U$ for all $\rho \in \mathbb{P}(E_Y)$.
b) The value functions can be computed recursively by the Bellman equation, i.e. for $(x, \rho) \in E_X \times \mathbb{P}(E_Y)$ it holds

$$J_0(x, \rho) = U(x),$$
$$J_n(x, \rho) = \sup_{a \in D(x)} \int\int J_{n-1}\Big((1+i)(x + a \cdot z), \Phi(\rho, z)\Big) Q^R(dz|y)\rho(dy).$$

c) The maximal value of problem (6.2) is given by $J_N(x, Q_0)$.
d) There exists a maximizer f_n^* of J_{n-1} and the portfolio strategy $(f_0, \ldots, f_{N-1})$ is optimal for the N-stage terminal wealth problem (6.2) where

$$f_n(h_n) := f_{N-n}^*(x_n, \mu_n(\cdot|h_n)), \; h_n = (x_0, a_0, z_1, x_1, \ldots z_n, x_n).$$

Proof. It is easily shown by using Assumption (FM) that the stationary filtered Markov Decision Model has an upper bounding function

$$b(x, \rho) := 1 + x, \quad (x, \rho) \in E_X \times \mathbb{P}(E_Y)$$

(cf. the proof of Proposition 4.2.1). Then the Structure Assumption (SA$_N$) is satisfied with

$$\mathbb{M}_n := \{v \in \mathbb{B}_b^+ \mid x \mapsto v(x, \rho) \text{ is strictly increasing, strictly concave and continuous for all } \rho \in \mathbb{P}(E_Y)\}$$

and $\Delta_n := F_n$. In particular the existence of maximizers follows along the same lines as the proof of Theorem 4.1.1 in view of (FM). Part c) and d) are implied by Theorem 5.3.2 and Theorem 5.3.3. □

In the next subsections we will specialize this result to power and logarithmic utility functions.

Power Utility

Here we assume that the utility function is of the form $U(x) = \frac{1}{\gamma}x^\gamma$ with $x \in [0,\infty)$ and $0 < \gamma < 1$. For convenience we define

$$\tilde{A} := \{\alpha \in \mathbb{R}^d \mid 1 + \alpha \cdot R(y) \geq 0 \ \mathbb{P}\text{-a.s.}\}.$$

Since by Assumption (FM) the support of $R(y)$ is independent of y, the set $\tilde{A}$ is independent of y. The proof of the next theorem is similar to the proof of Theorem 4.2.6.

Theorem 6.1.2. *Let U be the power utility with $0 < \gamma < 1$. Then it holds:*

a) The value functions are given by

$$J_n(x,\rho) = (xS_n^0)^\gamma d_n(\rho), \quad (x,\rho) \in E_X \times \mathbb{P}(E_Y)$$

where the sequence (d_n) satisfies the following recursion

$$d_0 \equiv \frac{1}{\gamma}$$

$$d_n(\rho) = \sup_{\alpha \in \tilde{A}} \int\int d_{n-1}(\Phi(\rho,z))\big(1 + \alpha \cdot z\big)^\gamma Q^R(dz|y)\rho(dy). \quad (6.3)$$

b) The optimal amounts which are invested in the stocks are given by

$$f_n^*(x,\rho) = \alpha_n^*(\rho)x, \quad x \geq 0$$

where $\alpha_n^*(\rho)$ is the solution of (6.3). The optimal portfolio strategy is given by $(f_0,\dots,f_{N-1})$ where

$$f_n(h_n) := f_{N-n}^*\big(x_n, \mu_n(\cdot|h_n)\big),\ h_n = (x_0,a_0,z_1,x_1,\dots,x_n).$$

Remark 6.1.3. In the case of a power utility function $U(x) = \frac{1}{\gamma}x^\gamma$ with $\gamma < 0$ we can proceed in the same way (see Remark 4.2.7) and obtain an analogous result. ◊

We consider now the special case of a binomial model with one stock and an unknown probability for an up movement, i.e. we consider a *Bayesian model* and assume

$$\mathbb{P}(\tilde{R} = \boldsymbol{u}|\vartheta = \theta) = \theta, \quad \mathbb{P}(\tilde{R} = \boldsymbol{d}|\vartheta = \theta) = 1 - \theta, \quad \theta \in (0,1) = \Theta$$

where $\boldsymbol{d} < 1+i < \boldsymbol{u}$. Obviously Assumption (FM) is satisfied in this market. Moreover, we exclude short-sellings now, i.e. we set $\tilde{A} := [0,1]$. Then we are able to compare the optimal investment strategy with the optimal investment strategy in the case of complete observation. More precisely the optimization problem in (6.3) reduces to one which is linear in θ:

$$d_n(\rho) = \sup_{0\le\alpha\le 1} \Big\{ d_{n-1}\big(\Phi(\rho,\bar{\boldsymbol{u}})\big) \int \theta\rho(d\theta)\,(1+i+\alpha(\boldsymbol{u}-1-i))^\gamma \tag{6.4}$$

$$+\, d_{n-1}\big(\Phi(\rho,\bar{\boldsymbol{d}})\big)\big(1-\int \theta\rho(d\theta)\big)\,(1+i+\alpha(\boldsymbol{d}-1-i))^\gamma \Big\}(1+i)^{-\gamma}$$

where $\bar{\boldsymbol{u}} := \frac{\boldsymbol{u}}{1+i} - 1$ and $\bar{\boldsymbol{d}} := \frac{\boldsymbol{d}}{1+i} - 1$. We can use Example 5.4.4 to show that $t_n(\tilde{h}_n) := (m,n)$ is a sufficient statistic where m is the number of ups. Then we obtain

$$\mu_n(C \mid \tilde{h}_n) = \hat{\mu}(C \mid t_n(\tilde{h}_n)) = \hat{\mu}(C \mid (m,n))$$
$$= \frac{\int_C \theta^m(1-\theta)^{n-m}Q_0(d\theta)}{\int_\Theta \theta'^m(1-\theta')^{n-m}Q_0(d\theta')}$$

where Q_0 is the given prior distribution of the unknown probability θ. We define by

$$\mathbb{D}_{Q_0} := \{\hat{\mu}(\cdot|m,n) \mid n \in \mathbb{N}_0, m \le n\}$$

the set of all possible posterior distributions. The next monotonicity result is important.

Lemma 6.1.4. *For $\rho \in \mathbb{D}_{Q_0}$ it holds:*

a) $\Phi(\rho,\bar{\boldsymbol{d}}) \le_{lr} \Phi(\rho,\bar{\boldsymbol{u}})$.
b) $\rho \le_{lr} \rho' \quad\Rightarrow\quad d_n(\rho) \le d_n(\rho')$.

Proof. a) Let $\theta \in (0,1)$ and $\rho,\rho' \in \mathbb{D}_{Q_0}$. We suppose that $\rho = \hat{\mu}(\cdot|m,n)$ and $\rho' = \hat{\mu}(\cdot|m',n')$. It holds that $\rho \le_{lr} \rho'$ if and only if $m' \ge m$ and $n-m \ge n'-m'$ (cf. equation (5.12) and Example B.3.8). This can be directly seen by inspecting the ratio

$$c \cdot \frac{\theta^{m'}(1-\theta)^{n'-m'}}{\theta^m(1-\theta)^{n-m}}.$$

In particular we have $\Phi(\rho,\bar{\boldsymbol{d}}) = \hat{\mu}(\cdot|m,n+1) \le_{lr} \hat{\mu}(\cdot|m+1,n+1) = \Phi(\rho,\bar{\boldsymbol{u}})$.
b) The proof is done by induction. For $n=0$ we have $d_0 \equiv \frac{1}{\gamma}$ and there is nothing to show. For convenience let us denote

$$c_u(\alpha) := (1 + i + \alpha(\boldsymbol{u} - 1 - i))^\gamma \quad \text{and} \quad c_d(\alpha) := (1 + i + \alpha(\boldsymbol{d} - 1 - i))^\gamma.$$

Suppose the statement is true for $k = 0, 1, \dots, n-1$. Since

$$\Phi(\rho, \bar{\boldsymbol{d}}) \leq_{lr} \Phi(\rho, \bar{\boldsymbol{u}}) \leq_{lr} \Phi(\rho', \bar{\boldsymbol{u}})$$

and $\Phi(\rho, \bar{\boldsymbol{d}}) \leq_{lr} \Phi(\rho', \bar{\boldsymbol{d}})$ and $\bar{\rho} := \int \theta \rho(d\theta) \leq \int \theta \rho'(d\theta) := \bar{\rho}'$ we obtain for arbitrary $\alpha \in (0, 1]$ (note that we have $c_u(\alpha) > c_d(\alpha)$ since $\alpha > 0$):

$$\begin{aligned}
& d_{n-1}\big(\Phi(\rho, \bar{\boldsymbol{u}})\big)\bar{\rho} c_u(\alpha) + d_{n-1}\big(\Phi(\rho, \bar{\boldsymbol{d}})\big)(1 - \bar{\rho}) c_d(\alpha) \\
&= \bar{\rho}\Big\{ d_{n-1}\big(\Phi(\rho, \bar{\boldsymbol{u}})\big) c_u(\alpha) - d_{n-1}\big(\Phi(\rho, \bar{\boldsymbol{d}})\big) c_d(\alpha) \Big\} + d_{n-1}\big(\Phi(\rho, \bar{\boldsymbol{d}})\big) c_d(\alpha) \\
&\leq \bar{\rho}'\Big\{ d_{n-1}\big(\Phi(\rho, \bar{\boldsymbol{u}})\big) c_u(\alpha) - d_{n-1}\big(\Phi(\rho, \bar{\boldsymbol{d}})\big) c_d(\alpha) \Big\} + d_{n-1}\big(\Phi(\rho, \bar{\boldsymbol{d}})\big) c_d(\alpha) \\
&\leq d_{n-1}\big(\Phi(\rho', \bar{\boldsymbol{u}})\big)\bar{\rho}' c_u(\alpha) + d_{n-1}\big(\Phi(\rho', \bar{\boldsymbol{d}})\big)(1 - \bar{\rho}') c_d(\alpha)
\end{aligned}$$

which implies the result. □

Recall that $\alpha_n^*(\rho)$ is the optimal fraction of wealth invested in the stock at time n when ρ is the posterior distribution of θ. The case of complete observation reduces to the special case $\rho := \delta_\theta$, and in this case we write $\alpha_n^*(\theta)$ instead of $\alpha_n^*(\delta_\theta)$. As before let us denote $\bar{\rho} = \int \theta \rho(d\theta) \in (0, 1)$.

Theorem 6.1.5. *For all $n \in \mathbb{N}_0$ and $\rho \in \mathbb{D}_{Q_0}$ it holds:*

$$\begin{aligned}
\alpha_n^*(\rho) &\geq \alpha_n^*(\bar{\rho}) \quad \text{if} \quad 0 < \gamma < 1, \\
\alpha_n^*(\rho) &\leq \alpha_n^*(\bar{\rho}) \quad \text{if} \quad \gamma < 0.
\end{aligned}$$

Proof. Let $0 < \gamma < 1$. By definition $\alpha_n^*(\rho)$ is a maximum point of (6.4). Then $\alpha_n^*(\rho)$ is also a solution of (we use the same notation as in the last proof)

$$\sup_{0 \leq \alpha \leq 1} \left\{ c_u(\alpha) + \frac{d_{n-1}\big(\Phi(\rho, \bar{\boldsymbol{d}})\big)}{d_{n-1}\big(\Phi(\rho, \bar{\boldsymbol{u}})\big)} \frac{1 - \bar{\rho}}{\bar{\rho}} c_d(\alpha) \right\}. \tag{6.5}$$

Let us define the function

$$h(\alpha) := c_u(\alpha) + \lambda c_d(\alpha)$$

and its maximum point (where we set $\delta := (1 - \gamma)^{-1}$)

$$\alpha^*(\lambda) = (1 + i) \frac{\lambda^{-\delta}(1 + i - \boldsymbol{d})^{-\delta} - (\boldsymbol{u} - 1 - i)^{-\delta}}{(\boldsymbol{u} - 1 - i)^{-\gamma\delta} + (1 + i - \boldsymbol{d})^{-\gamma\delta} \lambda^{-\delta}}$$

whenever this point is in the interval $(0, 1)$ and $\alpha^*(\lambda) = 1$ if the preceding expression is larger than 1 and $\alpha^*(\lambda) = 0$ if the preceding expression is smaller than 0.

Inserting

$$\lambda = \frac{d_{n-1}\big(\Phi(\rho, \bar{\boldsymbol{d}})\big)}{d_{n-1}\big(\Phi(\rho, \bar{\boldsymbol{u}})\big)} \frac{(1-\bar{\rho})}{\bar{\rho}}$$

gives us $\alpha_n^*(\rho)$ and inserting $(1-\bar{\rho})\bar{\rho}^{-1}$ gives us $\alpha_n^*(\bar{\rho})$. Hence we have to determine whether $\alpha^*(\lambda)$ is increasing or decreasing in λ. But this has been done in Lemma 4.2.9: $\alpha^*(\lambda)$ is decreasing in λ. Finally, it remains to show that

$$0 < \frac{d_{n-1}\big(\Phi(\rho, \bar{\boldsymbol{d}})\big)}{d_{n-1}\big(\Phi(\rho, \bar{\boldsymbol{u}})\big)} \leq 1.$$

But this follows by applying Lemma 6.1.4. For $\gamma < 0$ note that $d_n(\rho) < 0$ and hence we have

$$\frac{d_{n-1}\big(\Phi(\rho, \bar{\boldsymbol{d}})\big)}{d_{n-1}\big(\Phi(\rho, \bar{\boldsymbol{u}})\big)} \geq 1.$$

□

Remark 6.1.6. Theorem 6.1.5 tells us that we have to invest more in the stock in the case of an unobservable up probability, compared with the case where we know that $\bar{\rho} = \int \theta \rho(d\theta)$ is the up probability when $\gamma \in (0,1)$. If $\gamma < 0$ the situation is vice versa. A heuristic explanation of this fact is as follows. Though in all cases our investor is risk averse, the degree of risk aversion changes with γ: The risk aversion measured by the Arrow-Pratt absolute risk aversion coefficient is $\frac{1-\gamma}{x}$ and decreases for all wealth levels with γ. In particular if $\gamma \in (0,1)$, the investor is less risk averse than in the logarithmic utility case ($\gamma = 0$) and thus invests more in the stock (the logarithmic case is treated in the next subsection). ◊

In Figure 6.1 we have computed the optimal fractions $\alpha_0^*(\rho)$ for $\rho = U(0,1)$ and $\alpha_0^*(\frac{1}{2})$ in the case of partial and complete observation for the following data: $N = 2, i = 0, \boldsymbol{d} = 0.95925, \boldsymbol{u} = 1.04248$, i.e. we have a two-period problem and the prior distribution is the uniform distribution on $(0,1)$. The parameters belong to a stock with 30% volatility and zero interest rate. Figure 6.1 shows the optimal fractions in the stock in the observed case $\alpha_0^*(\frac{1}{2})$ (solid line) and in the unobserved case $\alpha_0^*(\rho)$, $\rho = U(0,1)$ (dotted line) as a function of γ. For $\gamma = 0$ both fractions coincide. Note that $\alpha_0^*(\gamma)$ is increasing in γ.

Logarithmic Utility

Here we assume that the utility function is of the form $U(x) = \log(x)$ with $x \in dom\, U = (0, \infty)$ and we define

$$\tilde{A} := \{\alpha \in \mathbb{R}^d \mid 1 + \alpha \cdot R(y) > 0 \ \mathbb{P}-\text{a.s.}\}.$$

We obtain the following statement:

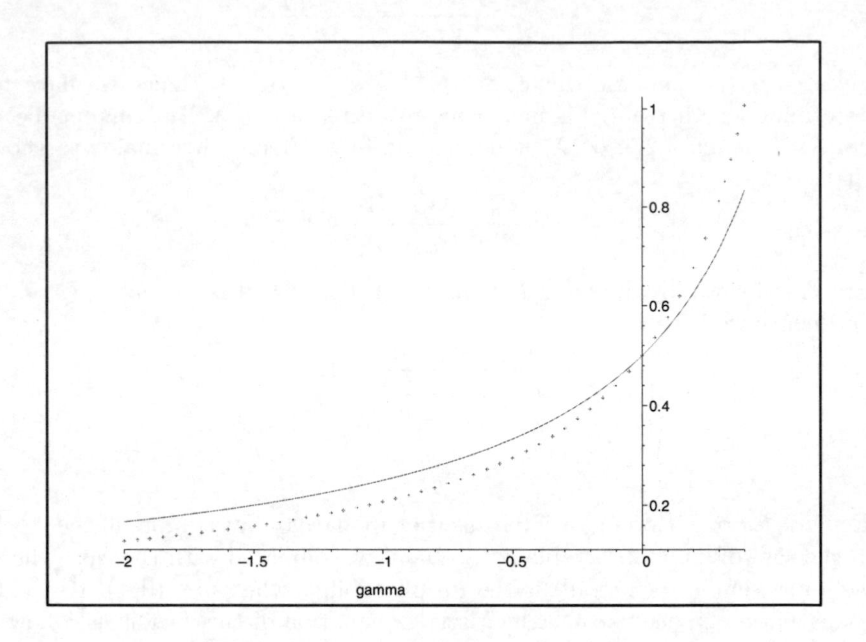

Fig. 6.1 Optimal fractions $\alpha_0^* = \alpha_0^*(\gamma)$ invested in the stock (cross=unobserved, solid=observed).

Theorem 6.1.7. *Let U be the logarithmic utility. Then it holds:*

a) The value functions are given by

$$J_n(x,\rho) = log(x) + d_n(\rho), \quad (x,\rho) \in E_X \times \mathbb{P}(E_Y)$$

where the (d_n) satisfy the following recursion

$$\begin{aligned} d_0(\rho) &= 0 \\ d_n(\rho) &= \log(1+i) + \sup_{\alpha \in \tilde{A}} \left\{ \int\int \log\big(1+\alpha \cdot z\big) Q^R(dz|y)\rho(dy) \right\} \\ &\quad + \int\int d_{n-1}(\Phi(\rho,z)) Q^R(dz|y)\rho(dy). \end{aligned} \tag{6.6}$$

b) The optimal amounts which are invested in the stocks are given by

$$f_n^*(x,\rho) = \alpha^*(\rho)x, \quad x \geq 0$$

where $\alpha^*(\rho)$ *is the solution of* (6.6). *The optimal portfolio strategy is given by* $(f_0, \ldots, f_{N-1})$ *where*

$$f_n(h_n) := f^*_{N-n}\big(x_n, \mu_n(\cdot|h_n)\big), \; h_n = (x_0, a_0, z_1, x_1, \ldots, x_n).$$

The proof is analogous to the proof of Theorem 4.2.13. Note that the optimal amounts which are invested do not depend on n.

We consider now the binomial model with one stock and unknown probability for an up movement, i.e. we consider the Bayesian model and assume

$$\mathbb{P}(\tilde{R} = \boldsymbol{u}|\vartheta = \theta) = \theta, \quad \mathbb{P}(\tilde{R} = \boldsymbol{d}|\vartheta = \theta) = 1 - \theta, \quad \theta \in (0,1) = \Theta$$

and $\boldsymbol{d} < 1 + i < \boldsymbol{u}$. The optimization problem in (6.6) reduces to one which is linear in θ.

$$\begin{aligned}
&\sup_{\alpha \in \tilde{A}} \Big\{ \int \theta \log\left(1 + i + \alpha(\boldsymbol{u} - 1 - i)\right) \rho(d\theta) \\
&\qquad\qquad + \int (1-\theta) \log\left(1 + i + \alpha(\boldsymbol{d} - 1 - i)\right) \rho(d\theta) \Big\} \\
= &\sup_{\alpha \in \tilde{A}} \Big\{ \int \theta \rho(d\theta) \log\left(1 + i + \alpha(\boldsymbol{u} - 1 - i)\right) \\
&\qquad\qquad + (1 - \int \theta\rho(d\theta)) \log\left(1 + i + \alpha(\boldsymbol{d} - 1 - i)\right) \Big\}.
\end{aligned}$$

The maximum point of the problem which represents the optimal fraction of the wealth invested in the stock is given by

$$\alpha^*(\rho) := (1+i) \left(\frac{\int \theta\rho(d\theta)}{1 + i - \boldsymbol{d}} - \frac{1 - \int \theta\rho(d\theta)}{\boldsymbol{u} - 1 - i} \right)$$

(cf. equation (4.12)). Thus, the optimal investment strategy is the same as in a model with complete observation and with probability $\int \theta\rho(d\theta)$ for an up-movement of the stock. Hence we simply have to replace the unknown probability in the formula (4.12) for the optimal invested fraction by its expectation, i.e. $\alpha^*_n(\rho) = \alpha^*_n(\bar{\rho})$ with $\bar{\rho} = \int \theta\rho(d\theta)$. This phenomenon is called the *certainty equivalence principle*.
Note that $\alpha^*(\rho)$ is the limit of the optimal fraction invested in the power utility case for $\gamma \to 0$.

6.2 Dynamic Mean-Variance Problems

Let us investigate the dynamic Markowitz problem of Section 4.6 where the relative risk distribution depends now on the Markov process (Y_n) as described at the beginning of this chapter. Since the discount factors $1 + i_n$ depend on n, the problem is not stationary. In Chapter 5 we restricted the presentation of Partially Observable Markov Decision Models to stationary models, however the extension to non-stationary models is straightforward and we will use appropriate modifications of the theorems here. More precisely we consider the following Partially Observable Markov Decision Model:

- $E_X := \mathbb{R}$ where $x \in E_X$ denotes the wealth,
- E_Y is the state space of (Y_n) where $y \in E_Y$ is the unobservable part of the state,
- $A := \mathbb{R}^d$ where $a \in A$ is the amount of money invested in the risky assets,
- $D_n(x) := A$,
- $\mathcal{Z} := [-1, \infty)^d$ where $z \in \mathcal{Z}$ denotes the relative risk,
- $T_{n,X}(x, a, z) := (1 + i_{n+1})(x + a \cdot z)$,
- $Q^{Z,Y}(B \times C|x, y, a) := Q^R(B|y)Q^Y(C|y)$ for $B \in \mathcal{B}(\mathcal{Z}), C \in \mathcal{B}(E_Y)$,
- Q_0 is the given initial (prior) distribution of Y_0.

The problem we want to solve is

$$(MV) \quad \begin{cases} \operatorname{Var}^\pi_{x_0}(X_N) \to \min \\ \mathbb{E}^\pi_{x_0}[X_N] \geq \mu \\ \pi \in \Pi_N \end{cases}$$

where $\mathbb{P}^\pi_{x_0}$ is the probability measure on $(E_X \times E_Y)^{N+1}$ which has been defined on page 150 and Π_N is the set of history-dependent policies, i.e. $\pi = (f_0, \ldots, f_{N-1}) \in \Pi_N$ with $f_n : H_n \to A$ depends on the observable history h_n until time n.
In order to obtain a well-defined problem we make the following assumptions throughout this section (cf. Section 4.6).

Assumption (FM):

(i) $\sup_y \mathbb{E}\,\|R(y)\| < \infty$.
(ii) *There exists a* $k \in \{1, \ldots, d\}$ *such that* $\mathbb{E}\,R^k(y) > 0$ *for all* $y \in E_Y$ *or* $\mathbb{E}\,R^k(y) < 0$ *for all* $y \in E_Y$.
(iii) *The covariance matrix of the relative risk process*

$$\Big(\operatorname{Cov}(R^j(y), R^k(y))\Big)_{1 \leq j,k \leq d}$$

is positive definite for all $y \in E_Y$.
(iv) $x_0 S^0_N < \mu$.

The second assumption implies that the discounted wealth process is not a martingale. Hence the set of admissible strategies for (MV) is not empty, and

(MV) is well-defined. For $|E_Y| = 1$ the Assumption (FM) reduces to the one in Section 4.6.

We proceed as in the case of complete observation. In particular the formulation as a Lagrange problem and the reduction to a Markov Decision Problem follow the same lines as for the classical mean-variance problem. We refer the reader to Section 4.6 for details. The key Markov Decision Problem we have to solve is

$$QP(b) \quad \begin{cases} \mathbb{E}_{x_0}^{\pi}\left[(X_N - b)^2\right] \to \min \\ \\ \pi \in \Pi_N. \end{cases}$$

In what follows define for $\rho \in I\!P(E_Y)$ and $n = 0, 1, \ldots, N$ the real-valued functions $d_n(\rho)$ by backwards recursion:

$$\begin{aligned} d_N(\rho) &= 1, \\ d_n(\rho) &= \underline{d}_{n+1}(\rho) - \ell_{n+1}(\rho)^\top C_{n+1}(\rho)^{-1} \ell_{n+1}(\rho), \end{aligned} \tag{6.7}$$

where

$$\begin{aligned} \underline{d}_n(\rho) &= \int\int d_n\big(\Phi(\rho, z)\big) Q^R(dz|y)\rho(dy), \\ \ell_n(\rho) &= \int\int d_n\big(\Phi(\rho, z)\big) z Q^R(dz|y)\rho(dy), \\ C_n(\rho) &= \int\int d_n\big(\Phi(\rho, z)\big) z z^\top Q^R(dz|y)\rho(dy). \end{aligned}$$

The following lemma will be crucial for Theorem 6.2.3.

Lemma 6.2.1. *For all $n = 0, 1, \ldots, N-1$ and for all $\rho \in I\!P(E_Y)$ it holds that $0 < d_n(\rho) < 1$.*

Proof. We proceed by induction. For $n = N - 1$ we obtain

$$d_{N-1}(\rho) = 1 - \ell_N(\rho)^\top C_N^{-1}(\rho) \ell_N(\rho).$$

First note that $\ell_N(\rho) \neq 0$ due to Assumption (FM) (ii). By definition the matrix

$$\Sigma_N := C_N(\rho) - \ell_N(\rho)\ell_N(\rho)^\top$$

is the covariance matrix of a random vector with distribution $\int Q^R(\cdot|y)\rho(dy)$. In view of Assumption (FM) (iii) Σ_N is positive definite. Following the proof of Lemma 4.6.4 it can now be shown that $0 < \ell_N(\rho)^\top C_N^{-1}(\rho)\ell_N(\rho) < 1$ and thus $0 < d_{N-1}(\rho) < 1$. Assume now that the statement is true for $N-1, \ldots, n+1$. We can write

$$d_n(\rho) = \underline{d}_{n+1}(\rho)\Big[1 - \frac{\ell_{n+1}(\rho)}{\underline{d}_{n+1}(\rho)}^\top \Big(\frac{C_{n+1}(\rho)}{\underline{d}_{n+1}(\rho)}\Big)^{-1} \frac{\ell_{n+1}(\rho)}{\underline{d}_{n+1}(\rho)}\Big]. \tag{6.8}$$

Our assumption implies that $0 < \underline{d}_{n+1}(\rho) < 1$. If $\ell_{n+1}(\rho) = 0$ the statement is shown. Now suppose $\ell_{n+1}(\rho) \neq 0$ and define the probability measure

$$Q_\rho(dz) := \frac{\int d_{n+1}\big(\Phi(\rho, z)\big) Q^R(dz|y)\rho(dy)}{\underline{d}_{n+1}(\rho)}.$$

Note that $d_{n+1}((\Phi(\rho, z)) > 0$ for all ρ, z by induction hypothesis. Thus we can interpret $\frac{\ell_{n+1}(\rho)}{\underline{d}_{n+1}(\rho)}$ as the expectation of a random vector with respect to $Q_\rho(dz)$. By definition the matrix

$$\Sigma_{n+1} := \frac{C_{n+1}(\rho)}{\underline{d}_{n+1}(\rho)} - \frac{\ell_{n+1}(\rho)}{\underline{d}_{n+1}(\rho)} \frac{\ell_{n+1}(\rho)}{\underline{d}_{n+1}(\rho)}^\top$$

is the covariance matrix of a random vector with distribution $Q_\rho(dz)$. Due to Assumption (FM)(iii) and since $d_{n+1} > 0$, the matrix Σ_{n+1} is positive definite. Following the proof of Lemma 4.6.4, the expression in brackets in (6.8) is in $(0, 1)$. Thus, the statement holds. □

Then we obtain the following result (this can be compared with Theorem 4.6.5 where the LQ-problem is treated without partial observation).

Theorem 6.2.2. *For the solution of the Markov Decision Problem $QP(b)$ it holds:*

a) The value functions are given by

$$V_n(x, \rho) = \Big(\frac{xS_N^0}{S_n^0} - b\Big)^2 d_n(\rho), \quad (x, \rho) \in E_X \times \mathbb{P}(E_Y)$$

where (d_n) is defined in (6.7). Then $V_0(x_0, Q_0)$ is the value of $QP(b)$.

b) Let

$$f_n^*(x, \rho) = \Big(\frac{bS_n^0}{S_N^0} - x\Big) C_{n+1}^{-1}(\rho)\ell_{n+1}(\rho), \quad (x, \rho) \in E_X \times \mathbb{P}(E_Y).$$

Then the portfolio strategy $\pi^* = (f_0, \dots, f_{N-1})$ is optimal for $QP(b)$ where

$$f_n(h_n) := f_n^*\big(x_n, \mu_n(\cdot|h_n)\big), \; h_n = (x_0, z_1, x_1, \dots, x_n).$$

c) The first and the second moment of X_N under π^* are given by

$$\mathbb{E}_{x_0}^{\pi^*}[X_N] = x_0 S_N^0 d_0(Q_0) + b(1 - d_0(Q_0))$$
$$\mathbb{E}_{x_0}^{\pi^*}[X_N^2] = (x_0 S_N^0)^2 d_0(Q_0) + b^2(1 - d_0(Q_0)).$$

Proof. For part a) and b) we proceed as in Theorem 4.6.5 by induction using similar sets for the structure assumption. We restrict here showing that the V_n satisfy the Bellman equation and that f_n^* is a minimizer. For $n = N$ we have $V_N(x, \rho) = (x - b)^2$ in which case the statement is true. Now suppose the statement is true for $k = n + 1, \ldots, N$. The Bellman equation gives

$$V_n(x, \rho) = \inf_{a \in \mathbb{R}^d} \int \int V_{n+1}\big((1 + i_{n+1})(x + a \cdot z), \Phi(\rho, z)\big) Q^R(dz|y)\rho(dy).$$

Inserting the induction hypothesis and after some algebra we obtain

$$V_n(x, \rho) = \inf_{a \in \mathbb{R}^d} \Big\{ \big(\frac{S_N^0}{S_n^0}\big)^2 \big(x^2 \underline{d}_{n+1}(\rho) + 2xa^\top \ell_{n+1}(\rho) + a^\top C_{n+1}(\rho) a\big) \\ - 2\big(\frac{S_N^0}{S_n^0}\big) b\big(x\underline{d}_{n+1}(\rho) + a\ell_{n+1}(\rho)\big) + b^2 \underline{d}_{n+1}(\rho) \Big\}.$$

The minimizer is given by

$$f_n^*(x, \rho) = \Big(\frac{bS_n^0}{S_N^0} - x\Big) C_{n+1}^{-1}(\rho) \ell_{n+1}(\rho).$$

Inserting the minimizer into the Bellman equation above gives after some lines of calculation

$$V_n(x, \rho) = \Big(\frac{xS_N^0}{S_n^0} - b\Big)^2 \big(\underline{d}_{n+1}(\rho) - \ell_{n+1}(\rho)^\top C_{n+1}^{-1}(\rho) \ell_{n+1}(\rho)\big).$$

Hence, part a) and b) are shown. For the proof of part c) we show by induction on n that

$$\mathbb{E}[X_N | X_n, \mu_n] = \frac{S_N^0}{S_n^0} X_n d_n(\mu_n) + b(1 - d_n(\mu_n)).$$

Since this is more cumbersome than in Theorem 4.6.5 we show the induction step here:

$$\begin{aligned}
&\mathbb{E}_{x_0}^{\pi^*}[X_N | X_n, \mu_n] = \mathbb{E}\left[\mathbb{E}[X_N | X_{n+1}, \mu_{n+1}] | X_n, \mu_n\right] \\
&= \mathbb{E}_{x_0}^{\pi^*}\left[\frac{S_N^0}{S_{n+1}^0} X_{n+1} d_{n+1}(\mu_{n+1}) + b(1 - d_{n+1}(\mu_{n+1})) | X_n, \mu_n\right] \\
&= \mathbb{E}_{x_0}^{\pi^*}\Big[\frac{S_N^0}{S_n^0}\Big(X_n + \big(\frac{S_n^0}{S_N^0} b - X_n\big) R_{n+1} \cdot C_{n+1}^{-1}(\mu_n) \ell_{n+1}(\mu_n)\Big) d_{n+1}(\mu_{n+1}) \\
&\qquad + b(1 - d_{n+1}(\mu_{n+1})) | X_n, \mu_n\Big] \\
&= \frac{S_N^0}{S_n^0} X_n d_n(\mu_n) + b(1 - d_n(\mu_n)).
\end{aligned}$$

The proof for the second moment can be done similarly. □

In order to solve the original mean-variance problem (MV) we proceed as in Section 4.6 and set

$$b^* := \mathbb{E}_{x_0}^{\pi^*}[X_N] + \lambda^* = \mu + \lambda^*$$

with

$$\lambda^* = (\mu - x_0 S_N^0)\frac{d_0(Q_0)}{1 - d_0(Q_0)}.$$

Note that $0 < d_0(Q_0) < 1$ due to Lemma 6.2.1. Altogether the solution of the mean-variance problem with partial information is given by:

Theorem 6.2.3. *For the mean-variance problem (MV) with partial observation it holds:*

a) The value of (MV) is given by

$$\text{Var}_{x_0}^{\pi^*}[X_N] = \frac{d_0(Q_0)}{1 - d_0(Q_0)}\Big(\mathbb{E}_{x_0}^{\pi^*}[X_N] - x_0 S_N^0\Big)^2$$

where $d_0(Q_0)$ is given in (6.7). Note that $\mathbb{E}_{x_0}^{\pi^*}[X_N] = \mu$.

b) For $(x, \rho) \in E_X \times \mathbb{P}(E_Y)$ let

$$f_n^*(x, \rho) = \left(\left(\frac{\mu - d_0(Q_0)x_0 S_N^0}{1 - d_0(Q_0)}\right)\frac{S_n^0}{S_N^0} - x\right) C_{n+1}^{-1}(\rho)\ell_{n+1}(\rho).$$

Then the portfolio strategy $\pi^* = (f_0, \dots, f_{N-1})$ is optimal for (MV) where

$$f_n(h_n) := f_n^*(x_n, \mu_n(\cdot|h_n)),\ h_n = (x_0, z_1, x_1, \dots, z_n, x_n).$$

6.3 Remarks and References

Partially observable portfolio problems in discrete time are sparsely studied in the literature. The models and results of this chapter seem to be new. Durst (1991) investigates some Bayesian nonparametric problems. Binomial models (with an unknown up-probability) have been considered by Runggaldier et al. (2002) and Favero (2001). Recently Taksar and Zeng (2007) treat a Hidden-Markov portfolio model with unknown drift and volatility. There are much more interesting applications in finance which can be investigated in a similar manner, e.g. mean-risk models, index-tracking and indifference pricing. It is left to the reader to extend the models of Chapter 4 to the partially observable setting.

Continuous-time portfolio problems with partial observation are studied e.g. in Sass and Haussmann (2004), Bäuerle and Rieder (2007) and Björk et al. (2010). The comparison result in Theorem 6.1.5 is the discrete-time analogue to a result in Rieder and Bäuerle (2005).

Part III
Infinite Horizon Optimization Problems

Chapter 7
Theory of Infinite Horizon Markov Decision Processes

In this chapter we consider Markov Decision Processes with an infinite time horizon. There are situations where problems with infinite time horizon arise in a natural way, e.g. when the random lifetime of the investor is considered. However more important is the fact that Markov Decision Models with finite but large horizon can be approximated by models with infinite time horizon. The latter one is often simpler to solve and admits mostly a (time) stationary optimal policy. On the other hand, the infinite time horizon makes it necessary to invoke some convergence assumptions. Moreover, for the theory it is necessary that properties of the finite horizon value functions carry over to the limit function.

The chapter is organized as follows. In Section 7.1 infinite horizon Markov Decision Models are introduced where we assume that the positive part of the rewards are bounded and converge. This formulation includes so-called *negative* and *discounted* Markov Decision Models. A first main result (Theorem 7.1.8) shows that the infinite horizon problem can indeed be seen as an approximation of the finite time horizon problems, given a structure assumption is satisfied. Moreover, the existence of an optimal stationary policy is shown. A characterization of the infinite horizon value function as the largest solution of the *Bellman equation* is also given. Analogously to Section 2.4, Section 7.2 provides some continuity and compactness conditions under which the structure assumption for the infinite horizon problem is satisfied. Section 7.3 deals with the favourable situation of *contracting* Markov Decision Models. In this case the maximal reward operator $\mathcal{T}$ is contracting on the Banach space $(\mathbb{B}_b, \|\cdot\|_b)$ and if a closed subspace $\mathbb{M}$ of $\mathbb{B}_b$ can be found with $\mathcal{T}: \mathbb{M} \to \mathbb{M}$ then Banach's fixed point theorem can be applied and the value function of the infinite horizon problem can be characterized as the unique fixed point of $\mathcal{T}$. Section 7.4 deals with *positive* Markov Decision Models. Here we make assumptions about the boundedness and convergence of the negative parts of the rewards. The main problem is to identify optimal policies in this case. A maximizer of the value function now does not necessarily define an optimal policy. In Section 7.5 some computational aspects of

N. Bäuerle and U. Rieder, *Markov Decision Processes with Applications to Finance*, Universitext, DOI 10.1007/978-3-642-18324-9_7,

infinite horizon problems are treated like *Howard's policy improvement algorithm*, solution via *linear programming* and *value iteration on a grid*. Finally, the last section contains four examples which explain the use of the developed theory. The first example treats a problem with *random horizon*. In the second example the *cash balance problem* of Section 2.6 is revisited and now treated as an infinite horizon Markov Decision Model. The third example deals with some classical *casino games* and the last example with the infinite horizon bandit.

7.1 Markov Decision Models with Infinite Horizon

Markov Decision Models with infinite horizon can be seen as an approximation of a model with finite but large horizon. Often the infinite horizon model is easier to solve and its optimal policy yields a reasonable policy for the model with finite horizon. This statement is formalized later in this section. In what follows we always assume that a (stationary) Markov Decision Model with infinite horizon is given.

Definition 7.1.1. A *stationary Markov Decision Model* with infinite horizon consists of a set of data (E, A, D, Q, r, β), where E, A, D, Q, r and β are given in Definition 2.1.1 (see also Remark 2.1.2 c) and Section 2.5). There is no terminal reward, i.e. $g \equiv 0$.

Let $\pi = (f_0, f_1, \dots) \in F^\infty$ be a policy for the infinite horizon Markov Decision Model. Then we define

$$J_{\infty\pi}(x) := \mathbb{E}_x^\pi \left[\sum_{k=0}^{\infty} \beta^k r\big(X_k, f_k(X_k)\big) \right], \quad x \in E$$

which gives the expected discounted reward under policy π over an infinite time horizon when we start in state x. The process (X_k) is the corresponding infinite horizon *Markov Decision Process*. The performance criterion is then

$$J_\infty(x) := \sup_\pi J_{\infty\pi}(x), \quad x \in E. \tag{7.1}$$

The function $J_\infty(x)$ gives the maximal expected discounted reward over an infinite time horizon when we start in state x. A policy $\pi^* \in F^\infty$ is called *optimal* if $J_{\infty\pi^*}(x) = J_\infty(x)$ for all $x \in E$.

In order to have a well-defined problem we assume throughout the

Integrability Assumption (A):

$$\delta(x) := \sup_\pi \mathbb{E}_x^\pi \left[\sum_{k=0}^\infty \beta^k r^+(X_k, f_k(X_k)) \right] < \infty, \quad x \in E.$$

Moreover, it is convenient to introduce the set

$$\mathbb{B} := \{v \in \mathbb{M}(E) \mid v(x) \le \delta(x) \text{ for all } x \in E\}.$$

Obviously, $J_{\infty\pi} \in \mathbb{B}$ for all policies π. In order to guarantee that the infinite horizon problem is an approximation of the finite horizon model, we make the following convergence assumption throughout this chapter.

Convergence Assumption (C):

$$\lim_{n\to\infty} \sup_\pi \mathbb{E}_x^\pi \left[\sum_{k=n}^\infty \beta^k r^+(X_k, f_k(X_k)) \right] = 0, \quad x \in E.$$

Below we discuss the convergence assumption and present some simple conditions which imply (C). For this purpose *upper bounding functions* will be important. An upper bounding function for a Markov Decision Model with infinite horizon is defined as follows (cf. Definition 2.4.1).

Definition 7.1.2. A measurable function $b : E \to \mathbb{R}_+$ is called an *upper bounding function* for the Markov Decision Model with infinite horizon, if there exist constants $c_r, \alpha_b \in \mathbb{R}_+$ such that

(i) $r^+(x, a) \le c_r b(x)$ for all $(x, a) \in D$.
(ii) $\int b(x')Q(dx'|x, a) \le \alpha_b b(x)$ for all $(x, a) \in D$.

Let us introduce the operator $\mathcal{T}_\circ$ by

$$\mathcal{T}_\circ v(x) := \sup_{a\in D(x)} \beta \int v(x')Q(dx'|x, a), \quad x \in E,$$

whenever the integral exists. The operator $\mathcal{T}_\circ$ causes a time-shift by one time step, i.e. $\mathcal{T}_\circ v(x)$ is the discounted maximal expectation of $v(X_1)$ when the current state is x. In particular it holds

$$\sup_\pi \mathbb{E}_x^\pi \left[\sum_{k=n}^\infty \beta^k r^+(X_k, f_k(X_k)) \right] = \beta^n \sup_\pi \mathbb{E}_x^\pi [\delta(X_n)] = \mathcal{T}_\circ^n \delta(x).$$

The last equation is true since δ is the value function of a positive Markov Decision Model (see Section 7.4) and Remark 2.3.14 applies. In particular if $v \in \mathbb{B}$ then $\mathcal{T}_\circ v \le \delta$. The Convergence Assumption (C) can also be expressed by the condition

$$\lim_{n\to\infty} \mathcal{T}_\circ^n \delta(x) = 0, \quad x \in E.$$

In particular suppose now that the infinite horizon Markov Decision Model has an upper bounding function b and recall the definition of the set

$$\mathbb{B}_b^+ := \Big\{ v \in \mathbb{M}(E) \mid v^+(x) \le cb(x) \text{ for some } c \in \mathbb{R}_+ \Big\}.$$

If $\|\delta\|_b < \infty$ and $\mathcal{T}_\circ^n b \to 0$, then the Integrability Assumption (A) and the Convergence Assumption (C) are obviously fulfilled. Next let us define α_b by

$$\alpha_b := \sup_{(x,a)\in D} \frac{\int b(x')Q(dx'|x,a)}{b(x)}.$$

Then we obtain

$$\mathcal{T}_\circ b(x) = \sup_{a\in D(x)} \beta \frac{\int b(x')Q(dx'|x,a)}{b(x)} b(x) \le \beta\alpha_b b(x)$$

and by induction

$$\mathcal{T}_\circ^n b(x) \le (\beta\alpha_b)^n b(x).$$

Moreover we get

$$\delta(x) \le c_r \sum_{k=0}^{\infty} (\beta\alpha_b)^k b(x).$$

These inequalities imply that (A) and (C) are satisfied when $\beta\alpha_b < 1$. Such a situation emerges in particular in the so-called *discounted case*, i.e. if the reward function is bounded from above and $\beta < 1$. In this case we have $\alpha_b \le 1$.

Assumptions (A) and (C) are also satisfied for so-called *negative* Markov Decision Models, i.e. if the reward function satisfies $r \le 0$.

Remark 7.1.3. It can be shown that history-dependent policies do not improve the maximal expected discounted reward. More precisely, let Π_∞ be the set of history-dependent policies for the infinite horizon Markov Decision Model. Then it holds:

$$J_\infty(x) = \sup_{\pi\in F^\infty} J_{\infty\pi}(x) = \sup_{\pi\in\Pi_\infty} J_{\infty\pi}(x), \quad x \in E.$$

For a proof see Hinderer (1970), Theorem 18.4 (see also Theorem 2.2.3). This result is also valid for positive Markov Decision Models which are investigated in Section 7.4. ◊

The convergence assumption implies the following (weak) monotonicity properties of the value functions

$$J_{n\pi} := \mathcal{T}_{f_0} \dots \mathcal{T}_{f_{n-1}} 0,$$
$$J_n := \sup_\pi J_{n\pi}.$$

Lemma 7.1.4. *For $n, m \in \mathbb{N}_0$ with $n \geq m$ it holds:*

a) $J_{n\pi} \leq J_{m\pi} + \mathcal{T}_\circ^m \delta$.
b) $J_n \leq J_m + \mathcal{T}_\circ^m \delta$.

Proof. For part a) let $\pi = (f_0, f_1 \dots)$ be an infinite-stage policy. By the reward iteration (see Theorem 2.5.3) we obtain for $n \geq m$

$$\begin{aligned} J_{n\pi} &= \mathcal{T}_{f_0} \dots \mathcal{T}_{f_{n-1}} 0 \\ &\leq \mathcal{T}_{f_0} \dots \mathcal{T}_{f_{m-1}} 0 + \sup_{\pi'} \mathbb{E}_x^{\pi'} \left[\sum_{k=m}^{n-1} \beta^k r^+(X_k, f_k'(X_k)) \right] \\ &\leq \mathcal{T}_{f_0} \dots \mathcal{T}_{f_{m-1}} 0 + \mathcal{T}_\circ^m \delta = J_{m\pi} + \mathcal{T}_\circ^m \delta. \end{aligned}$$

Part b) follows by taking the supremum over all policies in part a). □

In view of Lemma 7.1.4, the Convergence Assumption (C) implies in particular that the sequences $(J_{n\pi})$ and (J_n) are weakly decreasing. From Lemma A.1.4 we deduce that the limits $\lim_{n\to\infty} J_{n\pi}$ and $\lim_{n\to\infty} J_n$ exist. Moreover, for $\pi \in F^\infty$ we obtain from Theorem B.1.1

$$J_{\infty\pi} = \lim_{n\to\infty} J_{n\pi}.$$

We define the *limit value function* by

$$J(x) := \lim_{n\to\infty} J_n(x) \leq \delta(x), \quad x \in E.$$

By definition it obviously holds that $J_{n\pi} \leq J_n$ for all $n \in \mathbb{N}$, hence $J_{\infty\pi} \leq J$ for all policies π. Taking the supremum over all π implies

$$J_\infty(x) \leq J(x), \quad x \in E. \tag{7.2}$$

Note that in general $J \neq J_\infty$ (see Example 7.2.4) and the functions J_n, J and J_∞ are not in $\mathbb{B}$. However J_∞ and J are analytically measurable (see e.g. Bertsekas and Shreve (1978)). The following lemma shows that the L-operator and the limit can be interchanged.

Lemma 7.1.5. *Assume (C) and let $f \in F$. Then it holds:*

a) $\lim_{n\to\infty} LJ_n(x,a) = LJ(x,a)$ *for all* $(x,a) \in D$.
b) $\lim_{n\to\infty} \mathcal{T}_f J_n(x) = \mathcal{T}_f J(x)$ *for all* $x \in E$.

Proof. First recall the definition of the operator L

$$Lv(x,a) := r(x,a) + \beta \int v(x')Q(dx'|x,a), \quad (x,a) \in D,\ v \in \mathbb{B}.$$

Due to the definition of $\mathcal{T}_f$ it suffices to prove part a). If the sequence (J_n) is monotone decreasing, the statement follows directly from the monotone convergence theorem. In the general case let

$$w_m(x) := \sup_{n \geq m} J_n(x)$$

for $m \in \mathbb{N}$. Obviously the sequences (w_m) and (Lw_m) are monotone decreasing and hence $\lim_{m\to\infty} w_m$ and $\lim_{m\to\infty} Lw_m$ exist. Since by Lemma 7.1.4

$$w_m(x) \geq J_m(x) \geq w_m(x) - \mathcal{T}_\circ^m \delta(x)$$
$$Lw_m(x,a) \geq LJ_m(x,a) \geq Lw_m(x,a) - \mathcal{T}_\circ^{m+1}\delta(x)$$

for all $m \in \mathbb{N}$, we obtain due to Condition (C)

$$\lim_{m\to\infty} w_m(x) = \lim_{m\to\infty} J_m(x) \quad \text{and} \quad \lim_{m\to\infty} Lw_m(x,a) = \lim_{m\to\infty} LJ_m(x,a).$$

Finally since $w_m \leq \delta$, we get with monotone convergence

$$\lim_{m\to\infty} LJ_m(x,a) = \lim_{m\to\infty} Lw_m(x,a) = (L \lim_{m\to\infty} w_m)(x,a) = (L \lim_{m\to\infty} J_m)(x,a).$$

□

If $f^\infty := (f, f, \ldots) \in F^\infty$ is a so-called *stationary policy*, i.e. the same decision rule is used at each stage, then we write

$$J_f := J_{\infty f^\infty} = \lim_{n\to\infty} \mathcal{T}_f^n 0.$$

From Lemma 7.1.5 we obtain the infinite horizon *reward iteration.*

Theorem 7.1.6 (Reward Iteration).
Assume (C) and let $\pi = (f, \sigma) \in F \times F^\infty$. *Then it holds:*

a) $J_{\infty\pi} = \mathcal{T}_f J_{\infty\sigma}$.
b) $J_f \in \mathbb{B}$ *and* $J_f = \mathcal{T}_f J_f$.

As in Chapter 2 we formulate first a verification theorem in order to avoid the general measurability problems. It states that candidates for the optimal solution of problem (7.1) are given by fixed points of the maximal reward operator $\mathcal{T}$.

Theorem 7.1.7 (Verification Theorem). *Assume (C) and let* $v \in \mathbb{B}$ *be a fixed point of* $\mathcal{T}$ *such that* $v \geq J_\infty$. *If* f^* *is a maximizer of* v, *then* $v = J_\infty$ *and the stationary policy* $(f^*, f^*, \ldots)$ *is optimal for the infinite-stage Markov Decision Model.*

Proof. By assumption we obtain

$$v = \mathcal{T}_{f^*} v = \mathcal{T}_{f^*}^n v \leq \mathcal{T}_{f^*}^n 0 + \mathcal{T}_\circ^n \delta.$$

Letting $n \to \infty$ we conclude $v \leq J_{f^*} \leq J_\infty$. Thus, the statement follows since $J_\infty \leq v$. □

In what follows we want to solve the problem in (7.1) and at the same time would like to interpret it as a 'limit case', i.e. we want to have $J = J_\infty$. In order to guarantee this statement we require a structure assumption (with terminal reward $g \equiv 0$, see Section 2.5). Unfortunately the Structure Assumption (SA_N) of Chapter 2 is not enough (see Example 7.2.4). In addition we have to assume some properties of the limit value function $J = \lim_{n\to\infty} J_n$.

Structure Assumption (SA): *There exist sets* $\mathbb{M} \subset \mathbb{M}(E)$ *and* $\Delta \subset F$ *such that:*

(i) $0 \in \mathbb{M}$.
(ii) *If* $v \in \mathbb{M}$ *then*

$$\mathcal{T}v(x) := \sup_{a \in D(x)} \left\{ r(x,a) + \beta \int v(x') Q(dx'|x,a) \right\}, \quad x \in E$$

is well-defined and $\mathcal{T}v \in \mathbb{M}$.
(iii) *For all* $v \in \mathbb{M}$ *there exists a maximizer* $f \in \Delta$ *of* v.
(iv) $J \in \mathbb{M}$ *and* $J = \mathcal{T}J$.

Note that conditions (i)–(iii) together constitute the Structure Assumption of Section 2.5. Condition (iv) imposes additional properties on the limit value function. In Section 7.2 we will give important conditions which imply (SA).

Theorem 7.1.8 (Structure Theorem). *Let (C) and (SA) be satisfied. Then it holds:*

a) $J_\infty \in I\!M$, $J_\infty = \mathcal{T} J_\infty$ *and* $J_\infty = J = \lim_{n\to\infty} J_n$ **(Value iteration)**.
b) J_∞ *is the largest* r-subharmonic function v *in* $I\!M \cap I\!B$, *i.e* J_∞ *is the largest function* v *in* $I\!M$ *with* $v \le \mathcal{T} v$ *and* $v \le \delta$.
c) *There exists a maximizer* $f \in \Delta$ *of* J_∞, *and every maximizer* f^* *of* J_∞ *defines an optimal stationary policy* $(f^*, f^*, \dots)$ *for the infinite-stage Markov Decision Model.*

Proof. First recall that we always have $J_\infty \le J$. The Structure Assumption (SA) parts (iii) and (iv) imply that there exists an $f^* \in \Delta$ such that $\mathcal{T}_{f^*} J = \mathcal{T} J = J$. Thus we obtain by iterating the operator $\mathcal{T}_{f^*}$ in the same way as in Lemma 7.1.4:

$$J = \mathcal{T}_{f^*}^n J \le \mathcal{T}_{f^*}^n 0 + \mathcal{T}_\circ^n \delta$$

for all $n \in \mathbb{N}$. Taking the limit $n \to \infty$ yields

$$J \le J_{f^*} \le J_\infty,$$

since always $J_{f^*} \le J_\infty$. Therefore we get

$$J_\infty = J = J_{f^*}$$

and the stationary policy $(f^*, f^*, \dots)$ is optimal. This implies parts a) and c). For part b) note that $J_\infty \in I\!M$, $J_\infty = \mathcal{T} J_\infty$ and $J_\infty \le \delta$. Now let $v \in I\!M \cap I\!B$ be another r-subharmonic function. Then

$$v \le \mathcal{T}^n v \le \mathcal{T}^n 0 + \mathcal{T}_\circ^n \delta = J_n + \mathcal{T}_\circ^n \delta$$

for all $n \in \mathbb{N}$. For $n \to \infty$ we obtain: $v \le J = J_\infty$. □

The equation $J_\infty = \mathcal{T} J_\infty$ is called the *Bellman equation* for the infinite horizon Markov Decision Model. Often this fixed point equation is also called the *optimality equation.* Part a) of the preceding theorem shows that J_∞ is approximated by J_n for n large, i.e. the value of the infinite horizon Markov Decision Problem can be obtained by iterating the $\mathcal{T}$-operator. This justifies the notion *value iteration.*

Remark 7.1.9. a) Part c) of Theorem 7.1.8 shows that an optimal policy can be found among the stationary ones. Formally we have

$$J_\infty = \sup_{f \in F} J_f.$$

b) Under the assumptions of Theorem 7.1.8 the reverse statement of part c) is also true: if f^∞ is optimal, then f is a maximizer of J_∞. This can be seen as follows: Suppose f^∞ is optimal, i.e. $J_\infty = J_f$. Since $J_f = \mathcal{T}_f J_f$ by Theorem 7.1.6 it follows that $\mathcal{T}_f J_\infty = J_\infty = \mathcal{T} J_\infty$, i.e. f is a maximizer of J_∞. ◊

7.2 Semicontinuous Markov Decision Models

As in Section 2.4 we present some continuity and compactness conditions which imply the Structure Assumption (SA) and which are useful in applications. In what follows let E and A be Borel spaces, let D be a Borel subset of $E \times A$ and define

$$D_n^*(x) := \{a \in D(x) \mid a \text{ is a maximum point of } a \mapsto LJ_{n-1}(x,a)\}$$

for $n \in \mathbb{N} \cup \{\infty\}$ and $x \in E$ and define by

$$LsD_n^*(x) := \{a \in A \mid a \text{ is an accumulation point of a sequence } (a_n) \text{ with } a_n \in D_n^*(x) \text{ for } n \in \mathbb{N}\}$$

the *upper limit of the set sequence* $(D_n^*(x))$. The set $D_n^*(x)$ consists of the optimal actions in state x for an n-stage Markov Decision Problem and $D_\infty^*(x)$ are the optimal actions for the infinite horizon problem. We can state the following theorem (see also Theorem 2.4.6).

Theorem 7.2.1. *Suppose there exists an upper bounding function b with $\mathcal{T}_\circ^n b \to 0$ for $n \to \infty$, $\|\delta\|_b < \infty$ and the following conditions are satisfied:*

(i) *$D(x)$ is compact for all $x \in E$ and $x \mapsto D(x)$ is upper semicontinuous,*
(ii) *$(x,a) \mapsto \int v(x')Q(dx'|x,a)$ is upper semicontinuous for all upper semicontinuous $v \in \mathbb{B}_b^+$,*
(iii) *$(x,a) \mapsto r(x,a)$ is upper semicontinuous.*

Then it holds:

a) $J_\infty \in \mathbb{B}_b^+$, $J_\infty = \mathcal{T} J_\infty$ and $J_\infty = J$ **(Value Iteration)**.
b) If b is upper semicontinuous then J_∞ is upper semicontinuous.
c) $\emptyset \neq LsD_n^*(x) \subset D_\infty^*(x)$ for all $x \in E$ **(Policy Iteration)**.
d) There exists an $f^* \in F$ with $f^*(x) \in LsD_n^*(x)$ for all $x \in E$, and the stationary policy $(f^*, f^*, \dots)$ is optimal.

Proof. Assumption (C) is satisfied since $\delta \in I\!B_b^+$ and $\mathcal{T}_\circ^n b \to 0$ for $n \to \infty$. Then let

$$D^*(x) := \{a \in D(x) \mid a \text{ is a maximum point of } a \mapsto LJ(x,a)\}$$

for $x \in E$. We will first prove that

$$\emptyset \neq LsD_n^*(x) \subset D^*(x), \quad x \in E.$$

Let $x \in E$ be fixed and define for $n \in \mathbb{N}$ the functions $v_n : D(x) \to \mathbb{R} \cup \{-\infty\}$ by

$$v_n(a) := LJ_{n-1}(x,a).$$

With Lemma 7.1.5 it follows that $\lim_{n\to\infty} v_n(a) = LJ(x,a)$ for all $a \in D(x)$. Theorem 2.4.6 and assumptions (i)–(iii) imply that v_n is upper semicontinuous, since $J_{n-1} \in I\!B_b^+$ and J_{n-1} is upper semicontinuous. From Lemma 7.1.4 b) we obtain for $n \geq m$

$$\begin{aligned} v_n(a) &\leq L(J_{m-1} + \mathcal{T}_\circ^{m-1}\delta)(x,a) \leq LJ_{m-1}(x,a) + \mathcal{T}_\circ^m \delta(x) \\ &= v_m(a) + \mathcal{T}_\circ^m \delta(x). \end{aligned}$$

Thus we can apply Theorem A.1.5 and obtain $\emptyset \neq LsD_n^*(x) \subset D^*(x)$ and

$$\begin{aligned} \mathcal{T}J(x) &= \sup_{a \in D(x)} \lim_{n\to\infty} v_n(a) = \lim_{n\to\infty} \sup_{a\in D(x)} v_n(a) \\ &= \lim_{n\to\infty} \mathcal{T}J_{n-1}(x) = \lim_{n\to\infty} J_n(x) = J(x). \end{aligned}$$

The limit function J is measurable since J_n is measurable, thus $J \in I\!B$. By Theorem 2.4.6 there exist decision rules $f_n^* \in F$ with $f_n^*(x) \in D_n^*(x)$ for all $x \in E$ and $n \in \mathbb{N}$. In view of the measurable selection theorem (see Theorem A.2.3) there exists a decision rule $f^* \in F$ such that

$$f^*(x) \in Ls\{f_n^*(x)\} \subset LsD_n^*(x) \subset D^*(x).$$

Therefore f^* is a maximizer of J and we obtain

$$J(x) = \mathcal{T}_{f^*}^n J(x) \leq \mathcal{T}_{f^*}^n 0(x) + \mathcal{T}_\circ^n \delta(x)$$

for all $n \in \mathbb{N}$. Taking the limit $n \to \infty$ yields

$$J(x) \leq J_{f^*}(x) \leq J_\infty(x).$$

Thus we obtain $J(x) = J_\infty(x)$ and $D^*(x) = D_\infty^*(x)$ for all $x \in E$, and the stationary policy $(f^*, f^*, \dots)$ is optimal.
Moreover, we obtain for $n \geq m$

$$J_n \leq J_m + \mathcal{T}_\circ^m \delta \leq J_m + c\mathcal{T}_\circ^n b,$$

for some $c > 0$. Since J_m and $\mathcal{T}_\circ^m b$ are upper semicontinuous by Theorem 2.4.6, Lemma A.1.4 implies that J_∞ is upper semicontinuous. □

Suppose the assumptions of Theorem 7.2.1 are satisfied and the optimal stationary policy f^∞ is unique, i.e. in view of Remark 7.1.9 we obtain $D_\infty^*(x) = \{f(x)\}$. Now suppose (f_n^*) is a sequence of decision rules where f_n^* is a maximizer of J_{n-1}. According to Theorem 7.2.1 c) we must have $\lim_{n\to\infty} f_n^* = f$. This means that we can approximate the *optimal policy* for the infinite horizon Markov Decision Problem by a sequence of optimal policies for the finite-stage problems. This property is called *policy iteration.* If $A \subset \mathbb{R}$ and (f_n^*) is a sequence of decision rules where f_n^* is a maximizer of J_{n-1} then among others, the stationary policies $(\liminf_{n\to\infty} f_n^*)^\infty$ and $(\limsup_{n\to\infty} f_n^*)^\infty$ are optimal for the infinite horizon problem.

Corollary 7.2.2. *Suppose the Markov Decision Model has an upper semicontinuous upper bounding function b with $\beta\alpha_b < 1$. If the conditions (i),(ii) and (iii) of Theorem 7.2.1 are satisfied, then all statements a)–d) of Theorem 7.2.1 are valid.*

The proof follows from Theorem 7.2.1 since $\beta\alpha_b < 1$ implies $\|\delta\|_b < \infty$ and $\mathcal{T}_\circ^n b \to 0$ for $n \to \infty$. The convergence condition in Corollary 7.2.2 is true for the *discounted case* (i.e. r is bounded from above and $\beta < 1$) and also for the *negative case* (i.e. $r \le 0$).

Theorem 7.2.3. *Suppose there exists an upper bounding function b with $\mathcal{T}_\circ^n b \to 0$ for $n \to \infty$, $\|\delta\|_b < \infty$ and the following conditions are satisfied:*

(i) *$D(x)$ is compact for all $x \in E$,*
(ii) *$a \mapsto \int v(x')Q(dx'|x,a)$ is upper semicontinuous for all $v \in \mathbb{B}_b^+$ and for all $x \in E$,*
(iii) *$a \mapsto r(x,a)$ is upper semicontinuous for all $x \in E$.*

Then it holds:

a) $J_\infty \in \mathbb{B}_b^+$, $J_\infty = \mathcal{T} J_\infty$ and $J = J_\infty$ **(Value Iteration).**
b) $\emptyset \ne LsD_n^*(x) \subset D_\infty^*(x)$ for all $x \in E$ **(Policy Iteration).**
c) There exists an $f^* \in F$ with $f^*(x) \in LsD_n^*(x)$ for all $x \in E$, and the stationary policy $(f^*, f^*, \dots)$ is optimal.

Proof. The proof follows along the same lines as the proof of Theorem 7.2.1. Note that the v_n are again upper semicontinuous but we do not need to show here that J is upper semicontinuous. □

The preceding theorem gives conditions which guarantee that (SA) is satisfied for the set $\mathbb{B}_b^+$. It is also possible to derive another set of conditions involving continuous functions (using Theorem 2.4.10) which imply the value and policy iteration but since this is rather obvious we have skipped it here and will present these conditions only in the contracting case (cf. Theorem 7.3.6).

Example 7.2.4. This example shows that continuity and compactness conditions are necessary for the value and policy iteration stated in Theorem 7.2.1. We consider the following Markov Decision Model: Suppose that the state space is $E := \mathbb{N}$ and the action space is $A := \mathbb{N}$. Further let $D(1) := \{3, 4, \dots\}$ and $D(x) := A$ for $x \geq 2$ be the admissible actions. The transition probabilities are given by

$$\begin{aligned} q(a|1,a) &:= 1, \\ q(2|2,a) &:= 1, \\ q(x-1|x,a) &:= 1 \quad \text{for } x \geq 3. \end{aligned}$$

All other transition probabilities are zero. Note that state 2 is an absorbing state (see Figure 7.1). The discount factor is $\beta = 1$ and the one-stage reward function is given by

$$r(x,a) := -\delta_{x3}, \quad (x,a) \in D.$$

Since the reward is non-positive, assumptions (A) and (C) are satisfied. However, the set $D(x)$ is obviously not compact, but all other conditions of Theorem 7.2.1 are satisfied.

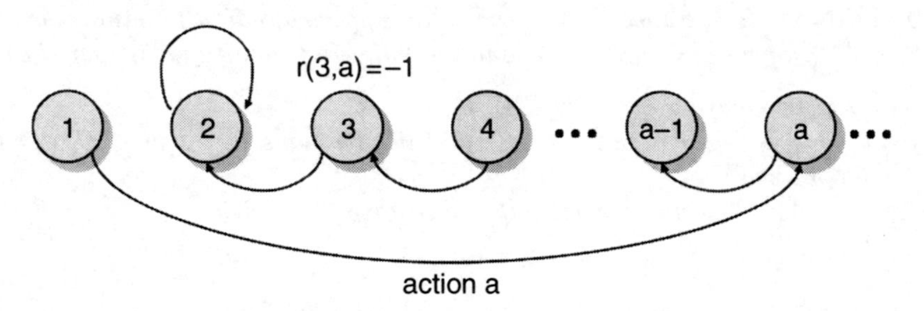

Fig. 7.1 Transition diagram.

We will compute now J and J_∞ and the corresponding sets of maximizers. Let us first consider J_∞. Since state 2 is absorbing, we obviously have $J_\infty(2) = 0$. In all other states it is not difficult to see that under any policy we will pass state 3 exactly once and will then be absorbed in state 2 which yields the maximal reward $J_\infty(x) = -1$ for $x \neq 2$. For state 1 the optimality equation reads

$$J_\infty(1) = \sup_{a \geq 3} J_\infty(a)$$

which yields the set $D^*_\infty(1) = \{3, 4, \ldots\}$. On the other hand we obtain for a finite horizon $n \in \mathbb{N}$ that

$$J_n(x) = \begin{cases} 0\ , & \text{for } x = 1, 2 \\ -1\ , & \text{for } 3 \le x \le n+2 \\ 0\ , & \text{for } x > n+2. \end{cases}$$

Since

$$J_n(1) = \sup_{a \ge 3} J_{n-1}(a)$$

the set of maximizers is given by $D^*_n(1) = \{n+2, n+3, \ldots\}$. Thus, taking the limit $n \to \infty$ we get $J(x) = 0$ for $x = 1, 2$ and $J(x) = -1$ for $x \ge 3$. Altogether, we have

$$J_\infty(1) = -1 < 0 = J(1),$$
$$LsD^*_n(1) = \emptyset \neq D^*_\infty(1).$$

We see that the value and policy iteration may no longer be true if $D(x)$ is not compact. Note that there exists an optimal stationary policy. ♦

7.3 Contracting Markov Decision Models

An advantageous and important situation arises when the operator $\mathcal{T}$ is contracting. Here we assume that the Markov Decision Model has a so-called *bounding function* instead of an upper bounding function which we have considered so far.

Definition 7.3.1. A measurable function $b : E \to \mathbb{R}_+$ is called a *bounding function* for the Markov Decision Model if there exist $c_r, \alpha_b \in \mathbb{R}_+$ such that

(i) $|r(x, a)| \le c_r b(x)$ for all $(x, a) \in D$.
(ii) $\int b(x')Q(dx'|x, a) \le \alpha_b b(x)$ for all $(x, a) \in D$.

If b is a bounding function then $(I\!B_b, \|\cdot\|_b)$ is a Banach space. The weighted supremum norm $\|\cdot\|_b$ has been defined in Section 2.4. We recall its definition here for $v \in I\!M(E)$

$$\|v\|_b := \sup_{x \in E} \frac{|v(x)|}{b(x)}.$$

Markov Decision Models with a bounding function b and $\beta\alpha_b < 1$ are called *contracting*. We will see in Lemma 7.3.3 that $\beta\alpha_b$ is the module of $\mathcal{T}$.

Remark 7.3.2. a) If r is bounded, then $b \equiv 1$ is a bounding function. If moreover $\beta < 1$, then the Markov Decision Model is contracting (the classical *discounted case*).

b) For any contracting Markov Decision Model the assumptions (C) and (A) are satisfied, since $\delta \in I\!\!B_b$ and there exists a constant $c > 0$ with

$$\lim_{n\to\infty} \sup_\pi \mathbb{E}_x^\pi \left[\sum_{k=n}^{\infty} \beta^k r^+(X_k, f_k(X_k)) \right] \le c \lim_{n\to\infty} (\beta\alpha_b)^n b(x) = 0. \qquad \diamondsuit$$

Lemma 7.3.3. *Suppose the Markov Decision Model has a bounding function b and let $f \in F$.*

a) For $v, w \in I\!\!B_b$ it holds:

$$\|\mathcal{T}_f v - \mathcal{T}_f w\|_b \le \beta\alpha_b \|v - w\|_b$$
$$\|\mathcal{T} v - \mathcal{T} w\|_b \le \beta\alpha_b \|v - w\|_b.$$

b) Let $\beta\alpha_b < 1$. Then $J_f = \lim_{n\to\infty} \mathcal{T}_f^n g$ for all $g \in I\!\!B_b$, and J_f is the unique fixed point of $\mathcal{T}_f$ in $I\!\!B_b$.

Proof. a) For $f \in F$ we obtain

$$\mathcal{T}_f v(x) - \mathcal{T}_f w(x) \le \beta \sup_{a\in D(x)} \int \Big(v(x') - w(x')\Big) Q(dx'|x,a)$$
$$\le \beta \|v - w\|_b \sup_{a\in D(x)} \int b(x') Q(dx'|x,a).$$

By interchanging v and w we conclude

$$\mathcal{T}_f w(x) - \mathcal{T}_f v(x) \le \beta \|v - w\|_b \sup_{a\in D(x)} \int b(x') Q(dx'|x,a).$$

Taking the weighted supremum norm yields

$$\|\mathcal{T}_f v - \mathcal{T}_f w\|_b \le \beta\alpha_b \|v - w\|_b.$$

For the second statement note that for functions $g, h : M \to \mathbb{R}$ it holds

$$\sup_{x\in M} g(x) - \sup_{x\in M} h(x) \le \sup_{x\in M} \Big(g(x) - h(x)\Big),$$

hence

$$\mathcal{T} w(x) - \mathcal{T} v(x) \le \beta \sup_{a\in D(x)} \int \Big(v(x') - w(x')\Big) Q(dx'|x,a).$$

Proceeding the same way as for $\mathcal{T}_f$ we derive the second inequality.

b) In view of part a) the operator $\mathcal{T}_f$ is contracting on $I\!\!B_b$, and the statements follow from Banach's fixed point theorem (Theorem A.3.5). $\square$

For contracting Markov Decision Models we are able to give the following slightly stronger version of the Verification Theorem 7.1.7.

Theorem 7.3.4 (Verification Theorem). *Let b be a bounding function, $\beta\alpha_b < 1$ and $v \in \mathbb{B}_b$ be a fixed point of $\mathcal{T} : \mathbb{B}_b \to \mathbb{B}_b$. If f^* is a maximizer of v, then $v = J_\infty = J$ and $(f^*, f^*, \ldots)$ is an optimal stationary policy.*

Proof. First we get $v = \mathcal{T}_{f^*}^n v \le \mathcal{T}_{f^*}^n 0 + \mathcal{T}_\circ^n \delta$ from which we conclude for $n \to \infty$ that $v \le J_{f^*}$. On the other hand, since by Theorem 2.3.7 and Banach's fixed point theorem $J \le \lim_{n\to\infty} \mathcal{T}^n 0 = v$, we obtain altogether

$$J_\infty \le J \le v \le J_{f^*} \le J_\infty. \qquad \square$$

Next we state the main result for contracting Markov Decision Models.

Theorem 7.3.5 (Structure Theorem). *Let b be a bounding function and $\beta\alpha_b < 1$. If there exists a closed subset $\mathbb{M} \subset \mathbb{B}_b$ and a set $\Delta \subset F$ such that*

(i) $0 \in \mathbb{M}$,
(ii) $\mathcal{T} : \mathbb{M} \to \mathbb{M}$,
(iii) *for all $v \in \mathbb{M}$ there exists a maximizer $f \in \Delta$ of v,*

then it holds:

a) $J_\infty \in \mathbb{M}$, $J_\infty = \mathcal{T} J_\infty$ and $J_\infty = J$ **(Value Iteration)**.
b) J_∞ is the unique fixed point of $\mathcal{T}$ in $\mathbb{M}$.
c) J_∞ is the smallest r-superharmonic function $v \in \mathbb{M}$, i.e. J_∞ is the smallest function $v \in \mathbb{M}$ with $v \ge \mathcal{T} v$.
d) Let $g \in \mathbb{M}$. Then

$$\|J_\infty - \mathcal{T}^n g\|_b \le \frac{(\beta\alpha_b)^n}{1 - \beta\alpha_b} \|\mathcal{T} g - g\|_b.$$

e) There exists a maximizer $f \in \Delta$ of J_∞, and every maximizer f^* of J_∞ defines an optimal stationary policy $(f^*, f^*, \ldots)$.

Proof. By Banach's fixed point theorem (Theorem A.3.5) there exists a function $v \in \mathbb{M}$ with $v = \mathcal{T} v$ and $v = \lim_{n\to\infty} \mathcal{T}^n 0$. Thus all assumptions of Theorem 7.1.8 are satisfied. Parts b) and d) follow directly from Banach's fixed point theorem. For c) note that if $v \in \mathbb{M}$ is a solution of $v \ge \mathcal{T} v$, then by iteration we obtain $v \ge \mathcal{T}^n v \to J_\infty$ for $n \to \infty$, and $v \ge J_\infty$ follows. $\square$

As in Section 7.2 we can impose continuity and compactness conditions which imply assumptions (i)–(iii) of Theorem 7.3.5. Here it is assumed that E and A are Borel spaces and D is a Borel subset of $E \times A$.

Theorem 7.3.6. *Let b be a continuous bounding function and $\beta\alpha_b < 1$. If the following assumptions are satisfied*

(i) *$D(x)$ is compact for all $x \in E$ and $x \mapsto D(x)$ is continuous,*
(ii) *$(x, a) \mapsto \int v(x')Q(dx'|x, a)$ is continuous for all continuous $v \in \mathbb{B}_b$,*
(iii) *$(x, a) \mapsto r(x, a)$ is continuous,*

then it holds:

a) J_∞ is continuous, $J_\infty \in \mathbb{B}_b$ and $J_\infty = J$ **(Value Iteration)**.
b) J_∞ is the unique continuous fixed point of $\mathcal{T}$ in $\mathbb{B}_b$.
c) $\emptyset \neq LsD_n^*(x) \subset D_\infty^*(x)$ for all $x \in E$ **(Policy Iteration)**.
d) There exists an $f^* \in F$ with $f^*(x) \in LsD_n^*(x)$ for all $x \in E$, and the stationary policy $(f^*, f^*, \dots)$ is optimal.

Proof. Theorem 2.4.10 implies that (i)–(iii) of Theorem 7.3.5 are satisfied with the set $\mathbb{M} := \{v \in \mathbb{B}_b \mid v \text{ is continuous}\}$. It remains to show that $\mathbb{M}$ is closed. Suppose that $(v_n) \subset \mathbb{M}$ with $\lim_{n\to\infty} \|v_n - v\|_b = 0$. By definition of $\|\cdot\|_b$ it follows that $\frac{v_n}{b}$ converges uniformly to $\frac{v}{b}$. Thus, since $\frac{v_n}{b}$ are continuous by assumption, the limit $\frac{v}{b}$ is continuous which implies that $\mathbb{M}$ is closed. The policy iteration follows as in Theorem 7.2.1. □

7.4 Positive Markov Decision Models

Now we suppose that an infinite horizon Markov Decision Model is given where the *negative* parts of the reward functions converge. This is in contrast to Section 7.1 where the positive parts are considered. Throughout this section we make use of the following assumptions:

Integrability Assumption (A):

$$\varepsilon(x) := \sup_\pi \mathbb{E}_x^\pi \left[\sum_{k=0}^{\infty} \beta^k r^-(X_k, f_k(X_k)) \right] < \infty, \quad x \in E.$$

Convergence Assumption (C$_-$): *For $x \in E$*

$$\lim_{n\to\infty} \sup_\pi \mathbb{E}_x^\pi \left[\sum_{k=n}^{\infty} \beta^k r^-(X_k, f_k(X_k)) \right] = 0.$$

Such a Markov Decision Model is called (generalized) *positive*. Obviously in the case $r \geq 0$ we have $\varepsilon(x) \equiv 0$ and (C$_-$) is satisfied. Note that (C$_-$) is not symmetric to (C) since we have a maximization problem in both cases. In particular $J_\infty(x) = \infty$ is possible for some $x \in E$.

If we define the operator $\mathcal{T}_\circ$ as in Section 7.1, then

$$\sup_\pi \mathbb{E}_x^\pi \left[\sum_{k=n}^{\infty} \beta^k r^- \big(X_k, f_k(X_k)\big) \right] = \beta^n \sup_\pi \mathbb{E}_x^\pi [\varepsilon(X_n)] = \mathcal{T}_\circ^n \varepsilon(x).$$

Hence the Convergence Assumption (C_-) can also be expressed as

$$\lim_{n\to\infty} \mathcal{T}_\circ^n \varepsilon(x) = 0 \text{ for } x \in E.$$

Analogously to Lemma 7.1.4 we obtain now that the value functions are weakly increasing.

Lemma 7.4.1. *For $n, m \in \mathbb{N}_0$ with $n \geq m$ it holds that*

a) $J_{n\pi} \geq J_{m\pi} - \mathcal{T}_\circ^m \varepsilon$.
b) $J_n \geq J_m - \mathcal{T}_\circ^m \varepsilon$.

This monotonicity of $(J_{n\pi})$ and (J_n) implies that

$$J_{\infty\pi} = \lim_{n\to\infty} J_{n\pi} \quad \text{for all } \pi \in F^\infty$$

and

$$J(x) := \lim_{n\to\infty} J_n(x), \quad x \in E$$

exist. Note that $J_{\infty\pi} \geq -\varepsilon$. Again as in Section 7.1 we obtain the following lemma by monotone convergence.

Lemma 7.4.2. *Assume (C_-) and let $f \in F$. Then it holds:*

a) $\lim_{n\to\infty} LJ_n(x,a) = LJ(x,a), \quad (x,a) \in D$.
b) $\lim_{n\to\infty} \mathcal{T}_f J_n(x) = \mathcal{T}_f J(x), \quad x \in E$.
c) $J_f = \mathcal{T}_f J_f$.

Theorem 7.4.3. *Let (C_-) be satisfied. Then it holds:*

a) $J_\infty = \mathcal{T} J_\infty$ *and* $J_\infty = J$ **(Value Iteration)**.
b) J_∞ *is the smallest r-superharmonic function v with* $v \geq -\varepsilon$, *i.e.* J_∞ *is the smallest function v with* $v \geq \mathcal{T} v$ *and* $v \geq -\varepsilon$.

Proof. a) We first prove that $J = \mathcal{T} J$. Fix $x \in E$ and define for $n \in \mathbb{N}$ the functions $v_n : D(x) \to \mathbb{R} \cup \{+\infty\}$ by

$$v_n(a) := LJ_{n-1}(x,a).$$

It is not difficult to show that for all $n \geq m$

$$v_n(a) \geq v_m(a) - \mathcal{T}_\circ^m \varepsilon(x).$$

Theorem A.1.6 then implies

$$\begin{aligned} J(x) &= \lim_{n\to\infty} J_n(x) = \lim_{n\to\infty} \mathcal{T} J_{n-1}(x) = \lim_{n\to\infty} \sup_{a\in D(x)} LJ_{n-1}(x,a) \\ &= \sup_{a\in D(x)} \lim_{n\to\infty} LJ_{n-1}(x,a) = \sup_{a\in D(x)} LJ(x,a) = \mathcal{T} J(x). \end{aligned}$$

Note that we do not assume (SA) here, hence it is not clear why $J_n = \mathcal{T} J_{n-1}$ holds. But this follows from general results of Markov Decision Process theory, see Remark 2.3.14.
Next we show that $J = J_\infty$. From the monotonicity it follows $J_\infty \geq J_{\infty\pi} \geq J_{m\pi} - \mathcal{T}_\circ^m \varepsilon$ for all $\pi \in F^\infty$ and therefore $J_\infty \geq \lim_{n\to\infty} J_n = J$. On the other hand, we have $J_{n\pi} \leq J_n$ for all n, hence $J_{\infty\pi} \leq J$ and $J_\infty \leq J$. Thus, $J_\infty = J$.

b) From part a) we know that J_∞ is a solution of $J_\infty \geq \mathcal{T} J_\infty$ and $J_\infty \geq -\varepsilon$. Let v be an arbitrary r-superharmonic function with $v \geq -\varepsilon$. Then we obtain for $n \in \mathbb{N}$ that $v \geq \mathcal{T}^n v \geq \mathcal{T}^n 0 - \mathcal{T}_\circ^n \varepsilon$ and for $n \to \infty$ we get $v \geq J_\infty$.

□

The next example shows that for positive Markov Decision Models a maximizer of J_∞ defines not necessarily an optimal stationary policy (cf. Theorem 7.1.8).

Example 7.4.4. We consider the following infinite horizon Markov Decision Model: Suppose that the state space is $E := \mathbb{N}_0$ and the action space is $A := \{1, 2\}$. Further let $D(x) := A$. The transition probabilities are given by

$$\begin{aligned} q(0|0,a) &:= 1 \\ q(x+1|x,1) &:= 1 \\ q(0|x,2) &:= 1 \quad \text{for } x \in \mathbb{N}. \end{aligned}$$

All other probabilities are zero (see Figure 7.2). The discount factor is $\beta := 1$ and the one-stage reward function is given by

$$r(x,1) := 0, \quad r(x,2) := 1 - \frac{1}{x+1}, \qquad x \in \mathbb{N}_0.$$

Assumption (C_-) is satisfied since the reward is non-negative. Moreover when we look directly at the problem, it is straightforward to show that $J_\infty(0) = 0$ and $J_\infty(x) = 1$ for $x \in \mathbb{N}$. Since

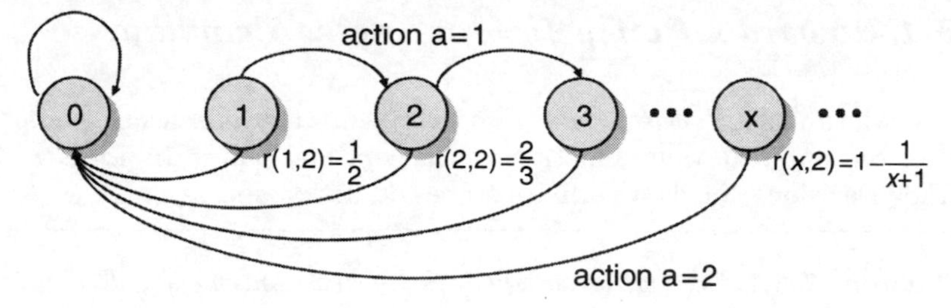

Fig. 7.2 Transition diagram.

$$J(x) = \max\left\{J(x+1), 1 - \frac{1}{x+1}\right\}, \quad x \in \mathbb{N}$$

it is easy to see that $f \equiv 1$ is a maximizer of J_∞, but $J_f \equiv 0$, i.e. f^∞ is not optimal. If the decision rule f is such that $f(x) = 1$ for $x = 0, 1, \ldots, x_0 - 1$ and $f(x_0) = 2$ for some $x_0 \in \mathbb{N}$, then $J_f(x_0) = 1 - \frac{1}{x_0+1}$. Therefore, an optimal stationary policy does not exist for this example. ♦

For positive Markov Decision Models we obtain the following optimality criterion which is different to the criterion in Section 7.1 (see Theorem 7.1.8 and Remark 7.1.9 c)).

Theorem 7.4.5. *Assume (C_-) and let $f \in F$. The following statements are equivalent:*

(i) *f^∞ is optimal.*
(ii) *J_f is an r-superharmonic function.*
(iii) *J_f is a fixed point of $\mathcal{T}$, i.e. $J_f = \mathcal{T} J_f$.*

Proof. *(i) ⇒ (iii)*: If f^∞ is optimal then we have $J_f = J_\infty$, and thus by Theorem 7.4.3 we obtain $J_f = \mathcal{T} J_f$.
(iii) ⇒ (ii): This follows by definition.
(ii) ⇒ (i): If J_f is an r-superharmonic function then Theorem 7.4.3 implies that $J_f \geq J_\infty$, i.e. f^∞ is optimal. □

7.5 Computational Aspects

From Theorem 7.1.8 we know that the value and an optimal policy of the infinite horizon Markov Decision Model can be obtained as limits from the finite horizon problem. This so-called *value iteration* already yields a first computational method to obtain a solution for the infinite horizon optimization problem. Other methods are discussed in this section.

7.5.1 Howard's Policy Improvement Algorithm

We next formulate *Howard's policy improvement algorithm* which is another tool to compute the value function and an optimal policy. It works well in Markov Decision Models with finite state and action spaces.

Theorem 7.5.1. *Let (C) be satisfied. For a decision rule $f \in F$ denote*

$$D(x, f) := \{a \in D(x) \mid LJ_f(x, a) > J_f(x)\}, \quad x \in E.$$

Then it holds:

a) If for some measurable $E_0 \subset E$ we define a decision rule h by

$$h(x) \in D(x, f) \quad \text{for } x \in E_0,$$
$$h(x) = f(x) \quad \text{for } x \notin E_0,$$

then $J_h \geq J_f$ and $J_h(x) > J_f(x)$ for $x \in E_0$. In this case the decision rule h is called an improvement *of f.*

b) If $D(x, f) = \emptyset$ for all $x \in E$, $J_f \geq 0$ and $\mathcal{T} : \mathbb{B} \to \mathbb{B}$ then $J_f = J_\infty$, i.e. the stationary policy $(f, f, \ldots) \in F^\infty$ is optimal.

c) Let b be a bounding function, $\beta\alpha_b < 1$ (i.e. the Markov Decision Model is contracting) and $\mathcal{T} : \mathbb{B}_b \to \mathbb{B}_b$. If $D(x, f) = \emptyset$ for all $x \in E$, then $J_f = J_\infty$.

Proof. a) From the definition of h we obtain

$$\mathcal{T}_h J_f(x) > J_f(x)$$

if $x \in E_0$ and $\mathcal{T}_h J_f(x) = J_f(x)$ if $x \notin E_0$. Thus by induction

$$J_f \leq \mathcal{T}_h J_f \leq \mathcal{T}_h^n J_f \leq \mathcal{T}_h^n 0 + \mathcal{T}_\circ^n \delta$$

where the first inequality is strict if we plug in $x \in E_0$. Letting $n \to \infty$ it follows by Assumption (C) that $J_f \leq J_h$ and in particular $J_f(x) < J_h(x)$ for $x \in E_0$.

b) The condition $D(x, f) = \emptyset$ for all $x \in E$ implies $\mathcal{T} J_f \leq J_f$. Since we always have $\mathcal{T} J_f \geq \mathcal{T}_f J_f = J_f$ we obtain $\mathcal{T} J_f = J_f$. Since $0 \leq J_f$ and $\mathcal{T} : \mathbb{B} \to \mathbb{B}$ it follows that

$$J_n \leq \mathcal{T}^n 0 \leq \mathcal{T}^n J_f = J_f$$

for $n \in \mathbb{N}$. Taking $n \to \infty$ yields $J \leq J_f$ and thus we obtain

$$J_\infty \leq J \leq J_f \leq J_\infty$$

which implies the result.

c) As in b) we get $\mathcal{T} J_f = J_f$. From Banach's fixed point theorem we know that $\mathcal{T}$ has a unique fixed point $v \in \mathbb{B}_b$ and $v = \lim_{n\to\infty} \mathcal{T}^n 0$. Since $J_\infty \le J \le v = J_f \le J_\infty$ the statement follows as in b). □

Remark 7.5.2. a) If F is finite (in particular if the state and action spaces are finite) then an optimal stationary policy can be obtained in a finite number of steps (see the algorithm below).

b) Obviously it holds that $f \in F$ defines an optimal stationary policy $(f, f, \dots)$ if and only if f cannot be improved by the algorithm.

Howard's Policy Improvement Algorithm.

1. Choose $f_0 \in F$ arbitrary and set $k = 0$.
2. Compute J_{f_k} as the largest solution $v \in \mathbb{B}$ of the equation $v = \mathcal{T}_{f_k} v$.
3. Compute f_{k+1} as a maximizer of J_{f_k} (where we set $f_{k+1}(x) = f_k(x)$ if possible). If $f_{k+1} = f_k$ and if $J_{f_k} \ge 0$ or the model is contracting then $J_{f_k} = J$ and $(f_k, f_k, \dots)$ is optimal. Else set $k := k + 1$ and go to step 2.

Corollary 7.5.3. *Let the assumptions of Theorem 7.5.1 be satisfied. In case the algorithm does not stop, it generates a sequence of decision rules (f_k) with $J_{f_k} \ge J_{f_{k-1}}$. If either $J_{f_k} \ge 0$ for some k or the Markov Decision Model is contracting then it holds*

$$\lim_{k\to\infty} J_{f_k} = J_\infty.$$

Proof. The proof follows along the same lines as the proof of Theorem 7.5.1 b), c). Let $\bar{J} := \lim_{k\to\infty} J_{f_k}$ then $\bar{J} \le J_\infty$. From the definition of the sequence of decision rules (f_k) it follows that

$$J_{f_{k+1}} \ge \mathcal{T} J_{f_k} \ge J_{f_k}.$$

Hence $\bar{J} = \mathcal{T}\bar{J}$. If $J_{f_k} \ge 0$ for some k, then $\bar{J} \ge 0$ and

$$\bar{J} = \mathcal{T}^n \bar{J} \ge \mathcal{T}^n 0 \to J, \text{ for } n \to \infty,$$

i.e. $\bar{J} = J = J_\infty$. If the Markov Decision Model is contracting, then $\mathcal{T}$ has a unique fixed point and $\bar{J} = J_\infty$. □

Example 7.5.4 (Howard's Toymaker). We revisit Example 2.5.5 to compute the optimal infinite horizon value function with Howard's policy improvement algorithm. Recall that $\beta \in (0, 1)$. Before we start with the recursion note that for arbitrary $v : E \to \mathbb{R}$ with $\Delta v := v(1) - v(2)$ it holds that

$$Lv(x, a) = r(x, a) + \beta q(1|x, a)\Delta v + \beta v(2)$$

and thus we obtain for $x \in E$:

$$Lv(x, 1) - Lv(x, 2) = 2 - 0.3\beta\Delta v.$$

Hence optimal stationary policies are either f_1^∞ with $f_1 \equiv 1$ or f_2^∞ with $f_2 \equiv 2$.
Let us start with $f := f_1$ and set $k = 0$. The first step is to compute J_f which can be done by solving $v = \mathcal{T}_f v$. This gives two equations:

$$v(x) = r(x, f(x)) + \beta \sum_{y \in E} q(y|x, f(x))v(y), \quad x \in \{1, 2\}.$$

The solution is given by

$$J_f = \frac{3}{(1-\beta)(10-\beta)} \begin{pmatrix} 20 - 17\beta \\ 13\beta - 10 \end{pmatrix}.$$

Now we compute a maximizer of J_f:

$$LJ_f(x, 1) - LJ_f(x, 2) = 2 - 0.3\beta\Delta J_f = 2 - \frac{27\beta}{10-\beta}.$$

Thus, $f := f_1$ is a maximizer of J_f if and only if $LJ_f(x, 1) - LJ_f(x, 2) \geq 0$, which is the case if and only if $\beta \leq \frac{20}{29}$.
Thus we obtain altogether: If $\beta \leq \frac{20}{29}$, then f_1^∞ is an optimal stationary policy and the value function is given by

$$J_\infty = J_{f_1} = \frac{3}{(1-\beta)(10-\beta)} \begin{pmatrix} 20 - 17\beta \\ 13\beta - 10 \end{pmatrix}.$$

If $\beta > \frac{20}{29}$, then f_2^∞ is an optimal stationary policy and the value function is given by

$$J_\infty = J_{f_2} = \frac{2}{(1-\beta)(10-\beta)} \begin{pmatrix} 20 - 11\beta \\ 34\beta - 25 \end{pmatrix}.$$

♦

7.5.2 Linear Programming

Markov Decision Problems can also be solved by linear programming. We restrict here to the contracting case. The key idea is to use the characterization of the value function as the smallest r-superharmonic function (see Theorem 7.3.5) and to express these properties as a linear programming problem. In order to explain the technique we consider a *contracting* Markov Decision Model with a *bounding function* $b : E \to \mathbb{R}_+$ satisfying

(i) $b(x) \geq 1$ *for all* $x \in E$.
(ii) $\int b(x)p(dx) < \infty$ *where* p *is the initial distribution of* X_0.

In the previous chapters we have assumed that p is concentrated on a certain state. We are now interested in computing so-called *p-optimal* policies by linear programming. A (Markov) policy $\pi = (f_n) \in F^\infty$ is called *p-optimal* if it maximizes the functional

$$\pi \mapsto J_\pi := \int J_{\infty\pi}(x)p(dx).$$

The value J_π is well-defined and finite, since by assumption

$$|J_\pi| \leq \frac{c_r}{1-\beta\alpha_b} \int b(x)p(dx) < \infty.$$

In what follows, let $I\!M$ be a closed linear subspace of $I\!B_b$ and we assume that $b \in I\!M$. Then we can formulate the following pair of linear programs:

$$(P) \begin{cases} \int v dp \to \min \\ v(x) - \beta \int v(x')Q(dx'|x,a) \geq r(x,a), \quad (x,a) \in D \\ v \in I\!M. \end{cases}$$

Note that the constraints in (P) are equivalent to $v \geq \mathcal{T}v$. For the dual program let $\mathcal{M}_b := \{\mu \text{ measure on } D \mid \int b d\mu < \infty\}$ and define

$$(D) \begin{cases} \int r d\mu \to \max \\ \int \big(v(x) - \beta \int v(x')Q(dx'|x,a)\big)\mu(d(x,a)) = \int v dp, \quad v \in I\!M \\ \mu \in \mathcal{M}_b. \end{cases}$$

Important measures in $\mathcal{M}_b$ are constructed as follows. For a policy $\pi \in F^\infty$ define the measure μ_π on D by

$$\mu_\pi(B) := I\!E_p^\pi \left[\sum_{k=0}^\infty \beta^k 1_B(X_k, A_k)\right] = \sum_{k=0}^\infty \beta^k \, I\!P_p^\pi \Big((X_k, A_k) \in B\Big)$$

for a measurable subset B of D. Then we obtain $\mu_\pi \in \mathcal{M}_b$, since

$$\mu_\pi(D) \le \sum_{k=0}^{\infty} \beta^k \, \mathbb{E}_p^\pi \Big[b(X_k) \Big] = \int b(x) \mu_\pi(d(x,a)) \le \frac{\int b dp}{1 - \beta \alpha_b} < \infty.$$

Note that we have used $b \geq 1$ for the first inequality. The measure μ_π is a (discounted) occupation measure. Later in Theorem 7.5.6 it will be shown that μ_π is indeed admissible for (D).

Remark 7.5.5. Problem (D) is not the dual program in the classical sense of linear programming. In general, we have to use finite additive set functions instead of the restricted set $\mathcal{M}_b$ (cf. Heilmann (1979)). For solving a contracting Markov Decision Problem however, the proposed program (D) is sufficient. ◊

In what follows we denote by

$$\mathcal{Z}_P := \Big\{ v \in I\!M \mid v(x) - Lv(x,a) \ge 0 \text{ for all } (x,a) \in D \Big\}$$

$$\mathcal{Z}_D := \Big\{ \mu \in \mathcal{M}_b \mid \int \big(v(x) - \beta \int v(x') Q(dx'|x,a)\big) \mu(d(x,a)) = \int v dp \text{ for all } v \in I\!M \Big\}$$

the feasible sets of the programs (P) and (D), and by $val(P)$ and $val(D)$ the minimal and maximal value respectively.

Theorem 7.5.6 (Weak Duality). *The feasible sets $\mathcal{Z}_P$ and $\mathcal{Z}_D$ are non-empty and it holds: $-\infty < val(D) \le val(P) < \infty$.*

Proof. a) Define

$$v(x) := \frac{c_r}{1 - \beta \alpha_b} b(x).$$

Then $v \in I\!M$ and we obtain

$$Lv(x,a) \le c_r b(x) + \frac{\beta \alpha_b}{1 - \beta \alpha_b} c_r b(x) = v(x).$$

Thus, it follows

$$v(x) - Lv(x,a) \ge 0$$

and $v \in \mathcal{Z}_P$.

b) We will show that the occupation measures μ_π satisfy the constraint in (D). For $v \in I\!M$ we obtain

$$\begin{aligned}
&\int \Big(v(x) - \beta \int v(x')Q(dx'|x,a)\Big)\mu_\pi(d(x,a)) \\
&= \sum_{k=0}^{\infty} \beta^k \Big(\mathbb{E}_p^\pi[v(X_k)] - \beta\, \mathbb{E}_p^\pi \Big[\int v(x')Q(dx'|X_k,A_k)\Big]\Big) \\
&= \sum_{k=0}^{\infty} \beta^k \Big(\mathbb{E}_p^\pi[v(X_k)] - \beta\, \mathbb{E}_p^\pi[v(X_{k+1})]\Big) \\
&= \int v(x)p(dx)
\end{aligned}$$

which implies that $\mu_\pi \in \mathcal{Z}_D$. Now let $v \in \mathcal{Z}_P$ and $\mu \in \mathcal{Z}_D$. Then

$$\begin{aligned}
\int v dp &= \int \Big(v(x) - \beta \int v(x')Q(dx'|x,a)\Big)\mu(d(x,a)) \\
&\geq \int r(x,a)\mu(d(x,a)) = \int r d\mu
\end{aligned}$$

which implies the last statement of the theorem. □

A direct consequence of *complementary slackness* in linear programming is the following result.

Theorem 7.5.7. *Let $v \in \mathcal{Z}_P$ and $\mu \in \mathcal{Z}_D$. Then the following statements are equivalent:*

(i) *v and μ are optimal and $val(P) = val(D)$.*
(ii) *$v(x) = Lv(x,a)$ for μ-almost all $(x,a) \in D$.*

Proof. First note that if $v \in \mathcal{Z}_P$ then (ii) is equivalent to

$$\int \Big(v(x) - Lv(x,a)\Big)\mu(d(x,a)) = 0,$$

since $v(x) - Lv(x,a)$ is non-negative.

(i) ⇒ (ii): By assumption we have $\int v dp = \int r d\mu$. Since $\mu \in \mathcal{Z}_D$ we obtain

$$\int \Big(v(x) - \beta \int v(x')Q(dx'|x,a)\Big)\mu(d(x,a)) = \int v dp = \int r d\mu.$$

This equation is equivalent to

$$\int \Big(v(x) - Lv(x,a)\Big)\mu(d(x,a)) = 0,$$

and the first part is shown.

$(ii) \Rightarrow (i)$ Similar to the first part, the equation

$$\int \Big(v(x) - Lv(x,a)\Big)\mu(d(x,a)) = 0$$

can be written as

$$\int \big(v(x) - \beta \int v(x')Q(dx'|x,a)\big)\mu(d(x,a)) = \int r d\mu.$$

Moreover, since $\mu \in \mathcal{Z}_D$ we obtain that the left-hand side equals $\int v dp$. Thus we have $\int v dp = \int r d\mu$, and the statement follows from Theorem 7.5.6. □

The next theorem indeed shows that the linear programs (P) and (D) help to find an optimal solution of the infinite horizon Markov Decision Problem. More precisely, the values of both linear programs coincide and yield the maximal value of the Markov Decision Problem.

Theorem 7.5.8 (Strong duality). *Suppose the assumptions of Theorem 7.3.5 are satisfied for the closed subspace $\mathbb{M} \subset \mathbb{B}_b$. Then the following statements hold:*

a) (P) has an optimal solution $v^* \in \mathbb{M}$, $v^* = J_\infty$ and

$$val(P) = \int J_\infty(x)p(dx) = val(D).$$

b) (D) has an optimal solution $\mu^* \in \mathcal{M}_b$ and there exists an $f^* \in F$ such that

$$val(D) = \int r d\mu^* = \int J_{f^*}(x)p(dx).$$

In particular, the stationary policy $(f^*, f^*, \ldots)$ is p-optimal.

Proof. a) By Theorem 7.3.5 the function J_∞ is the unique fixed point of $\mathcal{T}$ in $\mathbb{M}$ and thus by definition of (P) we have $J_\infty \in \mathcal{Z}_P$. Let $v \in \mathcal{Z}_P$ be arbitrary. Then $v \geq \mathcal{T}v$ and by iterating the operator $\mathcal{T}$ we obtain

$$v \geq \mathcal{T}^n v \to J_\infty, \quad \text{for } n \to \infty$$

i.e. $v \geq J_\infty$ and $\int v dp \geq \int J_\infty dp$. Hence J_∞ is an optimal solution of (P).

b) By Theorem 7.3.5 there exists a maximizer $f^* \in \Delta$ of J_∞. Let $\mu^* := \mu_{f^*}$ be the measure on D which is induced by the stationary policy $(f^*, f^*, \ldots)$ (see the proof of Theorem 7.5.6). Then $\mu^* \in \mathcal{Z}_D$, and since $J_\infty \in \mathcal{Z}_P$ (see part a)) and $J_\infty(x) = LJ_\infty(x, f^*(x))$, $x \in E$ we obtain by Theorem 7.5.7

$$\int J_\infty dp = \int r d\mu^* = val(D).$$

Moreover, we know from Theorem 7.3.5 that $J_\infty = J_{f^*}$. □

In what follows we consider the *special case* where the state and the action spaces are *finite*. Recall that we write in this case $q(y|x,a) := Q(\{y\}|x,a)$. Here the linear programs reduce to

$$(P)\begin{cases}\sum_{x\in E} v(x)p(x) \to \min\\ v(x) - \beta\sum_y q(y|x,a)v(y) \ge r(x,a), \quad (x,a)\in D,\\ v(x)\in\mathbb{R},\ x\in E.\end{cases}$$

$$(D)\begin{cases}\sum_{(x,a)\in D} r(x,a)\mu(x,a) \to \max\\ \sum_{(x,a)}\big(\delta_{xy} - \beta q(y|x,a)\big)\mu(x,a) = p(y), \quad y\in E,\\ \mu(x,a)\ge 0, (x,a)\in D.\end{cases}$$

We use for (P) the set $I\!M$ of all functions $v : E \to \mathbb{R}$ and it is sufficient for (D) to require the equation for the functions $v_y(x) = \delta_{xy}, x \in E$. Note that in this case (D) is indeed the dual program of (P) in the usual sense. In the finite case we get the following (somewhat stronger) result.

Theorem 7.5.9. *Let E and A be finite and $p(x) > 0$ for all $x \in E$. Then the following statements hold:*

a) (P) has an optimal solution v^* and $v^* = J_\infty$.
b) (D) has an optimal solution μ^*. Let μ^* be an optimal vertex. Then for all $x \in E$, there exists a unique $a_x \in D(x)$ such that $\mu^*(x,a_x) > 0$ and the stationary policy $(f^*, f^*, \ldots)$ with $f^*(x) := a_x,\ x \in E$, is optimal.

Proof. The existence of optimal solutions $v^* = J_\infty \in \mathcal{Z}_P$ and $\mu^* \in \mathcal{Z}_D$ follows from Theorem 7.5.8. In particular, $\mu^* \in \mathcal{Z}_D$ satisfies

$$\sum_{(x,a)\in D}\big(\delta_{xy} - \beta q(y|x,a)\big)\mu^*(x,a) = p(y)$$

which is equivalent to

$$\sum_{a\in D(y)}\mu^*(y,a) - \beta\sum_{(x,a)\in D} q(y|x,a)\mu^*(x,a) = p(y).$$

Since $p(x) > 0$ for all $x \in E$ we must have $\sum_{a\in D(x)}\mu^*(x,a) > 0$, i.e. for all x there exists at least one $a \in D(x)$ such that $\mu^*(x,a) > 0$. If μ^* is a vertex then there can be only $|E|$ positive entries in the vector $\mu^*(x,a)$ which implies that a_x^* with $\mu^*(x,a_x^*) > 0$ is unique. Now define $f^*(x) := a_x^*$. From Theorem 7.5.7 it follows that

$$v^*(x) = Lv(x, a_x^*) = Lv(x, f^*(x)), \quad x \in E$$

which means that $v^* = \mathcal{T}_{f^*} v^*$. On the other hand we know that $v^* = \mathcal{T} v^*$ which implies $\mathcal{T} v^* = \mathcal{T}_{f^*} v^*$. Thus, f^* is a maximizer of v^*, and the stationary policy $(f^*, f^*, \dots)$ is optimal. □

Example 7.5.10 (Howard's Toymaker). We revisit Example 2.5.5 to compute an optimal policy with the dual program. The dual program (D) has the following form

$$(D) \begin{cases} -6x_{11} - 4x_{12} + 3x_{21} + 5x_{22} \to \min \\ (1-0.5\beta)x_{11} + (1-0.8\beta)x_{12} - 0.4\beta x_{21} - 0.7\beta x_{22} = 1 \\ -0.5\beta x_{11} - 0.2\beta x_{12} + (1-0.6\beta)x_{21} + (1-0.3\beta)x_{22} = 1 \\ x_{11}, x_{21}, x_{12}, x_{22} \geq 0. \end{cases}$$

Using the well-known Simplex algorithm it follows that $(x_{11}^*, 0, x_{21}^*, 0)$ is an optimal vertex if $\beta < \frac{20}{29}$ with

$$x_{11}^* = \frac{10 - 2\beta}{(1-\beta)(10-\beta)}$$
$$x_{21}^* = \frac{(10}{(1-\beta)(10-\beta)}.$$

From Theorem 7.5.9 we conclude that $f^* \equiv 1$ determines the (unique) optimal stationary policy. In case $\beta > \frac{20}{29}$ then the vertex $(0, x_{12}^*, 0, x_{22}^*)$ is optimal with

$$x_{12}^* = \frac{10 + 4\beta}{(1-\beta)(10-\beta)}$$
$$x_{22}^* = \frac{10 - 6\beta}{(1-\beta)(10-\beta)}.$$

and thus $f^* \equiv 2$ determines the (unique) optimal stationary policy. In case $\beta = \frac{20}{29}$ the dual linear program has four optimal vertices, namely $(x_{11}^*, 0, x_{21}^*, 0)$, $(0, x_{12}^*, x_{21}^*, 0)$, $(x_{11}^*, 0, 0, x_{22}^*)$, $(0, x_{12}^*, 0, x_{22}^*)$ and thus all stationary policies are optimal. ♦

7.5.3 State Space Discretization

If the state space E is not discrete, then numerical results can only be obtained when the state space is approximated by a discrete grid and the value iteration is executed on the grid only. The value function in between has to be obtained by interpolation. In this section we consider a *contracting* Markov

Decision Problem with Borel spaces E and A, D a Borel subset of $E \times A$ and with a bounding function $b : E \to \mathbb{R}_+$ satisfying

(i) $b(x) \geq 1$ *for all* $x \in E$.
(ii) b *is uniformly continuous on* E.
(iii) *The Structure Assumption (SA) is satisfied for the set*

$$\mathbb{M}_c := \{v \in \mathbb{B}_b \mid v \text{ is uniformly continuous on } E\}.$$

We assume now that the state space is approximated by a grid $G \subset E$ and define the grid operator $\mathcal{T}_G$ on $\mathbb{M}_c$ by

$$\mathcal{T}_G v(x) := \begin{cases} \mathcal{T} v(x), & \text{for } x \in G \\ \sum_k \lambda_k \mathcal{T} v(x_k), & \text{for } x \notin G, x = \sum_k \lambda_k x_k, \\ & \text{and } x_k \in G, \lambda_k \geq 0, \sum_k \lambda_k = 1. \end{cases}$$

$\mathcal{T}_G$ coincides with $\mathcal{T}$ on G and $\mathcal{T}_G$ is a linear interpolation elsewhere. It is easy to see that $\mathcal{T}_G : \mathbb{M}_c \to \mathbb{M}_c$. Moreover, we define a different bounding function $b_G : E \to \mathbb{R}_+$ by

$$b_G(x) := \begin{cases} b(x), & \text{for } x \in G \\ \sum_k \lambda_k b(x_k), & \text{for } x \notin G, x = \sum_k \lambda_k x_k, \\ & \text{and } x_k \in G, \lambda_k \geq 0, \sum_k \lambda_k = 1. \end{cases}$$

Note that b_G is again a bounding function for our Markov Decision Model, thus for $v \in \mathbb{M}_c$ we may define:

$$\|v\|_G := \sup_{x \in E} \frac{|v(x)|}{b_G(x)}.$$

If the mesh size h of the grid is small enough, the operator $\mathcal{T}_G$ is again a contraction.

Proposition 7.5.11. *The module* α_G *is bounded by*

$$\alpha_G \leq \alpha_b m(h),$$

where $m(h) \to 1$ *if the mesh size* h *tends to zero.*

Proof. Let us define $\tilde{m}(h) := \|b - b_G\| = \sup_{x \in E} |b(x) - b_G(x)|$. Our assumption implies that $\tilde{m}(h) \to 0$ if the mesh size tends to zero. By definition we obtain

$$\begin{aligned}
\alpha_G &:= \sup_{(x,a)} \frac{\int b_G(y) Q(dy|x,a)}{b_G(x)} \\
&\le \sup_{(x,a)} \frac{\int b(y) Q(dy|x,a) + \tilde{m}(h)}{b(x) - \tilde{m}(h)} \\
&= \sup_{(x,a)} \frac{\int b(y) Q(dy|x,a)}{b(x)} \frac{b(x)}{b(x) - \tilde{m}(h)} + \sup_x \frac{\tilde{m}(h)}{b(x) - \tilde{m}(h)} \\
&\le \alpha_b \frac{1}{1 - \tilde{m}(h)} + \frac{\tilde{m}(h)}{1 - \tilde{m}(h)}
\end{aligned}$$

If we define

$$m(h) = \frac{\alpha_b + \tilde{m}(h)}{\alpha_b(1 - \tilde{m}(h))},$$

then $m(h) \to 1$ for $h \to 0$ and the statement follows. □

The next theorem states that our value function J_∞ can be approximated arbitrarily well by the iterates of the grid operator.

Theorem 7.5.12. *Suppose that $\beta\alpha_G < 1$. Then it holds for $g \in I\!M_c$*

$$\|J_\infty - \mathcal{T}_G^n g\|_G \le \frac{1}{1 - \beta\alpha_G} \Big((\beta\alpha_G)^n \|\mathcal{T}_G g - g\|_G + \|J_\infty - \mathcal{T}_G J_\infty\|_G \Big),$$

where $\|J_\infty - \mathcal{T}_G J_\infty\|_G \to 0$ if the mesh size h tends to zero.

Proof. Since $\beta\alpha_G < 1$ the $\mathcal{T}_G$-operator is contracting and has a unique fixed point which we denote by J_G, i.e. $J_G = \mathcal{T}_G J_G$. An application of the triangle inequality yields

$$\|J_\infty - \mathcal{T}_G^n g\|_G \le \|J_\infty - J_G\|_G + \|J_G - \mathcal{T}_G^n g\|_G$$

for all $n \in \mathbb{N}$. By Banach's fixed point theorem we obtain

$$\|J_G - \mathcal{T}_G^n g\|_G \le \frac{(\beta\alpha_G)^n}{1 - \beta\alpha_G} \|\mathcal{T}_G g - g\|_G.$$

Moreover, again by applying the triangle inequality we obtain

$$\begin{aligned}
\|J_\infty - J_G\|_G &\le \|J_\infty - \mathcal{T}_G J_\infty\|_G + \|\mathcal{T}_G J_\infty - \mathcal{T}_G J_G\|_G \\
&\le \|J_\infty - \mathcal{T}_G J_\infty\|_G + \beta c_G \|J_\infty - J_G\|_G.
\end{aligned}$$

Solving this for $\|J_\infty - J_G\|_G$ yields the desired result. Note that by definition of the $\mathcal{T}_G$-operator, $\mathcal{T}_G J_\infty(x) = J_\infty(x)$ for $x \in G$. Since J_∞ is uniformly continuous, we obtain $\|J_\infty - \mathcal{T}_G J_\infty\|_G \to 0$ for $h \to 0$. □

7.6 Applications and Examples

In this section we look at some simple applications to highlight the results of the preceding sections. More applications can be found in Chapter 9.

7.6.1 Markov Decision Models with Random Horizon

Sometimes Markov Decision Models with infinite horizon appear in a natural way. For example when the time horizon is random and not bounded. There may be situations where we do not know the horizon of the problem in advance. For example, there may occur events (like bankruptcy of certain companies) which lead to a termination of the project and these events occur only with a certain probability. In this case it is reasonable to model the horizon by a random variable $\tau : \Omega \to \mathbb{N}$.
Now suppose a stationary Markov Decision Model $(E, A, D, Q, r, g, \beta)$ is given (see Section 2.5). We *assume* that the random horizon τ is independent of the state process (X_n). For a fixed policy $\pi \in F^\infty$ let

$$V_\pi^\tau(x) := \mathbb{E}_x^\pi\left[\sum_{k=0}^{\tau-1} \beta^k r\big(X_k, f_k(X_k)\big) + \beta^\tau g(X_\tau)\right]. \tag{7.3}$$

The stochastic optimization problem is then given by

$$V^\tau(x) := \sup_{\pi\in F^\infty} V_\pi^\tau(x), \quad x \in E.$$

To simplify our analysis we assume that τ is geometrically distributed, i.e.

$$\mathbb{P}(\tau = n) = (1-p)p^{n-1},\ n \in \mathbb{N} \quad \text{and} \quad p \in (0,1).$$

In this case we claim that the given problem with random horizon is equivalent to an ordinary discounted Markov Decision Model with infinite horizon and with a modified reward function and a modified discount factor. To show this let us define the following infinite horizon Markov Decision Model $(E, A, D, Q, \tilde{r}, \tilde{\beta})$ where

- $\tilde{r}(x,a) := r(x,a) + \beta(1-p)\int g(x')Q(dx'|x,a)$, $(x,a) \in D$,
- $\tilde{\beta} := \beta p$.

We assume that assumptions (A) and (C) are satisfied. For a fixed policy $\pi = (f_0, f_1, \dots) \in F^\infty$ it holds

$$J_{\infty\pi}(x) = V_\pi^\tau(x)$$

and thus $J_\infty(x) = V^\tau(x)$ for $x \in E$. This is true since

$$
\begin{aligned}
J_{\infty\pi}(x) &= \mathbb{E}_x^\pi\left[\sum_{k=0}^{\infty}(\beta p)^k\Big(r\big(X_k, f_k(X_k)\big) + \beta(1-p)g(X_{k+1})\Big)\right] \\
&= \mathbb{E}_x^\pi\left[\sum_{k=0}^{\infty} 1_{[\tau>k]}\beta^k r\big(X_k, f_k(X_k)\big) + \sum_{k=1}^{\infty} 1_{[\tau=k]}\beta^k g(X_k)\right] \\
&= \mathbb{E}_x^\pi\left[\sum_{k=0}^{\tau-1}\beta^k r\big(X_k, f_k(X_k)\big) + \beta^\tau g(X_\tau)\right] = V_\pi^\tau(x)
\end{aligned}
$$

where we use the independence of τ from the state process (X_n) in the second equation. In case the distribution of τ is arbitrary, it is possible to proceed in the same way by using the time dependent discount factors $\beta_0 := 1$ and

$$
\beta_n := \frac{\beta\,\mathbb{P}(\tau > n)}{\mathbb{P}(\tau \geq n)}, \quad n \in \mathbb{N}
$$

instead of $\tilde{\beta}$.

7.6.2 A Cash Balance Problem with Infinite Horizon

In this section we reconsider the cash balance problem of Section 2.6.2, but now as an infinite horizon optimization problem. Recall the data of the cash balance problem:

- $E := \mathbb{R}$ where $x \in E$ denotes the cash level,
- $A := \mathbb{R}$ where $a \in A$ denotes the new cash level after transfer,
- $D(x) := A$,
- $\mathcal{Z} := \mathbb{R}$ where $z \in \mathcal{Z}$ denotes the cash change.
- $T(x, a, z) := a - z$,
- $Q^Z(\cdot|x, a) :=$ the distribution of the stochastic cash change Z_{n+1} (independent of (x, a)),
- $r(x, a) := -c(a - x) - L(a)$,
- $\beta \in (0, 1)$.

It is assumed (see Section 2.6.2) that $c(z) := c_u z^+ + c_d z^-$ with $c_u, c_d > 0$ and $L : \mathbb{R} \to \mathbb{R}_+$, $L(0) = 0$, $x \mapsto L(x)$ is convex and $\lim_{|x|\to\infty} \frac{L(x)}{|x|} = \infty$. Moreover we suppose that $Z := Z_1$ and $\mathbb{E}\, Z < \infty$. Note that assumptions (A) and (C) are fulfilled since $r \leq 0$.
We treat this problem as a cost minimization problem, i.e. J_∞ in the next theorem is the minimal cost function over an infinite horizon, and we obtain the following result.

Theorem 7.6.1. *For the cash balance problem with infinite horizon it holds:*

a) There exist critical levels S_- and S_+ such that

$$J_\infty(x) = \begin{cases} (S_- - x)c_u + L(S_-) + \beta\,\mathbb{E}\,J_\infty(S_- - Z) & \text{if } x < S_- \\ L(x) + \beta\,\mathbb{E}\,J_\infty(x - Z) & \text{if } S_- \le x \le S_+ \\ (x - S_+)c_d + L(S_+) + \beta\,\mathbb{E}\,J_\infty(S_+ - Z) & \text{if } x > S_+. \end{cases}$$

J_∞ is convex and $J_\infty = J = \lim_{n\to\infty} J_n$.

b) The stationary policy $(f^*, f^*, \ldots)$ is optimal with

$$f^*(x) := \begin{cases} S_- & \text{if } x < S_-, \\ x & \text{if } S_- \le x \le S_+, \\ S_+ & \text{if } x > S_+, \end{cases} \tag{7.4}$$

where S_- and S_+ are accumulation points of the sequences (S_{n-}) and (S_{n+}) given in Theorem 2.6.2.

Proof. a) We prove this part with the help of Theorem 7.1.8. Conditions (i)–(iii) of (SA) have already been shown in Section 2.6.2 with

$$I\!M := \{v : E \to \mathbb{R}_+ \mid v \text{ is convex and } v(x) \le c(-x) + d \text{ for some } d \in \mathbb{R}_+\}.$$

Indeed it is now crucial to see that $I\!M$ can be chosen as

$$I\!M := \{v : E \to \mathbb{R}_+ \mid v \text{ is convex and } v(x) \le c(-x) + d_0\}$$

with fixed $d_0 = \frac{\beta\,\mathbb{E}\,c(Z)}{1-\beta}$ since for $v \in I\!M$ with $v(x) \le c(-x) + d_0$ we get

$$\begin{aligned} \mathcal{T}v(x) &= c(-x) + \beta\,\mathbb{E}\,v(-Z) \\ &\le c(-x) + \beta\,\mathbb{E}\,c(Z) + \beta d_0 = c(-x) + d_0. \end{aligned}$$

So it remains to show condition (iv) of (SA). $J \in I\!M$ is obvious. To show that $J = \mathcal{T}J$ some work is required. Indeed it is crucial to see that the set of admissible actions can be restricted to a certain compact interval (cf. Remark 2.4.4). The arguments are as follows. Define

$$\tilde{A} := \left\{a \in \mathbb{R} \mid -\bar{c}|a| + L(a) \le d_0\right\}$$

where $\bar{c} := \max\{c_u, c_d\}$. Note that this set is a compact interval due to our assumptions on L and $0 \in \tilde{A}$. We claim now: If $a \notin \tilde{A}$ then for $v \in I\!M$ it holds

$$c(-x) + \beta\,\mathbb{E}\,v(-Z) \le c(a - x) + L(a) + \beta\,\mathbb{E}\,v(a - Z).$$

This means that when we consider the optimization problem $\mathcal{T}v$, action $a = 0$ is better than any action outside $\tilde{A}$. The proof of the claim is as follows: First note that $c(-x) \le c(a-x) + \bar{c}|a|$ for all $x, a \in \mathbb{R}$ and that $a \notin \tilde{A}$ implies

$$-\bar{c}|a| + L(a) > d_0 = \beta \mathbb{E}\, c(Z) + d_0\beta.$$

This yields

$$\begin{aligned} c(-x) + \beta \mathbb{E}\, v(-Z) &\le c(a-x) + \bar{c}|a| + \beta d_0 + \beta \mathbb{E}\, c(Z) \\ &\le c(a-x) + L(a) \\ &\le c(a-x) + L(a) + \beta \mathbb{E}\, v(a-Z), \end{aligned}$$

and thus we can get $\mathcal{T}v(x) = \inf_{a\in\tilde{A}}\{c(a-x) + L(a) + \beta \mathbb{E}\, v(a-Z)\}$ for $v \in I\!M$. Theorem A.1.5 then implies that $J = \mathcal{T}J$ and Theorem 7.1.8 finally yields the statement.

b) This part follows from the policy iteration in Theorem 7.2.1 d). Note that by a) we know that $D(x)$ can be replaced by the compact set $\tilde{A}$, thus conditions (i)–(iii) in Theorem 7.2.1 are certainly satisfied. Moreover, S_{n-} and $S_{n+} \in \tilde{A}$ for all $n \in \mathbb{N}$. Since $\tilde{A}$ is compact the sequences (S_{n-}) and (S_{n+}) have accumulation points in $\tilde{A}$, and any such accumulation point can be chosen for an optimal decision rule. □

7.6.3 Casino Games

Imagine a player who enters the casino and always plays the same game. The probability of winning one game is $p \in (0,1)$ and the games are independent. The player starts with an initial capital $x \in \mathbb{N}$ and can bet any non-negative integral amount less than or equal to the current fortune. When the player wins, she obtains twice her stake otherwise it is lost. The aim is to maximize the probability that the player reaches the amount $B \in \mathbb{N}$ before she goes bankrupt.

The formulation as a (substochastic) stationary Markov Decision Model is as follows. Since we want to maximize the probability of a certain event, we define the one-stage reward as an indicator of this event. When the game ends, i.e. if the player has lost everything or has reached a fortune of B, we have to make sure that the process ends and no further reward is obtained. Note that it cannot be optimal to bet more than $B - x$. We define

- $E := \{0, 1, \dots, B\}$ where $x \in E$ denotes the current fortune,
- $A := \mathbb{N}_0$ where $a \in A$ denotes the amount the player bets,
- $D(x) := \Big\{0, 1, \dots, \min\{x, B - x\}\Big\}$,

- $q(x+a|x,a) := p$ and $q(x-a|x,a) := 1-p$ for $0 < x < B$ and on the boundary $q(x|0,a) = q(x|B,a) := 0$ for all $x \in E$,
- $r(x,a) := 0$ for $x \neq B$ and $r(B,a) := 1$, $a \in D(x)$,
- $\beta := 1$.

If we choose an arbitrary policy $\pi = (f_0, f_1, \ldots) \in F^\infty$ it holds

$$\begin{aligned} J_{\infty\pi}(x) &= \mathbb{E}_x^\pi\Big[\sum_{k=0}^\infty r(X_k, f_k(X_k))\Big] = \mathbb{E}_x^\pi\Big[\sum_{k=0}^\infty 1_{[X_k=B]}\Big] \\ &= \mathbb{P}_x^\pi\Big(X_n = B \text{ for some } n \in \mathbb{N}_0\Big), \quad x \in E \end{aligned}$$

which is indeed the quantity we want to maximize. Obviously this is a positive Markov Decision Model with finite state and action spaces. Thus condition (C_-) is fulfilled and $\varepsilon(x) \equiv 0$.
Let us first consider the special policy where the player bets only one Euro per game. This policy is called *timid strategy.* Formally this stationary policy is defined by $(f_*, f_*, \ldots)$ with

$$f_*(x) := 1, \quad \text{for } x > 0 \quad \text{and} \quad f_*(0) = 0.$$

The value function $J_{f_*}(x)$ of this policy can be computed by solving the fixed point equation $J_{f_*} = \mathcal{T}_{f_*} J_{f_*}$, which gives the difference equation

$$J_{f_*}(x) = pJ_{f_*}(x+1) + (1-p)J_{f_*}(x-1), \quad x = 1, \ldots, B-1$$

with boundary conditions $J_{f_*}(0) = 0$ and $J_{f_*}(B) = 1$. The solution is given by

$$J_{f_*}(x) = \begin{cases} \dfrac{1-\left(\frac{1-p}{p}\right)^x}{1-\left(\frac{1-p}{p}\right)^B}, & p \neq \frac{1}{2} \\ \frac{x}{B}, & p = \frac{1}{2}. \end{cases}$$

It can be conjectured that if the game is favourable (i.e. $p \geq \frac{1}{2}$), then the player can be patient and make use of the strong law of large numbers which implies that in the long run she will win. We obtain the following theorem.

Theorem 7.6.2. *If $p \geq \frac{1}{2}$, the timid strategy is optimal, i.e. it maximizes the probability that the player will reach B before going bankrupt.*

Proof. According to Theorem 7.4.5 it suffices to show that $J_{f_*} \geq \mathcal{T} J_{f_*}$, i.e.

$$J_{f_*}(x) \geq pJ_{f_*}(x+a) + (1-p)J_{f_*}(x-a), \quad 0 < x < B,\ 0 \leq a \leq \min\{x, B-x\}.$$

Let us first consider the case $p = \frac{1}{2}$. Here we have $J_{f_*}(x) = \frac{x}{B}$ and the inequality is fulfilled. Now suppose $p > \frac{1}{2}$. Inserting the expression for $J_{f_*}(x)$

from above and rearranging terms, the inequality is equivalent to

$$1 \leq p\Big[\big(\frac{1-p}{p}\big)^a + \big(\frac{p}{1-p}\big)^{a-1}\Big], \quad a \in D(x).$$

For $a \in \{0,1\}$ this is obviously true. By inspecting the derivative, we see that

$$h(y) = \big(\frac{1-p}{p}\big)^y + \big(\frac{p}{1-p}\big)^{y-1}$$

is increasing in y for $y \geq 1$ if $p > \frac{1}{2}$. Thus, the statement follows. □

Let us next consider the special strategy where the player bets all the time her complete fortune or the part which is necessary to reach B. This strategy is called *bold strategy*. Formally it is defined by $(f_{**}, f_{**}, \ldots)$ with

$$f_{**}(x) := \min\{x, B-x\}, \quad \text{for } x \geq 0.$$

The value function $J_{f_{**}}(x)$ of this strategy can be computed by solving the fixed point equation $J_{f_{**}} = \mathcal{T}_{f_{**}} J_{f_{**}}$ which gives the difference equation

$$\begin{aligned} J_{f_{**}}(x) &= pJ_{f_{**}}(2x), \quad x < B-x \\ J_{f_{**}}(x) &= p + (1-p)J_{f_{**}}(2x-B), \quad x \geq B-x \end{aligned}$$

with boundary conditions $J_{f_{**}}(0) = 0$ and $J_{f_{**}}(B) = 1$. If the game is non-favourable (i.e. $p \leq \frac{1}{2}$), then the player has to try to reach B as fast as possible. We obtain the following theorem.

Theorem 7.6.3. *If $p \leq \frac{1}{2}$, the bold strategy is optimal. i.e. it maximizes the probability that the player will reach B before going bankrupt.*

Proof. Let us denote $W_n(x) := J_{nf_{**}}(x)$, which is the probability that B is reached before 0 until the n-th game under the bold strategy with initial capital x. It follows from the reward iteration (Theorem 2.5.3) that for $n \in \mathbb{N}$ and fortunes $0 < x < B$:

$$W_n(x) = \mathcal{T}_{f_{**}} W_{n-1}(x) = \begin{cases} pW_{n-1}(2x), & x \leq B-x \\ p + (1-p)W_{n-1}(2x-B), & x > B-x \end{cases}$$

and $W_n(0) = 0, W_n(B) = 1$. Moreover, we define $W_0 \equiv 0$. We claim now that for all $n \in \mathbb{N}_0$ and $x \in E, a \in D(x)$

$$W_{n+1}(x) \geq pW_n(x+a) + (1-p)W_n(x-a). \tag{7.5}$$

Since $J_{f_{**}}(x) = \lim_{n\to\infty} W_n(x)$ the statement follows from Theorem 7.4.5 as in the proof of Theorem 7.6.2.

We show (7.5) by induction. The case $n = 0$ is clear. Now suppose the statement holds for $n \in \mathbb{N}$. For the case $n+1$ we have to distinguish the following cases

Case 1: $x + a \le \frac{B}{2}$: Here we obtain with the reward iteration

$$\begin{aligned} & W_{n+1}(x) - pW_n(x+a) - (1-p)W_n(x-a) \\ &= p\Big[W_n(2x) - pW_{n-1}(2x+2a) - (1-p)W_{n-1}(2x-2a)\Big], \end{aligned}$$

which is non-negative by the induction hypothesis.

Case 2: $x \le \frac{B}{2} \le x + a$: Here we obtain with the reward iteration

$$\begin{aligned} & W_{n+1}(x) - pW_n(x+a) - (1-p)W_n(x-a) \\ &= p\Big[W_n(2x) - p - (1-p)W_{n-1}(2x+2a-B) - (1-p)W_{n-1}(2x-2a)\Big] \\ &= (1-p)\Big[pW_{n-1}(4x-B) - pW_{n-1}(2x+2a-B) - pW_{n-1}(2x-2a)\Big] \\ &= (1-p)\Big[W_n(2x - \frac{B}{2}) - pW_{n-1}(2x+2a-B) - pW_{n-1}(2x-2a)\Big] \\ &=: H(x,a) \end{aligned}$$

where we have used that $2x \ge x + a \ge \frac{B}{2}$ for the second equation and $2x - \frac{B}{2} \le \frac{B}{2}$ for the last equation. Now since $p \le \frac{1}{2}$ we have

$$H(x,a) \ge (1-p)\Big[W_n(2x-\frac{B}{2}) - pW_{n-1}(2x+2a-B) - (1-p)W_{n-1}(2x-2a)\Big]$$

which is non-negative when we use the induction hypothesis with $\hat{x} := 2x - \frac{B}{2}$ and $\hat{a} := 2a - \frac{B}{2}$. If $a < \frac{B}{4}$ we choose the inequality

$$H(x,a) \ge (1-p)\Big[W_n(2x-\frac{B}{2}) - (1-p)W_{n-1}(2x+2a-B) - pW_{n-1}(2x-2a)\Big]$$

which is non-negative when we use the induction hypothesis with $\hat{x} := 2x - \frac{B}{2}$ and $\hat{a} := \frac{B}{2} - 2a$.

Case 3: $x - a \le \frac{B}{2} \le x$: Follows analogously to the second case.

Case 4: $\frac{B}{2} \le x - a$: Follows analogously to the first case. □

If the game is fair (i.e. $p = \frac{1}{2}$), we know already from Theorem 7.6.2 and Theorem 7.6.3 that both the timid and the bold strategy are optimal. Indeed, in this case the wealth process is a martingale and every reasonable strategy yields the same probability to reach the amount B.

Theorem 7.6.4. *If $p = \frac{1}{2}$, then the maximal probability that the player will reach B before going bankrupt is given by $J_\infty(x) = \frac{x}{B}$, $x \in E$, and every stationary strategy $(f, f, \dots)$ with $f(x) > 0$ for $x > 0$ is optimal.*

Proof. From Theorem 7.6.2 we know already that the maximal value is given by $J_\infty(x) = \frac{x}{B}$. Now suppose $(f, f, \ldots)$ is an arbitrary stationary strategy with $f(x) > 0$ for $x > 0$. We obviously have

$$\mathbb{P}_x^f\Big(X_n = B \text{ for some } n \in \mathbb{N}_0\Big) + \mathbb{P}_x^f\Big(X_n = 0 \text{ for some } n \in \mathbb{N}_0\Big) = 1.$$

Now since $p = \frac{1}{2}$, we have due to symmetry arguments that

$$\mathbb{P}_x^f\Big(X_n = 0 \text{ for some } n \in \mathbb{N}_0\Big) = \mathbb{P}_{B-x}^f\Big(X_n = B \text{ for some } n \in \mathbb{N}_0\Big)$$

i.e. the probability of reaching 0 before B when we start in x is the same as the probability of reaching B before 0 when we start in $B-x$. Hence we have

$$\begin{aligned} J_f(x) &= \mathbb{P}_x^f\Big(X_n = B \text{ for some } n \in \mathbb{N}_0\Big) \\ &= 1 - \mathbb{P}_x^f\Big(X_n = 0 \text{ for some } n \in \mathbb{N}_0\Big) = 1 - J_f(B - x). \end{aligned}$$

On the other hand, we know that $J_f(x) \leq J(x) = \frac{x}{B}$ which implies that $J_f(x) = \frac{x}{B}$ for $x \in E$, and $(f, f, \ldots)$ is optimal. □

7.6.4 Bandit Problems with Infinite Horizon

Let us now reconsider the bandit problem of Section 5.5. For definitions and notations see Section 5.5. Here we investigate an infinite horizon bandit with $\beta < 1$ and assume that both success probabilities are unknown. In the infinite horizon case we get some more structural results about the optimal policy. In particular it can be shown that the optimal policy is a so-called *Index-policy.* For the model data we refer to Section 5.5.
First observe that since r is bounded (i.e. we can choose $b \equiv 1$) and $\beta < 1$ we have a *contracting Markov Decision Model.* Moreover, the assumptions of Theorem 7.3.5 are satisfied with $\mathbb{M} = \mathbb{B}_b$ and we obtain that the value function J_∞ of the infinite horizon Markov Decision Model is the unique solution of

$$J_\infty(x) = \max\Big\{p_1(x) + \beta Q_1 J_\infty(x),\ p_2(x) + \beta Q_2 J_\infty(x)\Big\}, \quad x \in \mathbb{N}_0^2 \times \mathbb{N}_0^2$$

and a maximizer f^* of J_∞ defines an optimal stationary policy $(f^*, f^*, \ldots)$.

Before we state the main result we have to do some preliminary work. A very helpful tool for the solution of the infinite horizon bandit are the so-called *K-stopping problems.* In a K-stopping problem only one arm of the bandit is considered and the decision maker can decide whether she pulls the arm and continues the game or whether she takes the reward K and quits. The maximal expected reward $J(m,n;K)$ of the K-stopping problem is then the unique solution of

$$v(m,n) = \max\Big\{K,\ p(m,n) + \beta(Pv)(m,n)\Big\}, \quad (m,n) \in \mathbb{N}_0^2$$

where $p(m,n)$ and Pv have been defined in Section 5.5. Obviously it holds that $J(\cdot;K) \geq K$ and if K is very large it will be optimal to quit the game, thus $J(m,n;K) = K$ for large K.

Definition 7.6.5. For $(m,n) \in \mathbb{N}_0^2$ we define the function

$$I(m,n) := \min\{K \in \mathbb{R} \mid J(m,n;K) = K\}$$

which is called *Gittins-index.*

An explicit representation of the index $I(m,n)$ is as follows.

Theorem 7.6.6. *Let $i_0 = (m_0,n_0) \in \mathbb{N}_0^2$ be fixed, $r(m,n) := p(m,n)$ and (X_k) be the state process of the K-stopping problem. Then*

$$I(m_0,n_0) = \frac{\mathbb{E}_{i_0}\Big[\sum_{k=0}^{\tau^*-1} \beta^k r(X_k)\Big]}{(1-\beta)\,\mathbb{E}_{i_0}\Big[\sum_{k=0}^{\tau^*-1} \beta^k\Big]} = \sup_{\tau\geq 1} \frac{\mathbb{E}_{i_0}\Big[\sum_{k=0}^{\tau-1} \beta^k r(X_k)\Big]}{(1-\beta)\,\mathbb{E}_{i_0}\Big[\sum_{k=0}^{\tau-1} \beta^k\Big]}$$

where $\tau^* := \inf\{n \in \mathbb{N} \mid I(X_n) \leq I(i_0)\}$ and the supremum is taken over all stopping times τ.

Proof. It holds for $i = (m,n)$ (cf. Section 10.2) that

$$J(m,n;K) = \sup_{\tau\geq 1} \mathbb{E}_i\Big[\sum_{k=0}^{\tau-1} \beta^k r(X_k) + K\beta^\tau\Big].$$

From the indifference property (see Corollary 7.6.8) we obtain

$$I(m_0,n_0) \geq \frac{\mathbb{E}_{i_0}\Big[\sum_{k=0}^{\tau-1} \beta^k r(X_k)\Big]}{(1-\beta)\,\mathbb{E}_{i_0}\Big[\sum_{k=0}^{\tau-1} \beta^k\Big]}$$

for all stopping times $\tau \geq 1$. For τ^* defined above the equality holds. □

The value function $J(\cdot;K)$ of the K-stopping problem has a number of important properties which we collect in the next proposition.

Proposition 7.6.7. *Let $(m,n) \in \mathbb{N}_0^2$ be fixed. Then it holds:*

a) $K \mapsto J(m,n;K)$ *is increasing, continuous and convex.*
b) $K \mapsto J(m,n;K) - K$ *is decreasing.*
c) The value function of the K-stopping problem can be written as

$$J(m,n;K) = \begin{cases} K, & K \geq I(m,n) \\ p(m,n) + \beta(PJ)(m,n;K), & K < I(m,n). \end{cases}$$

In particular, it holds: $K \leq J(m,n;K) \leq \max\{K, I(m,n)\}$.
d) $K \mapsto J(m,n;K)$ *is almost everywhere differentiable, the right- and left-hand side derivatives exist in each point and*

$$0 \leq \frac{\partial}{\partial K} J(m,n;K) \leq 1, \quad K \mapsto \frac{\partial}{\partial K} J(m,n;K) \text{ is increasing.}$$

Proof. The properties in part a) and b) follow by induction from the value iteration. For part c) note that by definition of the index $J(m,n,I(m,n)) = I(m,n)$. Now if $K \geq I(m,n)$, then we conclude from part b) that it holds $J(m,n;K) - K \leq 0$ which implies $J(m,n;K) = K$. If $K < I(m,n)$ we obtain $J(m,n;K) > K$ hence

$$J(m,n;K) = p(m,n) + \beta(PJ)(m,n;K) < I(m,n) = J\big(m,n;I(m,n)\big).$$

The last part follows from a), b) using properties of convex functions. □

Corollary 7.6.8. *The Gittins-indices have the following properties:*

a) The optimal stopping set for the K-stopping problem is given by

$$\{(m,n) \in \mathbb{N}_0^2 \mid J(m,n;K) = K\} = \{(m,n) \in \mathbb{N}_0^2 \mid I(m,n) \leq K\}.$$

b) The so-called indifference property *holds, i.e.*

$$I(m,n) = J\big(m,n;I(m,n)\big) = p(m,n) + \beta(PJ)\big(m,n;I(m,n)\big).$$

c) The indices are bounded by

$$\frac{p(m,n)}{1-\beta} \leq I(m,n) \leq \frac{1}{1-\beta}.$$

d) If the success probability p is known then

$$I(m,n) = \frac{p}{1-\beta} \quad \text{(independent of } (m,n)\text{)}.$$

Let us now consider the problem where we have three possible actions: either pull arm 1, pull arm 2 or quit the game and receive the reward K. For $K = 0$ this problem obviously reduces to our initial bandit problem. We denote the value function of this problem by $\tilde{J}(x; K)$ for $x \in \mathbb{N}_0^2 \times \mathbb{N}_0^2$ and obtain that it solves

$$\tilde{J}(x; K) = \max\Big\{K,\ \max_{a \in A}\{p_a(x) + \beta(Q_a\tilde{J})(x; K)\}\Big\}, \quad x \in \mathbb{N}_0^2 \times \mathbb{N}_0^2.$$

The value function $\tilde{J}(x; K)$ has similar properties as $J(m, n; K)$.

Proposition 7.6.9. *Let $x \in \mathbb{N}_0^2 \times \mathbb{N}_0^2$ be fixed. Then it holds:*

a) $K \mapsto \tilde{J}(x; K)$ is increasing, continuous and convex.
b) $K \mapsto \tilde{J}(x; K) - K$ is decreasing.
c) Let $I(x) := \max\{I(m_1, n_1), I(m_2, n_2)\}$ for $x = (m_1, n_1, m_2, n_2)$ and let $a^* \in \{1, 2\}$ be the number of the arm where the maximum is attained. Then

$$\tilde{J}(x; K) = \begin{cases} K, & K \geq I(x) \\ p_{a^*}(x) + \beta(Q_{a^*}\tilde{J})(x; K), & K < I(x) = I(m_{a^*}, n_{a^*}). \end{cases}$$

Proof. The statements in a) and b) follow from the value iteration. We have to show the statement in part c). The derivative of $\tilde{J}(x; y)$ with respect to y exists and it holds (for a proof see Tsitsiklis (1986))

$$\frac{\partial}{\partial y}\tilde{J}(x; y) = \frac{\partial}{\partial y}J(m_1, n_1; y) \cdot \frac{\partial}{\partial y}J(m_2, n_2; y).$$

By integration we obtain for $K \leq K_0$:

$$\tilde{J}(x; K) = \tilde{J}(x; K_0) - \int_K^{K_0} \frac{\partial}{\partial y}J(m_1, n_1; y) \cdot \frac{\partial}{\partial y}J(m_2, n_2; y)dy.$$

For $K_0 \geq \frac{1}{1-\beta}$ it holds $\tilde{J}(x; K_0) = K_0$. Thus, we define the function

$$W(x; K) := K_0 - \int_K^{K_0} \frac{\partial}{\partial y}J(m_1, n_1; y) \cdot \frac{\partial}{\partial y}J(m_2, n_2; y)dy, \quad x \in E, K \leq K_0.$$

By partial integration we obtain with the notation $P_a(x; y) := \frac{\partial}{\partial y}J(m_b, n_b; y)$ for $b \neq a$

$$W(x; K) = K_0 - P_a(x; y)J(m_a, n_a; y)\Big|_K^{K_0} + \int_K^{K_0} J(m_a, n_a; y)dP_a(x; y).$$

According to Proposition 7.6.7 $P_a(x; y)$ has the following properties:

- $0 \le P_a(x;y) \le 1$,
- $x \mapsto P_a(x;y)$ is increasing in x,
- $P_a(x;y) = 1$ for $y \ge I(x)$.

For $K_0 \to \infty$ we obtain

$$W(x;K) = P_a(x;K)J(m_a,n_a;K) + \int_K^\infty J(m_a,n_a;y)dP_a(x;y).$$

Let us introduce the following functions

$$\Delta_a(x;K) := W(x;K) - p_a(x) - \beta(Q_aW)(x;K)$$
$$\delta(m,n;K) := J(m,n;K) - p(m,n) - \beta(PJ)(m,n;K).$$

It can be shown that they are related as follows:

$$\Delta_a(x;K) = \delta(m_a,n_a;K)P_a(x;K) + \int_K^\infty \delta(m_a,n_a;y)dP_a(x;y).$$

From Proposition 7.6.7 we conclude that $\delta(m_a,n_a;K) \ge 0$ and that for $K < I(m_a,n_a)$ it holds $\delta(m_a,n_a;K) = 0$. From these relations we obtain

$$K \ge I(x) \Rightarrow W(x;K) = K$$
$$K < I(m_{a^*},n_{a^*}) = I(x) \Rightarrow W(x;K) = p_{a^*}(x) + \beta(Q_{a^*}W)(x;K).$$

Further $\Delta_a(x;K) \ge 0$ and $W(x;K) \ge K$ and

$$K \ge I(m_{a^*},n_{a^*}) = I(x) \Rightarrow K \ge \max_{a\in A}\{p_a(x) + \beta(Q_aW)(x;K)\}$$
$$K < I(m_{a^*},n_{a^*}) = I(x) \Rightarrow K \le p_{a^*}(x) + \beta(Q_{a^*}W)(x;K).$$

From these conclusions we derive

$$W(x;K) = \max\Big\{K,\ \max_{a\in A}\{p_a(x) + \beta(Q_aW)(x;K)\}\Big\}, \quad x \in \mathbb{N}_0^2 \times \mathbb{N}_0^2, \quad (7.6)$$

i.e. $W(x;K)$ is a fixed point of (7.6). Since the fixed point is unique we get $W(x;K) = \tilde{J}(x;K)$ which implies the statements. □

The main result of this section is the optimality of the Gittins-index policy.

Theorem 7.6.10. *The stationary Index-policy $(f^*, f^*, \ldots)$ is optimal for the infinite horizon bandit problem where for $x = (m_1,n_1,m_2,n_2)$*

$$f^*(x) := \begin{cases} 2 & \text{if } I(m_2,n_2) \ge I(m_1,n_1) \\ 1 & \text{if } I(m_2,n_2) < I(m_1,n_1). \end{cases}$$

Remarkable about this policy is that we can compute for each arm separately its own index (which depends only on the model data of this arm) and choose the arm with the higher index. This is of interest for the computation of the optimal solution since the dimension of the state space for the separate problems is reduced dramatically.

Proof. From Proposition 7.6.9 c) it follows directly that a maximizer f_K^* of $\tilde{J}(x; K)$ is given by

$$f_K^*(x) = \begin{cases} \text{stop} & \text{if } I(x) \leq K \\ 2 & \text{if } I(m_2, n_2) = I(x) > K \\ 1 & \text{if } I(m_1, n_1) = I(x) > K. \end{cases}$$

Letting $K \to 0$ the statement follows. □

Using the partial order relation

$$(m, n) \leq (m', n') :\Longleftrightarrow m \leq m', \ n \geq n'$$

it follows that $x \mapsto \tilde{J}(x; K)$ is increasing. From the definition of the index we can thus conclude that the index is increasing, i.e.

$$(m, n) \leq (m', n') \Longrightarrow I(m, n) \leq I(m', n').$$

This now implies that the stationary Index-policy $(f^*, f^*, \ldots)$ has the *stay-on-a-winner* property, i.e.

$$(m_1, n_1) \leq (m_1', n_1') \text{ and } f^*(m_1, n_1, m_2, n_2) = 1 \Rightarrow f^*(m_1', n_1', m_2, n_2) = 1.$$

Of course the analogous statement holds for the second arm. For corresponding finite horizon results see Theorem 5.5.1. Finally we can characterize the Gittins-indices as the unique fixed point of certain equations. These properties are very useful for numerical computations.

Proposition 7.6.11. *Let* $i_0 := (m_0, n_0) \in \mathbb{N}_0^2$ *be fixed and define* $J^0(m, n) := J(m, n; I(i_0))$ *for* $(m, n) \in \mathbb{N}_0^2$. *Then* J^0 *is the unique solution of*

$$v(m, n) = \max\{p(i_0) + \beta(Pv)(i_0), \ p(m, n) + \beta(Pv)(m, n)\}, \quad (m, n) \in \mathbb{N}_0^2$$

and it holds: $I(m_0, n_0) = J^0(m_0, n_0)$.

Proof. We know already that

$$J^0(m, n) = J(m, n; I(i_0)) = \max\{I(i_0); \ p(m, n) + \beta(PJ)(m, n; I(i_0))\}$$

and $I(i_0) = p(i_0) + \beta(PJ)(i_0; I(i_0))$. Hence J^0 is a fixed point of the preceding equation. Since $\beta < 1$ uniqueness follows from Banach's fixed point theorem (see Theorem A.3.5). □

7.7 Exercises

Exercise 7.7.1 (Blackwell Optimality). Suppose a Markov Decision Model with finite state and action spaces and $\beta < 1$ is given. In order to stress the dependence of the value function on β we write J_∞^β.

a) Show that for fixed $x \in E$ the function $\beta \mapsto J_\infty^\beta(x)$ is continuous and

$$J_\infty^\beta(x) = \max_{f \in F} \frac{P_{f,x}(\beta)}{P(\beta)}$$

where $P_{f,x}(\beta)$ and $P(\beta)$ are polynomials in β.
b) Denote $M^\beta := \{f \in F | f^\infty \text{ is optimal for } \beta\}$. Show that the function $\beta \mapsto M^\beta$ is piecewise constant, i.e. there exists a $k \in \mathbb{N}_0$ and constants $\beta_0 := 0 < \beta_1 < \ldots < \beta_k < 1 =: \beta_{k+1}$ such that $\beta \mapsto M^\beta$ is constant on the intervals (β_m, β_{m+1}).
c) Show that there exists a $\beta_0 < 1$ and a stationary policy $\pi^* = (f, f, \ldots)$ such that π^* is optimal for all $\beta \in (\beta_0, 1)$. Such a policy is called *Blackwell-optimal*.

Exercise 7.7.2 (Coin Game). In a game of chance you can decide in every round how many of maximal 10 fair coins you would like to toss. The aim is to get (at least) four times head (the 'head'-tosses are summed up) with minimal cost. If you decide to toss a coins in one round you have to pay $a+1$ Euro.

a) Formulate this problem as a Markov Decision Model.
b) Show that (SA) can be satisfied.
c) Show that the minimal expected cost $J(x)$ when still $x \in \{0, 1, 2, 3, 4\}$ 'head'-tosses are needed satisfies: $J(0) = 0$ and for $x > 0$

$$J(x) = \min_{1 \le a \le 10} \Big\{ (2^a - 1)^{-1} \Big[(1 + a)2^a + \sum_{y=1}^{x-1} \binom{a}{x-y} J(y) \Big] \Big\}.$$

d) Determine an optimal stationary strategy.
e) Is the game favourable when you receive 11 Euros after completing the task?

Exercise 7.7.3 (Controlled Queue). Suppose $x \in \mathbb{N}_0$ customers wait in front of a queue. A server can be activated at cost $d \ge 0$. When the server is active, the customer at the head of the queue will be served with probability

$p \in (0, 1]$ until the next time slot. Each waiting customer produces a cost $c > 0$ in every time slot. The costs are discounted by a factor $\beta < 1$. What is the optimal service policy that minimizes the expected discounted cost of the system over an infinite horizon? The data of the Markov Decision Model is given by:

- $E = \mathbb{N}_0$,
- $A = \{0, 1\}$,
- $D = E \times A$,
- $q(y|0, \cdot) = \delta_{0y}, q(y|x, 0) = \delta_{xy}, q(y|x, 1) = p\delta_{x-1,y} + (1 - p)\delta_{xy}$, for $x \in \mathbb{N}, y \in \mathbb{N}_0$ (see Figure 7.3),
- $r(x, a) = -cx - d\delta_{1a}$,
- $\beta < 1$.

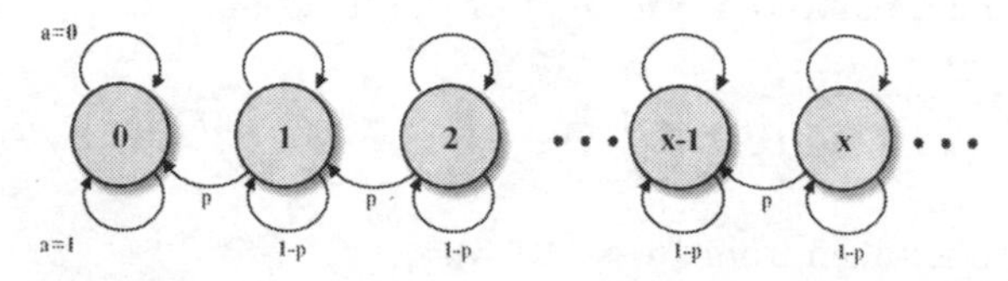

Fig. 7.3 Transition diagram.

Let $f \equiv 1$ and $g \equiv 0$. Show that

a) $\frac{d}{p} \leq \frac{\beta c}{1-\beta}$ implies f^∞ is optimal.
b) $\frac{d}{p} \geq \frac{\beta c}{1-\beta}$ implies g^∞ is optimal.

Hint: Compare $\mathcal{T}_f \mathcal{T}_g v$ and $\mathcal{T}_g \mathcal{T}_f v$, $v \in I\!B$ and show in part a) by induction $\mathcal{T}_f J_n \geq \mathcal{T}_g J_n$ and in part b) $\mathcal{T}_g J_n \geq \mathcal{T}_f J_n$. This is a so-called *interchange argument*

Exercise 7.7.4 (Moving Particle). A particle is moving randomly on the set $\{0, 1, \ldots, M\}$ where the states 0 and M are absorbing. If Y_n is the place where the particle is at time n, we assume that (Y_n) is a Markov chain and for $0 < i < M$:

$$\mathbb{P}(Y_{n+1} = i + 1 \mid Y_n = i) = p$$
$$\mathbb{P}(Y_{n+1} = i - 1 \mid Y_n = i) = 1 - p$$

where $p \in (0, 1)$. Now you are allowed to move the particle. You have initially a 'fuel reserve' of $T \in \mathbb{N}$ and can decide at each time point to use the amount $t \in \mathbb{N}_0$ of your reserve to move the particle t units to the right. The aim is to maximize the probability that the particle gets to the state M.

a) Set this up as a Markov Decision Model.

b) Show that every reasonable policy is optimal. (We call a policy reasonable when it moves the particle to M if there is enough fuel and if in state 0 at least one fuel unit is used to escape the trap.)

Exercise 7.7.5. Suppose we have a positive Markov Decision Model and (C_-) is satisfied. Let f be a maximizer of J_∞. We know already that the stationary policy $(f, f, \ldots)$ is not necessarily optimal. Show that $(f, f, \ldots)$ is optimal if and only if $\lim_{n\to\infty} \mathbb{E}_x^f[J_\infty(X_n)] = 0$.

Exercise 7.7.6. Let an infinite horizon Markov Decision Model be given which satisfies (C) and (SA) and let $f, g \in F$ be two decision rules. Show that $J_h(x) \geq \max\{J_f(x), J_g(x)\}$ if $h \in F$ is defined by

a) $h(x) := \begin{cases} f(x), & \text{if } J_f(x) \geq J_g(x) \\ g(x), & \text{if } J_f(x) < J_g(x), \end{cases} \qquad x \in E.$

b) $h(x)$ is a maximum point on $D(x)$ of the function

$$a \mapsto r(x, a) + \beta \int \max\{J_f(x'), J_g(x')\} Q(dx'|x, a).$$

(This exercise is taken from Ross (1983).)

Exercise 7.7.7 (Howard's Policy Improvement). Consider Example 7.4.4. This is a positive Markov Decision Model and Howard's policy improvement does not work here in general. Find out what happens if you start the algorithm with $f_0 \equiv 0$.

Exercise 7.7.8 (Howard's Policy Improvement). Consider the following Markov Decision Model:

- $S = \{1, 2\}$,
- $A = \{1, 2\}$,
- $D(x) = A$ for all x,
- transition probabilities are given in Figure 7.4,
- $r(1, 1) = r(2, \cdot) = 0, r(1, 2) = 1$,
- $\beta = 1$.

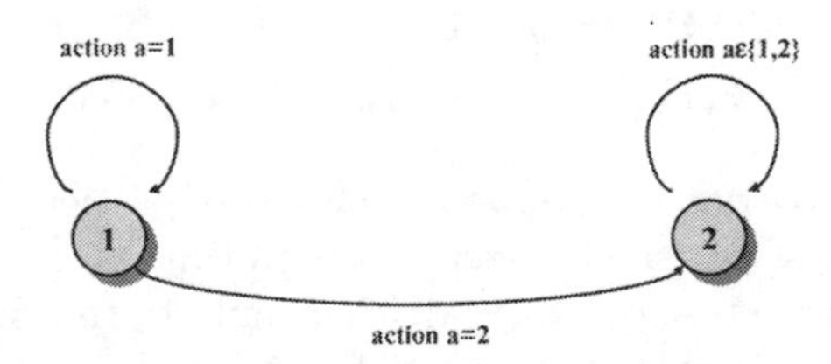

Fig. 7.4 Transition diagram.

a) Show that (C) is satisfied.
b) Let $f \equiv 1$ and $g \equiv 2$. Compute J_f and J_g.
c) Start Howard's policy improvement with $f_0 = f$ and show that it is important to choose $f_k(x) = f_{k-1}(x)$ if possible.

Exercise 7.7.9 (Howard's Policy Improvement). Suppose a Markov Decision Model with finite state and action spaces and $\beta < 1$ is given. We suppose that $|E| = n$, hence $\mathcal{T} : \mathbb{R}^n \to \mathbb{R}^n$. Define $F(x) := \mathcal{T}x - x$ for $x \in \mathbb{R}^n$.

a) Suppose f is the unique maximizer of J. Show that

$$\frac{\partial F(x)}{\partial x} = \beta q_f - I$$

where $q_f = \big(q(y|x, f(x))\big) \in \mathbb{R}^{n \times n}$ and I is the n-dimensional identity matrix.

b) Suppose Howard's policy improvement algorithm yields a unique maximizer throughout. Show that the algorithm computes the same sequence J_{f_k} as the Newton method for solving $F(x) = 0$ with initial point $x_0 := J_{f_0}$.

Exercise 7.7.10. Consider the following Markov Decision Model:

- $S = \{1, 2\}$,
- $A = \{1, 2\}$,
- $D(x) = A$ for all x,
- transition probabilities are given in Figure 7.5,
- $r(1,1) = 2, r(1,2) = \frac{1}{2}, r(2,1) = 1, r(2,2) = 3$,
- $\beta = 0.9$.

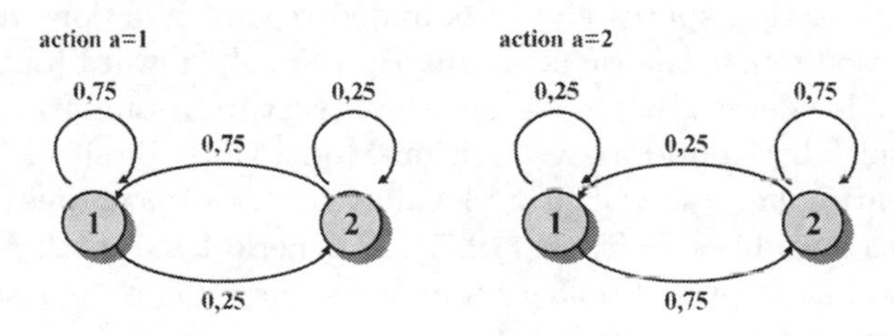

Fig. 7.5 Transition diagram.

Compute an optimal stationary policy by using

a) Howard's policy improvement algorithm
b) Linear programming.

Exercise 7.7.11 (K-Stopping Problem). Let us consider a K-stopping problem with bounded reward function and $\beta < 1$. The value function satisfies:

$$J(x;K) = \max\left\{K,\ r(x) + \beta \int Q(dy|x)J(y;K)\right\}.$$

a) Show that $K \mapsto J(x;K)$ is increasing and convex for fixed x.
b) Show that $K \mapsto J(x;K) - K$ is decreasing for fixed x.
c) Let τ^* be the optimal stopping time, i.e. the first time one chooses K and quits. Show that

$$\frac{\partial}{\partial K} J(x;K) = \mathbb{E}_x\, \beta^{\tau^*}.$$

Exercise 7.7.12 (K-Stopping Problem). Let us consider a K-stopping problem with $\beta < 1$. Now we assume that the reward is a random variable and its density $q(\cdot|\theta)$ depends on an unknown parameter θ (the Bernoulli bandit is a special case). It is also assumed that $q(z|\theta)$ is MTP_2. The value function satisfies:

$$J(\rho;K) = \max\left\{K,\ \int\int \Big(z + \beta J\big(\Phi(\rho,z);K\big)\Big) q(z|\theta)dz\rho(d\theta)\right\}.$$

Show:

a) It is optimal to stop if $K \ge I(\rho)$ where $I(\rho) := \min\{K | J(\rho;K) = K\}$.
b) $\rho \le_{lr} \rho'$ implies $I(\rho) \le I(\rho')$.

7.8 Remarks and References

In this chapter we consider infinite horizon Markov Decision Models with Borel state and action spaces and unbounded reward functions under weak convergence conditions. The weak assumptions on the reward functions were introduced by Hinderer (1971). We use the Structure Assumption (SA) and solve the infinite horizon Markov Decision Model as the limit of finite-stage models. In particular, the policy and value iteration techniques are valid. This approach goes back to Schäl (1975). A general framework for deriving structural results of optimal policies and the value function is also provided in Schäl (1990) and Puterman (1994).
We note that there exists an interesting martingale characterization of optimal policies. A policy is optimal if and only if it is value conserving (or thrifty) and value equalizing (see Sudderth (1971) and Rieder (1976)). Thrifty policies are maximizers of the value function. Under (C) every policy is value equalizing, but this is not true under (C_-).

Section 7.5: The value iteration suffers from the curse of dimensionality. There are a lot of different ways to speed up this procedure, to approximate it or to exclude suboptimal policies in advance. Accelerated Jacobi and

Gauss-Seidel procedures as well as parallelization of the value iteration are discussed e.g. in Kushner and Dupuis (2001) and Bertsekas (2001). Tests of suboptimality and turnpike theorems can be found in Hernández-Lerma and Lasserre (1999) and Puterman (1994). Large Markov Decison Problems can be solved by *neuro-dynamic programming* or *reinforcement learning*. The idea here is essentially to step forward in time and use iterative algorithms to approximate the value function (e.g. by simulation). For these methods see e.g. Bertsekas and Tsitsiklis (1996), Sutton and Barto (1998), Van Roy (2002), Chang et al. (2007) and Powell (2007). The policy improvement technique for Markov Decision Processes with finite state and action spaces was introduced by Howard (1960). There is a strong relationship to the linear programming approach, see Derman (1970), Kallenberg (1983), Filar and Vrieze (1997) and Altman (1999). The linear programming formulation for solving Markov Decision Processes with Borel state and action spaces is based on Heilmann (1979) and also discussed in Klein-Haneveld (1986), Piunovskiy (1997) and Hernández-Lerma and Lasserre (1996). For a recent survey on this topic see Hernández-Lerma and Lasserre (2002).

Section 7.6: Optimization problems with a random horizon are considered e.g. in Ross (1970, 1983), Puterman (1994) and Iida and Mori (1996). For remarks and references concerning the cash balance problem see the end of Chapter 2. The classical casino games which have been presented here, were extensively studied for the first time in Dubins and Savage (1965). Recently, the problem was investigated under the presence of inflation. Surprisingly Chen et al. (2004) found that the bold strategy is not necessarily optimal for subfair casino games with inflation. For a recent positive result see Chen et al. (2005).
The Bernoulli bandit with infinite horizon is a special case of the multiproject bandit. In a multiproject bandit problem m projects are available which are all in some states. One project has to be selected to work on or one chooses to retire. The project which is selected then changes its state whereas the other projects remain unchanged. Gittins (1979) was the first to show that multiproject bandits can be solved by considering single projects and that the optimal policy is an index-policy, see also Berry and Fristedt (1985), Gittins (1989). The method of proof we have used here is due to Whittle (1980). For an extension to open bandit processes see Friis et al. (1993). Alternative proofs can be found in Weber (1992), Varaiya et al. (1985) and Kaspi and Mandelbaum (1998). Bertsimas and Niño Mora (1996) used a significantly different proof via an achievable region approach, see also Bäuerle and Stidham (2001) for applications in fluid networks. Computational results for the indices of the multi-armed bandit are given in Katehakis and Veinott (1987). Further extensions are *restless bandits* where the other projects can change their state too (see e.g. Whittle (1988), Weber and Weiss (1990), Glazebrook et al. (2002)), bandits with availability constraints (see Dayanik et al. (2008)) and bandits in continuous-time (see e.g. Karatzas (1984), El Karoui

and Karatzas (1994)). Bandit models with applications in finance are e.g. treated in Bank and Föllmer (2003).

Chapter 8
Piecewise Deterministic Markov Decision Processes

In this chapter we deal with optimization problems where the state process is a Piecewise Deterministic Markov Process. These processes evolve through random jumps at random time points while the behavior between jumps is governed by an ordinary differential equation. They form a general and important class of non-diffusions. It is known that every strong Markov process with continuous paths of bounded variation is necessarily deterministic. We assume that both the jump behavior as well as the drift behavior between jumps can be controlled. Hence this leads to a control problem in continuous-time which can be tackled for example via the Hamilton-Jacobi-Bellman equation. However, since the evolution between jumps is deterministic these problems can also be reduced to a discrete-time Markov Decision Process where however the action space is now a function space. We can treat these problems with the methods we have established in the previous chapters. More precisely we will restrict the presentation to problems with infinite horizon, thus we will use the results of Chapter 7. We show that under some continuity and compactness conditions the value function of the Piecewise Deterministic Markov Decision Process is a fixed point of the Bellman equation (Theorem 8.2.6) and the computational methods of Chapter 7 apply. In Section 8.3 the important special class of continuous-time Markov Decision Chains is investigated, in particular for problems with finite time horizon.

8.1 Piecewise Deterministic Markov Decision Models

First we introduce the ingredients of a Piecewise Deterministic Markov Decision Model where we restrict to a stationary model with infinite horizon.

Definition 8.1.1. A *Piecewise Deterministic Markov Decision Model* consists of the data $(E, \mathcal{U}, \mu, \lambda, Q, r, \beta)$ with the following meaning:

N. Bäuerle and U. Rieder, *Markov Decision Processes with Applications to Finance*, Universitext, DOI 10.1007/978-3-642-18324-9_8,

- E is the *state space*. We assume that E is a Borel subset of $\mathbb{R}^d$. The elements (states) are denoted by $x \in E$.
- $\mathcal{U}$ is the *control action space* and is assumed to be a Borel subset of a Polish space. Let

$$A := \{\alpha : \mathbb{R}_+ \to \mathcal{U} \text{ measurable } \} \tag{8.1}$$

  be the set of *control functions*. We write $\alpha_t = \alpha(t)$. We will not restrict the set of control actions available at state x.
- The stochastic evolution is given by a marked point process (T_n, Z_n), where (T_n) is the increasing sequence of jump time points of a Poisson process with rate $\lambda > 0$ and the marks (Z_n) are the post jump states. We set $T_0 := 0$. Between the jump times T_n and T_{n+1} the process is described by a *deterministic flow*. More precisely, let $\mu(x, u) \in \mathbb{R}^d$ be the deterministic drift between the jumps if the state is $x \in E$ and control action $u \in \mathcal{U}$ is taken. We assume that for all $\alpha \in A$ there exists a unique solution $\phi_t^\alpha(x) \in E$ of the following initial value problem:

$$dx_t = \mu(x_t, \alpha_t)dt, \quad x_0 = x \in E.$$

  Then $\phi_t^\alpha(x)$ is the state of the piecewise deterministic process at time $T_n + t < T_{n+1}$ if $Z_n = x$. It is assumed that $\phi_t^\alpha(x)$ is measurable in (x, α) (see below for a definition of a σ-algebra in A) and continuous in t.
- Q is a stochastic kernel from $E \times \mathcal{U}$ to E which describes the distribution of the jump goals, i.e. $Q(B|x, u)$ is the probability that the process jumps in the set B given the state $x \in E$ immediately before the jump and the control action $u \in \mathcal{U}$ at the jump time.
- $r : E \times \mathcal{U} \to \mathbb{R}$ is a measurable function, where $r(x, u)$ gives the reward rate in state x if control action u is taken.
- $\beta \geq 0$ is the discount rate.

At time T_n the evolution of the process up to time T_{n+1} is known to the decision maker who can therefore fix the control action $\alpha(t)$ for $T_n + t \leq T_{n+1}$ by some $\alpha \in A$. This leads to the idea of treating the continuous-time control problem as a discrete-time Markov Decision Process where one now looks on α as the action at time T_n. But then the action space is a function space! It is known that A becomes a Borel space if A is endowed with the coarsest σ-algebra such that

$$\alpha \mapsto \int_0^\infty e^{-t} w(t, \alpha_t) dt$$

is measurable for all bounded and measurable functions $w : \mathbb{R}_+ \times \mathcal{U} \to \mathbb{R}$ (see e.g. Yushkevich (1980)). Then $f : E \to A$ is measurable if and only if there exists a measurable function $\tilde{f} : \mathbb{R}_+ \times E \to \mathcal{U}$ such that

$$f(x)(t) = \tilde{f}(t, x) \quad \text{for } t \in \mathbb{R}_+, x \in E.$$

In the sequel we will not distinguish between f and $\tilde{f}$.
A *Markov policy* (or piecewise open loop policy) $\pi = (\pi_t)$ is defined by a sequence of measurable functions $f_n : E \to A$ such that

$$\pi_t = f_n(Z_n)(t - T_n) \text{ for } t \in (T_n, T_{n+1}].$$

In the sequel we restrict to Markov policies and write $\pi = (\pi_t) = (f_n)$ and

$$\phi^\pi_{t-T_n}(Z_n) := \phi^{f_n(Z_n)}_{t-T_n}(Z_n) \text{ for } t \in [T_n, T_{n+1}).$$

Remark 8.1.2. The process (π_t) is predictable (cf. Definition B.2.9). The most general form of a predictable control process (π_t) is given by

$$\pi_t = f_n(T_0, Z_0, \ldots, T_n, Z_n, t - T_n) \text{ for } t \in (T_n, T_{n+1}]$$

for measurable functions (f_n). Due to the Markovian structure of the state process, this larger class of policies does not increase the value of the control problem (see Section 2.2). ◊

The piecewise deterministic process (X_t) is given by

$$X_t = \phi^\pi_{t-T_n}(Z_n) \text{ for } t \in [T_n, T_{n+1}).$$

Note that $Z_n = X_{T_n}$.
Given a policy π and an initial state $x \in E$ there is a probability space $(\Omega, \mathcal{F}, \mathbb{P}^\pi_x)$ on which the random variables T_n and Z_n are defined such that $X_0 = Z_0 = x$ and for all Borel sets $B \subset E$

$$\begin{aligned}
&\mathbb{P}^\pi_x \left(T_{n+1} - T_n \le t, Z_{n+1} \in B \mid T_0, Z_0, \ldots, T_n, Z_n\right) \\
&= \lambda \int_0^t e^{-\lambda s} Q\big(B | X_{T_n+s}, \pi_{T_n+s}\big) ds \\
&= \lambda \int_0^t e^{-\lambda s} Q\big(B | \phi^\pi_s(Z_n), f_n(Z_n)(s)\big) ds.
\end{aligned}$$

We impose now the following

Integrability Assumption (A):

$$\sup_\pi \mathbb{E}^\pi_x \left[\int_0^\infty e^{-\beta t} r^+(X_t, \pi_t) dt \right] < \infty, \quad x \in E.$$

Then the expected discounted total reward is well-defined for all π by

$$V_\pi(x) := \mathbb{E}^\pi_x \left[\int_0^\infty e^{-\beta t} r(X_t, \pi_t) dt \right], \quad x \in E. \tag{8.2}$$

Moreover, the value function of the Piecewise Deterministic Markov Decision Model is given by

$$V_\infty(x) := \sup_\pi V_\pi(x), \quad x \in E, \tag{8.3}$$

where the supremum is taken over all Markov policies.

Remark 8.1.3. We restrict the presentation here to the problem of maximizing the expected discounted reward. This is usually done for Piecewise Deterministic Markov Processes. However, the theory also allows to include instantaneous rewards at the jump time points, i.e. to look at the objective

$$\mathbb{E}_x^\pi \left[\int_0^\infty e^{-\beta t} r(X_t, \pi_t) dt \right] + \mathbb{E}_x^\pi \left[\sum_{k=1}^\infty e^{-\beta T_k} \bar{r}(X_{T_k}, \pi_{T_k}) \right]$$

where the measurable function $\bar{r} : E \times \mathcal{U} \to \mathbb{R}$ gives the reward for each jump. Moreover, it is of interest (in particular in finance applications) to consider optimization problems with a finite time horizon, i.e. to maximize the objective

$$\pi \mapsto \mathbb{E}_x^\pi \left[\int_0^T r(X_t, \pi_t) dt + g(X_T) \right]$$

where $g : E \to \mathbb{R}$ is the measurable terminal reward function. For a treatment of these problems see Bäuerle and Rieder (2010) and Sections 9.3 and 9.4. $\Diamond$

The optimization problem (8.3) is a continuous-time control problem. However, we can show that the value function $V_\infty(x)$ can be obtained by a discrete-time Markov Decision Problem. This point of view implies a number of interesting results. The first one is that under some conditions the value function is a fixed point of the Bellman equation. Differentiability or continuity of the value function is not needed in contrast to the classical continuous-time stochastic control approach. Second, the existence of an optimal policy is rather easy to prove. Moreover, several different computational approaches arise. Value iteration or Howard's policy improvement algorithm can be used to solve the continuous-time problem.

Remark 8.1.4. In order to outline the approach via discrete-time Markov Decision Processes we have chosen a fairly simple model. The Piecewise Deterministic Markov Decision Model can be extended in various ways.

- It is possible to extend the constant jump intensity to a state and action dependent intensity $\lambda(x, u)$ (see e.g. Davis (1993)).
- As in Section 4.4 we can consider a regime switching model, i.e. the parameters are allowed to vary randomly. Suppose (I_t) is a continuous-time Markov chain with finite state space. We may assume that the jump time

points (T_n) are generated by a Cox-process (N_t) with (stochastic) intensity $\lambda(I_t)$. Of course, in this case the discrete-time Markov Decision Process is more complicated.

- We may assume that the Markov chain (I_t) is 'hidden' and cannot be observed by the controller. Using the methods of Chapter 5, it is again possible to solve this problem as a discrete-time Partially Observable Markov Decision Process (see also Rieder and Winter (2009)). $\diamond$

8.2 Solution via a Discrete-Time Markov Decision Process

We introduce here a discrete-time Markov Decision Process which is equivalent to the control problem of the previous section. The idea is to look at the time points (T_n) and choose actions $\alpha \in A$ at time T_n, since the evolution of the state process between jumps is deterministic.
Now suppose a Piecewise Deterministic Markov Decision Model is given as described in the previous section. Let us define the following stationary infinite-stage Markov Decision Model (E, A, D, Q', r').

- E is the *state space*. A state x describes the state of the process directly after a jump.
- A is given by (8.1). Recall that the function space A is a Borel space.
- $D := E \times A$.
- For all Borel sets $B \subset E$, $x \in E$ and $\alpha \in A$, the stochastic kernel Q' is given by

$$Q'(B|x, \alpha) := \lambda \int_0^\infty e^{-(\lambda+\beta)t} Q\big(B|\phi_t^\alpha(x), \alpha_t\big) dt. \tag{8.4}$$

This is obviously a substochastic transition law. In order to make it stochastic, we may add an artificial cemetery state $\Delta \notin E$ to the state space and define

$$Q'\big(\{\Delta\}|x, \alpha\big) := 1 - \frac{\lambda}{\beta + \lambda}, \quad Q'\big(\{\Delta\}|\Delta, \alpha\big) := 1.$$

- The reward function r' is defined by

$$r'(x, \alpha) := \int_0^\infty e^{-(\beta+\lambda)t} r\big(\phi_t^\alpha(x), \alpha_t\big) dt, \tag{8.5}$$
$$r'(x, \Delta) := 0.$$

- The discrete-time discount factor is given by $\beta' = 1$.

In what follows we treat the Markov Decision Process as a substochastic problem and skip the state Δ (see e.g. the discussion in Example 2.3.13). For this discrete-time Markov Decision Model we define for a policy (f_n):

$$J_{\infty(f_n)}(x) = \mathbb{E}_x^{(f_n)}\left[\sum_{n=0}^{\infty} r'(Z_n', f_n(Z_n'))\right]$$
$$J_\infty(x) = \sup_{(f_n)} J_{\infty(f_n)}(x), \quad x \in E$$

where (Z_n') is the corresponding state process of the Markov Decision Process up to absorption in Δ. Note that $Z_n' = Z_n$ as long as the process is not absorbed (cp. Exercise 8.4.2).

Theorem 8.2.1. *For a Markov policy $\pi = (f_n)$ we have*

$$V_\pi(x) = J_{\infty(f_n)}(x), \quad x \in E.$$

Moreover, it holds: $V = J_\infty$.

Proof. Let $H_n := (T_0, Z_0, \dots, T_n, Z_n)$. Then we obtain with Theorem B.1.1:

$$\begin{aligned}
V_\pi(x) &= \sum_{n=0}^{\infty} \mathbb{E}_x^\pi\left[\int_{T_n}^{T_{n+1}} e^{-\beta t} r(X_t, \pi_t)dt\right] \\
&= \sum_{n=0}^{\infty} \mathbb{E}_x^\pi\left[e^{-\beta T_n}\, \mathbb{E}_x^\pi\left[\int_{T_n}^{T_{n+1}} e^{-\beta(t-T_n)} r(X_t, \pi_t)dt \Big| H_n\right]\right] \\
&= \sum_{n=0}^{\infty} \mathbb{E}_x^\pi\left[e^{-\beta T_n} r'(Z_n, f_n(Z_n))\right] \\
&= \sum_{n=0}^{\infty} \mathbb{E}_x^\pi\left[\prod_{k=1}^{n} e^{-\beta(T_k - T_{k-1})} r'(Z_n, f_n(Z_n))\right] \\
&= \sum_{n=0}^{\infty} \mathbb{E}_x^{(f_n)}\left[r'(Z_n', f_n(Z_n'))\right] = J_{\infty(f_n)}(x)
\end{aligned}$$

since the transition kernel of (Z_n') is given by (8.4) and r' by (8.5). The statement $V = J_\infty$ follows directly from the definition. □

Remark 8.2.2. There is another equivalent discrete-time Markov Decision Model with a state and action dependent discount factor $\beta(x, \alpha, x')$ where $\beta(x, \alpha, x')$ is defined by

$$\beta(Z_n, \alpha_n, Z_{n+1}) := \mathbb{E}_x^\pi\left[e^{-\beta(T_{n+1}-T_n)} \Big| Z_n, Z_{n+1}\right]$$

with $\alpha_n = f_n(Z_n)$ (see Forwick et al. (2004)). If $\phi_t^\alpha(x) = x$ then we have $\beta(Z_n, \alpha_n, Z_{n+1}) = \frac{\lambda}{\beta+\lambda}$. $\diamond$

Theorem 8.2.1 implies that $V_\infty(x)$ can also be seen as the value function of the discrete-time Markov Decision Model. Note that the integrability assumption for the Piecewise Deterministic Markov Decision Process implies the integrability assumption for the discrete-time Markov Decision Model. The maximal reward operator $\mathcal{T}$ is given by

$$(\mathcal{T}v)(x) = \sup_{\alpha \in A} \left\{ \int_0^\infty e^{-(\beta+\lambda)t} \Big[r\big(\phi_t^\alpha(x), \alpha_t\big) + \lambda \int v(z) Q\big(dz|\phi_t^\alpha(x), \alpha_t\big) \Big] dt \right\}.$$

From now on we assume that $\mathcal{U}$ is compact. In order to prove the existence of optimal controls we need certain continuity and compactness assumptions. To achieve this, we have to enlarge the action space and we introduce

$$\mathcal{R} := \{\alpha : \mathbb{R}_+ \to \mathbb{P}(\mathcal{U}) \text{ measurable}\}, \tag{8.6}$$

the set of *relaxed controls* where $\mathbb{P}(\mathcal{U})$ is the set of all probability measures on $\mathcal{U}$ equipped with the σ-algebra of the Borel sets, i.e. α_t can be seen as a randomized action. The problem is to define a topology on A which allows for a compact action space and a continuous target function – two competing aims. The set A of deterministic controls is a measurable subset of $\mathcal{R}$ in the sense that for $\alpha \in A$ the measures α_t are one-point measures on $\mathcal{U}$. A suitable topology on $\mathcal{R}$ is given by the so-called *Young topology*. The definition and important properties of this topology are summarized in the following remark. It can be shown that the set A of deterministic controls is dense in $\mathcal{R}$ with respect to the Young topology. This means that a relaxed control can be approximated by a deterministic control and given some continuity properties of the value function, this carries over to the values of the controls. This statement is also known as the *Chattering Theorem* (see e.g. Kushner and Dupuis (2001), Section 4).

Remark 8.2.3 (Young Topology). The Young topology on $\mathcal{R}$ is the coarsest topology such that all mappings of the form

$$\mathcal{R} \ni \alpha \mapsto \int_0^\infty \int_{\mathcal{U}} w(t,u) \alpha_t(du) dt$$

are continuous for all functions $w : [0,\infty] \times \mathcal{U} \to \mathbb{R}$ which are continuous in the second argument and measurable in the first argument and satisfy

$$\int_0^\infty \max_{u \in \mathcal{U}} |w(t,u)| dt < \infty.$$

We denote this class by $Car(\mathbb{R}_+ \times \mathcal{U})$, the class of so-called strong Carathéodory functions. With respect to the Young topology $\mathcal{R}$ is a separable metric and compact Borel space. In order to have well-defined integrals the following characterizations of measurability are important:

(i) A function $\alpha : \mathbb{R}_+ \to \mathbb{P}(\mathcal{U})$ is measurable if and only if

$$t \mapsto \int_{\mathcal{U}} v(u)\alpha_t(du)$$

is measurable for all bounded and continuous $v : \mathcal{U} \to \mathbb{R}_+$.

(ii) A function $f : E \to \mathcal{R}$ is measurable if and only if

$$x \mapsto \int_{\mathbb{R}_+} \int_{\mathcal{U}} w(s,u) f(s,x;du)ds$$

is measurable for all $w \in Car(\mathbb{R}_+ \times \mathcal{U})$.

Moreover, the following characterization of convergence in $\mathcal{R}$ is crucial for our applications. Suppose $(\alpha_n) \subset \mathcal{R}$ and $\alpha \in \mathcal{R}$. Then $\lim_{n\to\infty} \alpha_n = \alpha$ if and only if

$$\lim_{n\to\infty} \int_0^\infty \int_{\mathcal{U}} w(t,u)\alpha_t^n(du)dt = \int_0^\infty \int_{\mathcal{U}} w(t,u)\alpha_t(du)dt$$

for all $w \in Car(\mathbb{R}_+ \times \mathcal{U})$. ◊

Now we have to extend the domain of functions already defined on A. In particular we define for $\alpha \in \mathcal{R}$

$$\begin{aligned} d\phi_t^\alpha(x) &= \int \mu(\phi_t^\alpha(x), u)\alpha_t(du)dt, \quad \phi_0^\alpha(x) = x, \qquad (8.7)\\ r'(x,\alpha) &= \int_0^\infty e^{-(\beta+\lambda)t} \int r(\phi_t^\alpha(x), u)\alpha_t(du)dt, \\ Q'(B|x,\alpha) &= \lambda \int_0^\infty e^{-(\beta+\lambda)t} \int Q(B|\phi_t^\alpha(x), u)\alpha_t(du)dt \end{aligned}$$

where we again assume that a unique solution of (8.7) exists (according to the Theorem of Carathéodory this is the case if e.g. $\mu(x,u)$ is Lipschitz-continuous in x uniformly in u). If $\alpha \in A$ then the definitions of ϕ_t^α, r' and Q' coincide with those we have used so far. In case α is a relaxed control there is no physical interpretation of the model. The operator $\mathcal{T}$ has the following form:

$$\begin{aligned} (\mathcal{T}v)(x) = \sup_{\alpha\in\mathcal{R}} \Big\{ \int_0^\infty e^{-(\beta+\lambda)t} \int \Big[r\big(\phi_t^\alpha(x), u\big) \\ + \lambda \int v(z) Q\big(dz|\phi_t^\alpha(x), u\big)\Big] \alpha_t(du)dt \Big\}. \end{aligned}$$

In the Markov Decision Model with relaxed controls the decision maker can thus do at least as well as in the case without relaxed controls. When we denote by J_∞^{rel} the corresponding value function we obtain

$$J_\infty^{rel}(x) \geq J_\infty(x) = V_\infty(x), \quad x \in E.$$

We will show that these value functions are equal under some conditions (cp. Theorem 8.2.7). Next we introduce the notion of an upper bounding function for the given Piecewise Deterministic Markov Decision Model.

Definition 8.2.4. A measurable function $b : E \to \mathbb{R}_+$ is called an *upper bounding function* for the Piecewise Deterministic Markov Decision Model, if there exist constants $c_r, c_Q, c_\phi \in \mathbb{R}_+$ such that

(i) $r^+(x,u) \leq c_r b(x)$ for all $(x,u) \in E \times \mathcal{U}$.
(ii) $\int b(z)Q(dz|x,u) \leq c_Q b(x)$ for all $(x,u) \in E \times \mathcal{U}$.
(iii) $\lambda \int_0^\infty e^{-(\lambda+\beta)t} b(\phi_t^\alpha(x))dt \leq c_\phi b(x)$ for all $x \in E, \alpha \in \mathcal{R}$.

If r is bounded from above then $b \equiv 1$ is an upper bounding function and $c_Q = 1, c_\phi = \frac{\lambda}{\beta+\lambda}$. From properties (ii) and (iii) it follows

$$\int b(z)Q'(dz|x,\alpha) \leq \lambda c_Q \int_0^\infty e^{-(\lambda+\beta)t} b(\phi_t^\alpha(x))dt \leq c_Q c_\phi b(x).$$

Thus if b is an upper bounding function for the Piecewise Deterministic Markov Model, then b is also an upper bounding function for the discrete-time Markov Decision Model (with and without relaxed controls) and

$$\alpha_b \leq c_Q c_\phi.$$

The Integrability Assumption (A) and the Convergence Assumption (C) are satisfied for the discrete-time Markov Decision Process (with and without relaxed controls) if b is an upper bounding function and $\alpha_b < 1$. Throughout this section we make use of the following

Continuity and Compactness Assumptions:

(i) $\mathcal{U}$ is compact,
(ii) $(t,x,\alpha) \mapsto \phi_t^\alpha(x)$ is continuous on $\mathbb{R}_+ \times E \times \mathcal{R}$,
(iii) $(x,\alpha) \mapsto \int_0^\infty e^{-(\lambda+\beta)t} b(\phi_t^\alpha(x))dt$ is continuous on $E \times \mathcal{R}$,
(iv) $(x,u) \mapsto \int v(z)Q(dz|x,u)$ is upper semicontinuous for all upper semicontinuous $v \in \mathbb{B}_b^+$,
(v) $(x,u) \mapsto r(x,u)$ is upper semicontinuous.

Lemma 8.2.5. *Let b be a continuous upper bounding function and let (ii) and (iii) of the continuity assumptions be satisfied. Let $w : E \times \mathcal{U} \to \mathbb{R}$ be upper semicontinuous with $w(x,u) \leq c_w b(x)$ for some $c_w > 0$. Then*

$$(x, \alpha) \mapsto \int_0^\infty e^{-(\beta+\lambda)t} \Big(\int w(\phi_t^\alpha(x), u)\alpha_t(du) \Big) dt$$

is upper semicontinuous on $E \times \mathcal{R}$.

Proof. First we prove that the function

$$W(x, \alpha) := \int_0^\infty e^{-(\beta+\lambda)t} \Big(\int w(\phi_t^\alpha(x), u)\alpha_t(du) \Big) dt$$

is bounded and continuous if w is bounded and continuous. Boundedness is obvious. Now suppose $(x^n, \alpha^n) \to (x, \alpha)$. Let $\phi_t^n := \phi_t^{\alpha^n}(x_n)$ and $\phi_t := \phi_t^\alpha(x)$. We consider

$$\begin{aligned}
&|W(x^n, \alpha^n) - W(x, \alpha)| \\
&= \Big| \int_0^\infty e^{-(\beta+\lambda)t} \Big(\int_{\mathcal{U}} w(\phi_t^n, u)\alpha_t^n(du) - \int_{\mathcal{U}} w(\phi_t, u)\alpha_t(du) \Big) dt \Big| \\
&\le \int_0^\infty e^{-(\beta+\lambda)t} \int_{\mathcal{U}} |w(\phi_t^n, u) - w(\phi_t, u)| \alpha_t^n(du) dt + \\
&\quad + \Big| \int_0^\infty e^{-(\beta+\lambda)t} \Big(\int_{\mathcal{U}} w(\phi_t, u)\alpha_t^n(du) - \int_{\mathcal{U}} w(\phi_t, u)\alpha_t(du) \Big) dt \Big|.
\end{aligned}$$

The first term on the right-hand side can be further bounded by

$$\int_0^\infty e^{-(\beta+\lambda)t} \sup_{u \in \mathcal{U}} |w(\phi_t^n, u) - w(\phi_t, u)| dt$$

which converges to zero for $n \to \infty$ due to dominated convergence and the continuity of ϕ and w. The second term converges to zero in view of the definition of convergence w.r.t. the Young topology and the fact that w is continuous. Now let w be upper semicontinuous with $w \le c_w b$. Then $w^b(x, u) := w(x, u) - c_w b(x) \le 0$ and is upper semicontinuous. According to Lemma A.1.3 b), there exists a sequence (w_n^b) of bounded and continuous functions with $(w_n^b) \downarrow w^b$. From the first part of the proof we know that

$$W_n(x, \alpha) := \int_0^\infty e^{-(\beta+\lambda)t} \Big(\int w_n^b(\phi_t^\alpha(x), u)\alpha_t(du) \Big)$$

is bounded and continuous and decreases for $n \to \infty$ against

$$W(x, \alpha) - c_w \int_0^\infty e^{-(\beta+\lambda)t} b(\phi_t^\alpha(x)) dt$$

which is thus an upper semicontinuous function. In view of (iii) the function W is upper semicontinuous. □

From Theorem 7.2.1 we directly obtain the main results for the control problem (with relaxed controls). Let us introduce the set

$$\mathbb{M}_{usc} := \{v \in \mathbb{B}_b^+ \mid v \text{ is upper semicontinuous }\}.$$

Theorem 8.2.6. *Suppose the Piecewise Deterministic Markov Decision Process has a continuous upper bounding function b with $\alpha_b < 1$ and the continuity and compactness assumptions are satisfied. Then it holds:*

a) $J_\infty^{rel} \in \mathbb{M}_{usc}$ and $J_\infty^{rel} = \mathcal{T} J_\infty^{rel}$.
b) There exists an optimal relaxed Markov policy $\pi^* = (\pi_t^*)$ such that

$$\pi_t^* = f(Z_n)(t - T_n), \quad t \in (T_n, T_{n+1}]$$

for a decision rule $f : E \to \mathcal{R}$.

Proof. Recall from Remark 8.2.3 that $\mathcal{R}$ is compact. Then it follows from Lemma 8.2.5 that

$$(x, \alpha) \mapsto r'(x, \alpha)$$

is upper semicontinuous and for $v \in \mathbb{M}_{usc}$

$$(x, \alpha) \mapsto \int v(z) Q'(dz|x, \alpha)$$

is upper semicontinuous. Hence the statement follows from Theorem 7.2.1. □

Note that the optimal π_t^* takes values in $\mathbb{P}(\mathcal{U})$. In applications the existence of optimal *nonrelaxed* controls is more interesting. Here we are able to prove the following result.

Theorem 8.2.7. *Suppose the Piecewise Deterministic Markov Decision Process has a continuous upper bounding function b with $\alpha_b < 1$ and the continuity and compactness assumptions are satisfied. If $\phi_t^\alpha(x)$ is independent of α (uncontrolled flow) or if $\mathcal{U}$ is convex, $\mu(x, u)$ is linear in u and*

$$u \mapsto r(x, u) + \lambda \int J_\infty^{rel}(z) Q(dz|x, u)$$

is concave on $\mathcal{U}$, then there exists an optimal nonrelaxed *Markov policy $\pi^* = (\pi_t^*)$ such that*

$$\pi_t^* = f(Z_n)(t - T_n), \quad t \in (T_n, T_{n+1}]$$

for a decision rule $f : E \to A$. Note that π_t^* takes values in $\mathcal{U}$ and that $V_\infty = J_\infty = J_\infty^{rel}$. In particular, V_∞ is a fixed point of $\mathcal{T}$.

Proof. For $v \in I\!M_{usc}$ define

$$w(x,u) := r(x,u) + \lambda \int v(z)Q(dz|x,u), \quad x \in E, u \in \mathcal{U}.$$

Then

$$(Lv)(x,\alpha) = \int e^{-(\beta+\lambda)t} \int w\big(\phi_t^\alpha(x),u\big)\alpha_t(du)dt$$

and

$$(\mathcal{T}v)(x) = \sup_{\alpha\in\mathcal{R}} (Lv)(x,\alpha).$$

a) Let $\phi_t^\alpha(x)$ be independent of α (uncontrolled flow). There exists a measurable function $\tilde{f} : E \to \mathcal{U}$ such that

$$w\big(x,\tilde{f}(x)\big) = \sup_{u\in\mathcal{U}} w(x,u), \quad x \in E.$$

Define $f(x)(t) := \tilde{f}\big(\phi_t(x)\big)$ for $t \geq 0$. Then $f : E \to A$ is measurable and it is easily shown (by a pointwise maximization) that for $\alpha \in \mathcal{R}$

$$\begin{aligned}(Lv)(x,\alpha) &\leq \int e^{-(\beta+\lambda)t} w\big(\phi_t(x), f(x)(t)\big)dt \\ &= (Lv)\big(x,f(x)\big), \quad x \in E.\end{aligned}$$

Hence the statement follows as in the proof of Theorem 8.2.6.

b) Let $u \mapsto w(x,u)$ be concave on $\mathcal{U}$. There exists a measurable function $f^{rel} : E \to \mathcal{R}$ such that (see Theorem A.2.4)

$$\sup_{\alpha\in\mathcal{R}} (Lv)(x,\alpha) = (Lv)\big(x,f^{rel}(x)\big), \quad x \in E.$$

Define $f(x) := \int_{\mathcal{U}} u f^{rel}(x)(du)$ for $x \in E$. Then $f(x) \in A$ since $\mathcal{U}$ is convex, and $f : E \to A$ is measurable. Moreover, since $\mu(x,u)$ is linear in u we obtain $\phi_t^\alpha = \phi_t^{\bar{\alpha}}$ where $\bar{\alpha}_t = \int u\alpha_t(du)$. From the concavity of $w(x,\cdot)$ we conclude

$$(Lv)(x,\alpha) \leq \int e^{-(\beta+\lambda)t} w\big(\phi_t^{\bar{\alpha}}(x),\bar{\alpha}_t\big)dt = (Lv)(x,\bar{\alpha})$$

and hence

$$\sup_{\alpha\in\mathcal{R}} (Lv)(x,\alpha) = (Lv)\big(x,f(x)\big), \quad x \in E.$$

For $v = J_\infty^{rel}$ the (nonrelaxed) control function f is a maximizer of J_∞^{rel}, hence optimal and $V_\infty(x) = J_\infty^{rel}(x)$ (see Theorem 7.1.7). □

Sufficient conditions for the concavity assumption can be formulated by using the results of Section 2.4 (in particular Proposition 2.4.18). Often it is easier to check the concavity of

$$u \mapsto r(x,u) + \lambda \int v(z) Q(dz|x,u)$$

for all functions v in some function class which contains J_∞^{rel} (cf. the applications in Sections 9.3 and 9.4). The fixed point equation is called the Bellman equation. It is related to the Hamilton–Jacobi–Bellman equation in stochastic control which can be seen as a local form of the Bellman equation.

The Hamilton-Jacobi-Bellman equation

We will briefly outline the classical verification technique. We assume that the Piecewise Deterministic Markov Decision Process has a *bounding* function b, i.e. there exists an upper bounding function b with r replaced by $|r|$. Note that the generator of the controlled state process is given by

$$\mathcal{A}v(x,u) := \mu(x,u)v_x + \lambda \int \big(v(y) - v(x)\big) Q(dy|x,u)$$

for a continuously differentiable $v : E \to \mathbb{R}$ and $x \in E, u \in \mathcal{U}$. The Hamilton-Jacobi-Bellman equation (HJB equation) for problem (8.3) has the form:

$$\sup_{u \in \mathcal{U}} \{\mathcal{A}v(x,u) + r(x,u)\} = \beta v(x), \quad x \in E. \tag{8.8}$$

We call a decision rule $f^* : E \to \mathcal{U}$ a maximizer of the HJB equation when $f^*(x)$ is a maximum point of

$$u \mapsto \mathcal{A}v(x,u) + r(x,u), \quad u \in \mathcal{U}.$$

A decision rule $f : E \to \mathcal{U}$ is called a *feedback* control function when f defines a state process $x_t = \phi_t^f(x)$ between two jumps as a solution of

$$dx_t = \mu\big(x_t, f(x_t)\big)dt, \quad x_0 = x. \tag{8.9}$$

We obtain the following result.

Theorem 8.2.8 (Verification Theorem). *Let a Piecewise Deterministic Markov Decision Process be given with a bounding function b, $\alpha_b < 1$ and $\mathbb{E}_x^\pi[e^{-\beta t} b(X_t)] \to 0$ for $t \to \infty$ for all π, x. Suppose that $v \in C^1(E) \cap \mathbb{B}_b$ is a solution of the HJB equation and that f^* is a maximizer of the HJB equation and defines a state process (X_t^*). Then $v = V_\infty$ and $\pi_t^* := f^*(X_{t-}^*)$ is an optimal Markov policy (in feedback form).*

Proof. Let $\pi = (\pi_t)$ be an arbitrary predictable control process. We denote by $\tilde{N}^\pi$ the corresponding compensated counting measure, i.e.

$$\tilde{N}^\pi(t,B) := \sum_{n=0}^\infty 1_{[T_n \le t]} 1_{[Z_n \in B]} - \lambda \int_0^t Q(B|X_s,\pi_s)ds.$$

The Itô formula for jump processes implies for $v \in C^1(E) \cap \mathbb{B}_b$ that

$$\begin{aligned} e^{-\beta t} v(X_t) &= v(x) + \int_0^t e^{-\beta s}\Big(\mathcal{A}v(X_s,\pi_s) - \beta v(X_s)\Big)ds \\ &+ \int_0^t e^{-\beta s} \int (v(y) - v(X_{s-}))\tilde{N}^\pi(ds,dy). \end{aligned}$$

Since

$$\begin{aligned} &\mathbb{E}_x^\pi\Big[\sum_{n=1}^\infty e^{-\beta T_n} |v(X_{T_n}) - v(X_{T_n-})|\Big] \\ &\le \sum_{n=1}^\infty \mathbb{E}_x^\pi\Big[e^{-\beta T_n} b(Z_n)\Big] + \sum_{n=1}^\infty \mathbb{E}_x^\pi\Big[e^{-\beta T_{n-1}} b(\phi^\pi_{T_n - T_{n-1}}(Z_{n-1}))\Big] \\ &\le \sum_{n=1}^\infty \alpha_b^n b(x) + \sum_{n=1}^\infty c_\phi \alpha_b^{n-1} b(x) < \infty, \end{aligned}$$

the second integral is a martingale. Thus taking the conditional expectation $\mathbb{E}_x^\pi$ and using the fact that v satisfies the HJB equation yields:

$$\begin{aligned} \mathbb{E}_x^\pi\left[e^{-\beta t} v(X_t)\right] &= v(x) + \mathbb{E}_x^\pi\Big[\int_0^t e^{-\beta s}\Big(\mathcal{A}v(X_s,\pi_s) - \beta v(X_s)\Big)ds\Big] \\ &\le v(x) - \mathbb{E}_x^\pi\Big[\int_0^t e^{-\beta s} r(X_s,\pi_s)ds\Big]. \end{aligned}$$

Taking $t \to \infty$ this implies by our transversality assumption that

$$v(x) \ge \mathbb{E}_x^\pi\Big[\int_0^\infty e^{-\beta s} r(X_s,\pi_s)ds\Big]$$

for arbitrary π. Inserting π^* yields equality and the statement is shown. □

8.3 Continuous-Time Markov Decision Chains

Continuous-time Markov Decision Chains with a countable state space E are an important special case of Piecewise Deterministic Markov Decision Processes. For a recent book on this topic see Guo and Hernández-Lerma (2009).

Instead of giving transition probabilities, continuous-time Markov Chains are often described by transition rates which are assumed to be controllable here, i.e. $q_{xy}(u), u \in \mathcal{U}$. In what follows we assume that the rates $q_{xy}(u) \in \mathbb{R}$ are *conservative*, i.e.

$$\sum_{y \in E} q_{xy}(u) = 0, \quad x \in E, u \in \mathcal{U}$$

and *bounded*, i.e. $\lambda \geq -q_{xx}(u)$ for all $x \in E, u \in \mathcal{U}$. A continuous-time Markov Decision Processes (X_t) with such transition rates can be constructed from a Poisson process (N_t) with intensity λ and a discrete-time Markov Decision Process (Z_n) with transition probabilities

$$Q(\{y\}|x,u) := \begin{cases} \frac{1}{\lambda} q_{xy}(u), & y \neq x \\ 1 + \frac{1}{\lambda} q_{xx}(u), & y = x. \end{cases}$$

It holds that $X_t = Z_{N_t}$. This representation is called a *uniformized* Markov Decision Chain. In this case we have $\phi_t^\alpha(x) = x$, i.e. the flow between two jumps is independent of α. This implies that we can restrict to control functions α which are constant, i.e. $\alpha_t \equiv u$. The discrete-time Markov Decision Model of the last section has the form $(E, \mathcal{U}, Q', r')$ with

- $Q'(\{y\}|x,u) := \frac{\lambda}{\beta+\lambda} Q(\{y\}|x,u)$,
- $r'(x,u) := \frac{1}{\beta+\lambda} r(x,u)$.

The maximal reward operator $\mathcal{T}$ is given by

$$(\mathcal{T}v)(x) = \frac{1}{\beta+\lambda} \sup_{u \in \mathcal{U}} \Big\{ r(x,u) + \sum_{y \in E} q_{xy}(u) v(y) \Big\} + \frac{\lambda}{\beta+\lambda} v(x).$$

The next result follows now directly from Theorem 8.2.6 and Theorem 8.2.7. Note that E is countable. Moreover, we have here

$$\alpha_b \leq \frac{\lambda}{\beta+\lambda} c_Q.$$

Theorem 8.3.1. *Suppose the continuous-time Markov Decision Chain has an upper bounding function b with $\alpha_b < 1$ and the following continuity and compactness assumptions are satisfied:*

(i) *$\mathcal{U}$ is compact,*
(ii) *$u \mapsto q_{xy}(u)$ is continuous for all $x, y \in E$,*
(iii) *$u \mapsto \sum_{y \in E} b(y) q_{xy}(u)$ is continuous for all $x \in E$,*
(iv) *$u \mapsto r(x,u)$ is upper semicontinuous for all $x \in E$.*

Then it holds:

a) $V_\infty \in \mathbb{B}_b^+$ and V_∞ is a fixed point of $\mathcal{T}$, i.e.

$$\beta V_\infty(x) = \sup_{u \in \mathcal{U}} \Big\{ r(x,u) + \sum_{y \in E} q_{xy}(u) V_\infty(y) \Big\}, \quad x \in E. \tag{8.10}$$

b) There exists an optimal Markov policy $\pi^* = (\pi_t^*)$ *such that*

$$\pi_t^* = f^*(X_{t-}), \quad t \geq 0$$

for a decision rule $f^* : E \to \mathcal{U}$*, where* $f^*(x)$ *is a maximum point of*

$$u \mapsto r(x,u) + \sum_{y \in E} q_{xy}(u) V_\infty(y), \quad u \in \mathcal{U}.$$

Note that the fixed point equation (8.10) coincides with the Hamilton-Jacobi-Bellman equation.

Problems with Finite Time Horizon

Continuous-time Markov Decision Chains with finite time horizon T are more interesting, in particular as far as applications in finance and insurance are concerned. Since the time horizon is finite we have to consider also the jumps times T_n besides the marks (post-jump states) Z_n. The function space A is here defined by

$$A := \{\alpha : [0,T] \to \mathcal{U} \text{ measurable } \}. \tag{8.11}$$

Hence, a Markov policy (or piecewise open loop policy) $\pi = (\pi_t)$ is defined by a sequence of measurable functions $f_n : [0,T] \times E \to A$ such that

$$\pi_t = f_n(T_n, Z_n)(t - T_n) \text{ for } t \in (T_n, T_{n+1}].$$

Since we consider here control problems with a finite time period $[0,T]$ we have a measurable terminal reward $g : E \to \mathbb{R}$. We impose now the following

Integrability Assumption (A):

$$\sup_\pi \mathbb{E}_x^\pi \left[\int_0^T r^+(X_s, \pi_s) ds + g^+(X_T) \right] < \infty, \quad x \in E.$$

Then the expected total reward when we start at time t in state x is well-defined for all π by

$$V_\pi(t,x) := \mathbb{E}_{tx}^\pi \left[\int_t^T r(X_s, \pi_s) ds + g(X_T) \right], \quad x \in E, t \in [0,T]$$

where $\mathbb{E}_{tx}^\pi$ denotes the conditional expectation given that $X_t = x$. The value function of the continuous-time Markov Decision Chain is given by

$$V(t,x) := \sup_\pi V_\pi(t,x), \quad x \in E, t \in [0,T] \tag{8.12}$$

where the supremum is taken over all Markov policies. It holds that $V_\pi(T,x) = g(x) = V(T,x)$.
Using a similar embedding procedure as in Section 8.2 the value function $V(t,x)$ can be obtained by the following discrete-time infinite-stage Markov Decision Model (E', A, Q', r'):

- $E' = [0,T] \times E$ is the *state space.* A state (t,x) describes the time point of a jump and the state of the process directly after the jump.
- A is given by (8.11). Recall that the function space A is a Borel space.
- For all Borel sets $B \subset [0,T], y \in E$ and $(t,x) \in E'$, $\alpha \in A$, the stochastic kernel Q' is given by

$$Q'\big(B \times \{y\}|t,x,\alpha\big) := \lambda \int_0^{T-t} e^{-\lambda s} 1_B(t+s) Q\big(\{y\}|x,\alpha_s\big) ds. \quad (8.13)$$

  This is obviously a substochastic transition law. In order to make it stochastic, we may add a cemetery state $\Delta \notin E'$ to the state space and define

$$Q'\big(\{\Delta\}|t,x,\alpha\big) := e^{-\lambda(T-t)}, \quad Q'\big(\{\Delta\}|\Delta,\alpha\big) := 1.$$

- The reward function $r' : E' \times A \to \mathbb{R}$ is defined by

$$r'(t,x,\alpha) := \int_0^{T-t} e^{-\lambda s} r\big(x,\alpha_s\big) ds + e^{-\lambda(T-t)} g(x), \quad (8.14)$$
$$r'(x,\Delta) := 0.$$

In what follows we treat the problem as a substochastic problem and skip the state Δ. For this discrete-time Markov Decision Model we define for a policy $\pi = (f_n)$:

$$J_{\infty(f_n)}(t,x) = \mathbb{E}_{tx}^{(f_n)} \left[\sum_{n=0}^{\infty} r'\big(T_n', Z_n', f_n(T_n', Z_n')\big) \right]$$
$$J_\infty(t,x) = \sup_{(f_n)} J_{\infty(f_n)}(t,x), \quad (t,x) \in E'$$

where (T_n', Z_n') is the corresponding state process of the Markov Decision Process up to absorption in Δ. Note that $(T_n', Z_n') = (T_n, Z_n)$ as long as $T_n \le T$.

Theorem 8.3.2. *For a Markov policy $\pi = (f_n)$ we have*

$$V_\pi(t,x) = J_{\infty(f_n)}(t,x), \quad (t,x) \in E'.$$

Moreover, it holds: $V = J_\infty$.

Proof. Let $H_n := (T_0, Z_0, \dots, T_n, Z_n)$ and $T_n \le T$. We consider only the time point $t = 0$. Arbitrary time points can be treated similarly by adjusting the notation. We obtain:

$$\begin{aligned}
V_\pi(0,x) &= \mathbb{E}_x^\pi \left[\sum_{n=0}^\infty \Big(\int_{T_n \wedge T}^{T_{n+1}\wedge T} r(X_s, \pi_s) ds + 1_{[T_n \le T < T_{n+1}]} g(X_T) \Big) \right] \\
&= \sum_{n=0}^\infty \mathbb{E}_x^\pi \left[\mathbb{E}_x^\pi \left[\int_{T_n}^{T_{n+1}\wedge T} r(X_s, \pi_s) ds \Big| H_n \right] \right] \\
&\quad + \sum_{n=0}^\infty \mathbb{E}_x^\pi \left[\mathbb{E}_x^\pi \left[1_{[T_n \le T < T_{n+1}]} g(X_T) | H_n \right] \right] \\
&= \sum_{n=0}^\infty \mathbb{E}_{0x}^{(f_n)} \left[r'(T_n', Z_n', f_n(T_n', Z_n')) \right] = J_{\infty (f_n)}(0,x)
\end{aligned}$$

since the transition kernel of (T_n', Z_n') is given by (8.13) and r' by (8.14). □

The maximal reward operator $\mathcal{T}$ for the discrete-time model is given by

$$\begin{aligned}
(\mathcal{T} v)(t,x) &= \sup_{\alpha \in A} \Big\{ e^{-\lambda(T-t)} g(x) \\
&\quad + \int_0^{T-t} e^{-\lambda s} \Big[r(x, \alpha_s) + \lambda \sum_{y \in E} v(t+s, y) Q(\{y\}|x, \alpha_s) \Big] ds \Big\} \\
&= e^{-\lambda(T-t)} g(x) \\
&\quad + \int_0^{T-t} e^{-\lambda s} \sup_{u \in \mathcal{U}} \Big[r(x,u) + \lambda \sum_{y \in E} v(t+s, y) Q(\{y\}|x,u) \Big] ds.
\end{aligned}$$

The last equality follows by pointwise maximization. If $b : E \to \mathbb{R}_+$ is an *upper bounding function* for the continuous-time Markov Decision Chain, i.e. there exist $c_r, c_g, c_Q \in \mathbb{R}_+$ such that

(i) $r^+(x,u) \le c_r b(x)$ for all $(x,u) \in E \times \mathcal{U}$,
(ii) $g^+(x) \le c_g b(x)$ for all $x \in E$,
(iii) $\sum_{y \in E} b(y) Q(\{y\}|x,u) \le c_Q b(x)$ for all $(x,u) \in E \times \mathcal{U}$,

then it is easily shown that

$$b(t,x) = e^{\gamma(T-t)} b(x), \quad \text{for } \gamma \ge 0$$

is an upper bounding function for the discrete-time model and

$$\alpha_b \le c_Q \frac{\lambda}{\lambda + \gamma} \big(1 - e^{-(\lambda+\gamma)T} \big).$$

Hence we always have $\alpha_b < 1$ for γ large.

Theorem 8.3.3. *Suppose the continuous-time Markov Decision Chain has an upper bounding function b and the continuity and compactness assumptions (i)–(iv) of Theorem 8.3.1 are satisfied. Then there exists an optimal Markov policy $\pi^* = (\pi_t^*)$ such that*

$$\pi_t^* = f^*(t, X_{t-}), \quad t \in [0, T]$$

for a decision rule $f^* : E' \to \mathcal{U}$, where $f^*(t,x)$ is a maximum point of

$$u \mapsto r(x,u) + \sum_{y \in E} q_{xy}(u) V(t,y), \quad u \in \mathcal{U}.$$

Moreover, $V(t,x)$ is a fixed point of the operator $\mathcal{T}$ in $I\!B_b^+$, i.e.

$$V(t,x) = e^{-\lambda(T-t)} g(x) + \int_0^{T-t} e^{-\lambda s} \sup_{u \in \mathcal{U}} \Big[r(x,u) + \lambda \sum_{y \in E} V(t+s,y) Q\big(\{y\}|x,u\big) \Big] ds.$$

Proof. The proof is similar to the proof of Theorem 8.2.7 a). For $v \in I\!B_b^+$ define

$$w(t,x,u) := r(x,u) + \lambda \sum_{y \in E} v(t,y) Q(\{y\}|x,u), \quad x \in E, u \in \mathcal{U}.$$

Then there exists a measurable function $f^* : E' \to \mathcal{U}$ such that (see Theorem A.2.4)

$$w\big(t,x,f^*(t,x)\big) = \sup_{u \in \mathcal{U}} w(t,x,u), \quad (t,x) \in E'.$$

Define $f(t,x)(s) := f^*(t+s,x)$ for $s \geq 0$. Then $f : E' \to A$ is measurable and by pointwise maximization

$$\begin{aligned}\sup_{\alpha \in A} (Lv)(t,x,\alpha) &= e^{-\lambda(T-t)} g(x) + \int_0^{T-t} e^{-\lambda s} w\big(t+s,x,f(t,x)(s)\big) ds \\ &= (Lv)\big(t,x,f(t,x)\big).\end{aligned}$$

For $v = V$ the decision rule f is a maximizer of V. Then the policy $\pi^* = (\pi_t^*)$ is optimal, where for $t \in (T_n, T_{n+1}]$

$$\pi_t^* := f(T_n, Z_n)(t - T_n) = f^*(t, Z_n) = f^*(t, X_{t-})$$

(see also the Verification Theorem 7.1.7). □

Remark 8.3.4. If there exists a *bounding* function, then $V(t,x)$ is the unique fixed point of $\mathcal{T}$ in $\mathbb{B}_b$. In this case, $\mathcal{T}$ is a contraction. Moreover, $V(t,x)$ is continuous in t for all $x \in E$. ◊

Remark 8.3.5. The Hamilton-Jacobi-Bellman equation for the continuous-time Markov Decision Chain with finite horizon T has the form

$$\begin{aligned} g(x) &= V(T,x) \\ 0 &= V_t(t,x) + \sup_{u\in\mathcal{U}} \Big\{ r(x,u) + \sum_{y\in E} V(t,y) q_{xy}(u) \Big\} \end{aligned}$$

where V_t denotes the derivative w.r.t. time. A verification theorem can be formulated as in the last section (see Theorem 8.2.8). ◊

8.4 Exercises

Exercise 8.4.1 (Relaxed Controls). Consider the controlled Piecewise Deterministic Markov process which is given by: $E = \mathbb{R}$, $\mathcal{U} = \{-1,+1\}$ and $\mu(x,u) = u$, $Q(B|x,u) = \delta_x(B)$, i.e. jumps cannot be controlled and are only 'virtual' jumps. For a relaxed control $\pi = (f_n)$ we have

$$\phi_t^\pi(x) = x + \int_0^t \int_{\mathcal{U}} u f_n(s,x)(du) ds, \quad t \in [T_n, T_{n+1}).$$

The aim is to minimize

$$\mathbb{E}_x \left[\int_0^\infty e^{-\beta t} X_t^2 dt \right].$$

a) Let $x_0 = 0$. Compute an optimal relaxed control.
b) Show that there does not exist an optimal nonrelaxed control.

Exercise 8.4.2 (Killing at Rate β). Another way of looking at the reformulation as a discrete-time problem in Theorem 8.2.1 is to interpret the discount factor as a *killing rate*. This enables us to write a modified problem without discount factor which is then straightforward to describe as a Markov Decision Problem. In order to explain this let U be a random variable independent of the state process (X_t) which is uniformly distributed over $[0,1]$. Let

$$\tau := \inf \{ t > 0 \mid e^{-\beta t} \le U \}$$

and define the killed process (X_t') by

$$X'_t = \begin{cases} X_t, t < \tau \\ \Delta, \ t \geq \tau. \end{cases}$$

If we set $r(\Delta, \cdot) = 0$, show that

$$\begin{aligned} \mathbb{E}_x^\pi \left[\int_0^\infty e^{-\beta t} r(X_t, \pi_t) dt \right] &= \mathbb{E}_x^\pi \left[\int_0^\infty r(X'_t, \pi_t) dt \right] \\ &= \sum_{n=0}^\infty \mathbb{E}_x^{(f_n)} \left[r'(Z'_n, f_n(Z'_n)) \right]. \end{aligned}$$

Exercise 8.4.3. Suppose a controlled Piecewise Deterministic Markov Process as in Section 8.1 is given where the aim is now to maximize

$$V_\pi(t,x) := \mathbb{E}_{tx}^\pi \left[\int_t^T r(X_s, \pi_s) ds + g(X_T) \right], \quad x \in E, \qquad (8.15)$$

the expected reward over a finite time horizon. Find an equivalent discrete-time Markov Decision Model to solve this problem.

Exercise 8.4.4. Formulate a Markov Decision Model which is equivalent to the optimization problem in (8.3) with a state and action dependent discount factor less than 1 as indicated in Remark 8.2.2.

Exercise 8.4.5. Consider the financial market introduced in Section 3.2 and solve the problem of maximizing the expected utility of terminal wealth $\mathbb{E}_x^\pi U(X_T)$, where $U(x) = -e^{-\gamma x}, x \in \mathbb{R}$ and $\gamma > 0$ and $U(x) = \log x, x > 0$.

Exercise 8.4.6 (μc-rule). Consider a queueing system with two queues and one server. Customers arrive at queue i according to a Poisson process with rate λ_i and need exponential service times with rates μ_i, $i = 1, 2$. After each departure of a customer, the server has to be assigned to the queues, where the capacity of the server can be divided, i.e. if we assign the fraction $a \in [0, 1]$ to queue 1, then the first customer in this queue is served with rate $\mu_1 a$ and the first customer in queue 2 is served with rate $\mu_2(1-a)$ (see Figure 8.1). Customers produce linear holding cost at rate c_i, $i = 1, 2$. The aim is to minimize the expected discounted holding cost over an infinite time horizon.

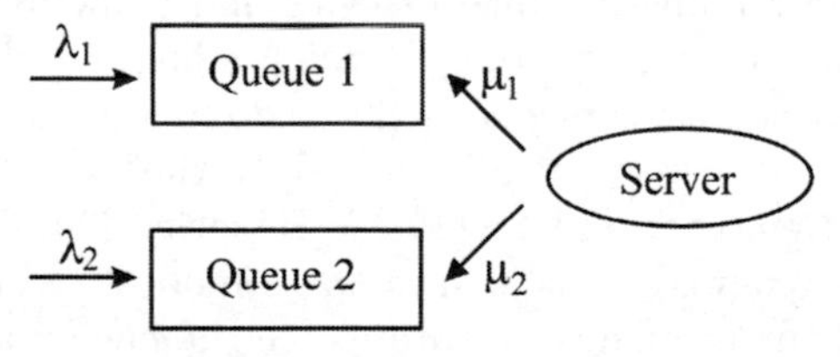

Fig. 8.1 Queueing system of Exercise 8.4.6.

a) Formulate this queueing problem as a continuous-time Markov Decision Chain.
b) Suppose the discount factor is $\beta > 0$ and let $\lambda = \lambda_1 + \lambda_2 + \mu_1 + \mu_2$ be the uniformization rate. Show that the Bellman equation has the form

$$\begin{aligned}(\lambda + \beta)v(x_1, x_2) = c_1 x_1 + c_2 x_2 + \lambda_1 v(x_1 + 1, x_2) + \lambda_2 v(x_1, x_2 + 1)\\ + \min\big\{\mu_1 v\big((x_1 - 1)^+, x_2\big) + \mu_2 v(x_1, x_2);\\ \mu_2 v\big(x_1, (x_2 - 1)^+\big) + \mu_1 v(x_1, x_2)\big\},\ x_1, x_2 \in \mathbb{N}_0.\end{aligned}$$

c) Assume that $\mu_1 c_1 \geq \mu_2 c_2$. Consider the embedded discrete-time Markov Decision Process with a finite horizon N. Show by induction that the stationary policy $\pi^* = (f^*, \dots, f^*)$ with

$$f^*(x_1, x_2) = 1_{[x_1 > 0]}$$

is optimal. This means that queue 1 with the greater $\mu_1 c_1$ has priority over queue 2 and will always be served as long as a customer is waiting there. This decision rule is called the μc-rule.
d) Use part c) to prove that the μc-rule is optimal for the infinite horizon problem.

8.5 Remarks and References

A first systematic study of controlled Piecewise Deterministic Markov Processes is done in Hordijk and van der Duyn Schouten (1984, 1985). The idea of reducing the control problems of this type to a discrete-time Markov Decision Problem is due to Yushkevich (1980). For a recent paper on this topic see Almudevar (2001). In Yushkevich (1987, 1989) optimality conditions are given in a weak form based on a continuous-time approach. Davis introduced the name *Piecewise Deterministic Markov Process* (see e.g. Davis (1984)) and summarizes the state of the art in his book Davis (1993). Schäl (1998) and Forwick et al. (2004) extend the existing results to unbounded reward problems. They impose certain assumptions on the drift which imply (using a time transformation) the existence of nonrelaxed controls. Relaxed controls are known from deterministic control theory and allow to define a topology on the action space (Young topology) which simplifies the task to have a compact action space and continuous target functions at the same time. It is well known that concavity conditions imply the existence of nonrelaxed controls (see e.g. Dempster and Ye (1992), Bäuerle (2001)).
There has been a renewed interest into these models recently, in particular as far as applications in finance, insurance and queueing are concerned (for references see Chapter 9).

Continuous-time Markov Decision processes are investigated in Guo and Hernández-Lerma (2009). They also consider problems with unbounded reward and transition rates.

Chapter 9
Optimization Problems in Finance and Insurance

We will now apply the theory of infinite horizon Markov Decision Models to solve some optimization problems in finance. In Section 9.1 we consider a *consumption and investment problem with random horizon* which leads to a contracting Markov Decision Model with infinite horizon as explained in Section 7.6.1. Explicit solutions in the case of a power utility are given. In Section 9.2 a classical *dividend pay-out problem* for an insurance company is investigated. In this example the state and action space are both discrete which implies that all functions on $E \times A$ are continuous and we can work with Theorem 7.2.1. Here the Markov Decision Model is not contracting. The main part of this section is to show that there exists an optimal stationary policy which is a so-called *band-policy*. In special cases this band-policy reduces to a barrier-policy, i.e. it is optimal to pay out all the money which is above a certain threshold. In Section 9.3 we consider a utility maximization problem in a financial market where the stock prices are *Piecewise Deterministic Markov Processes*. This optimization problem is contracting and our results from Chapters 7 and 8 allow a characterization of the value function and some computational approaches which complement the classical stochastic control approach via the *Hamilton-Jacobi-Bellman* equation. Some numerical results are also given. In Section 9.4 we study the liquidation of a large amount of shares in so-called dark pools. This is a continuous-time Markov Decision Chain with finite time horizon (see Section 8.3). Using the discrete-time solution approach we are able to derive some interesting properties of the optimal liquidation policy.

9.1 Consumption-Investment Problems with Random Horizon

In this section we reconsider the consumption and investment problem of Section 4.3. However, this time we assume that the investment horizon of the

N. Bäuerle and U. Rieder, *Markov Decision Processes with Applications to Finance*, Universitext, DOI 10.1007/978-3-642-18324-9_9,

agent is random (cf. Section 7.6.1). This is reasonable since there may be a drastic change of the agent's plan in the future with some probability. For example the agent may need all her money because she was disabled due to an accident.
A financial market with d risky asset and one riskless bond (with interest rate $i_n = 0$) is given as introduced in Section 4.2. Recall that $\mathcal{F}_n := \mathcal{F}_n^S$. Here we assume that $R_1, R_2, \dots$ are independent and identically distributed random vectors and that the following Assumption (FM) holds:

Assumption (FM):

(i) *There are no arbitrage opportunities.*
(ii) $\mathbb{E}\,\|R_1\| < \infty$.

The first assumption means that there is no arbitrage opportunity for any finite horizon. The random horizon is here described by a *geometrically distributed* random variable τ with parameter $p \in (0,1)$, i.e.

$$\mathbb{P}(\tau = n) = (1-p)p^{n-1},\ n \in \mathbb{N}.$$

It is assumed that the random horizon τ is independent of $(\mathcal{F}_n)$. The aim is to maximize the expected discounted utility from consumption and investment until time τ. The initial wealth is given by $x > 0$. In what follows suppose that $U_c, U_p : E \to \mathbb{R}_+$ are two continuous utility functions with $dom\, U_c = dom\, U_p := [0, \infty)$ which are used to evaluate the consumption and the terminal wealth. The wealth process (X_n) evolves as follows

$$X_{n+1} = X_n - c_n + \phi_n \cdot R_{n+1},$$

where $(c, \phi) = (c_n, \phi_n)$ is a consumption-investment strategy i.e. ϕ_n and c_n are $(\mathcal{F}_n)$-adapted and $0 \le c_n \le X_n$, for all $n \in \mathbb{N}$. The optimization problem is then given by

$$\begin{cases} \mathbb{E}_x \left[\sum_{n=0}^{\tau-1} \beta^n U_c(c_n) + \beta^\tau U_p(X_\tau^{c,\phi}) \right] \to \max \\ (c,\phi) \text{ is a consumption-investment strategy and} \\ \qquad X_\tau^{c,\phi} \in dom\, U_p \ \mathbb{P}\text{-a.s.} \end{cases} \tag{9.1}$$

According to Section 7.6.1 we can formulate this optimization problem with random horizon as a stationary Markov Decision Model with infinite horizon:

- $E := [0, \infty)$ where $x \in E$ denotes the wealth,
- $A := \mathbb{R}_+ \times \mathbb{R}^d$ where $a \in \mathbb{R}^d$ is the amount of money invested in the risky assets and $c \in \mathbb{R}_+$ the amount which is consumed,
- $D(x)$ is given by

$$D(x) := \{(c, a) \in A \mid 0 \le c \le x \text{ and } x - c + a \cdot R_1 \ge 0\, \mathbb{P}\text{-a.s.}\},$$

- $\mathcal{Z} := [-1, \infty)^d$ where $z \in \mathcal{Z}$ denotes the relative risk,
- $T(x, c, a, z) := x - c + a \cdot z$,
- $Q^Z(\cdot|x, c, a) :=$ distribution of R_{n+1} (independent of (x, c, a)),
- $r(x, c, a) := U_c(c) + \beta(1-p)\,\mathbb{E}[U_p(T(x, c, a, R_1))]$,
- $\tilde{\beta} := \beta p$ with $\beta \in (0, 1]$.

The value $J_\infty(x)$ is then the maximal value of (9.1) and an optimal policy for the Markov Decision Models defines an optimal consumption-investment strategy.

Next define $\tilde{A} := \{\alpha \in \mathbb{R}^d \mid 1 + \alpha \cdot R_1 \geq 0\, \mathbb{P}\text{-a.s.}\}$. The optimization problem

$$v_0 := \sup_{\alpha \in \tilde{A}} \mathbb{E}(1 + \alpha \cdot R_1)$$

has a finite value v_0 since $\tilde{A}$ is bounded in view of (FM). It is not difficult to show that $b(x) := 1 + x$ is a bounding function for the stationary Markov Decision Model, and we obtain

$$\begin{aligned} \alpha_b &:= \sup_{(x,c,a)} \frac{1 + x - c + a \cdot \mathbb{E} R_1}{1 + x} \\ &= \sup_{x \in E, \alpha \in \tilde{A}} \frac{1 + x + x\alpha \cdot \mathbb{E} R_1}{1 + x} = \sup_{x \in E} \frac{1 + x v_0}{1 + x} = \max\{1, v_0\}. \end{aligned}$$

If $\tilde{\beta}\alpha_b < 1$ then the Markov Decision Model is contracting. The operator $\mathcal{T}$ is given by

$$\mathcal{T}v(x) = \sup_{(c,a) \in D(x)} \left\{ r(x, c, a) + \beta p\, \mathbb{E}\, v\big(x - c + a \cdot R_1\big) \right\}, \; x \in E$$

for $v \in \mathbb{B}_b$. We obtain:

Theorem 9.1.1. *For the consumption-investment problem with random horizon and $\beta p \alpha_b < 1$ it holds:*

a) J_∞ is increasing, concave and continuous.
b) $J_\infty = \lim_{n\to\infty} J_n$ and J_∞ is the unique fixed point of $\mathcal{T}$ in $\mathbb{B}_b$.
c) There exists a maximizer f^* of J_∞, and the stationary strategy $(f^*, f^*, \ldots)$ is optimal for the consumption-investment problem with random horizon.
d) The policy iteration holds.

Proof. We show that the assumptions of Theorem 7.3.5 are satisfied with

$$\mathbb{M} := \{v \in \mathbb{B}_b \mid v \text{ is non-negative, increasing, concave and continuous}\}.$$

This in turn implies our statements. That $0 \in \mathbb{M}$ is obvious and the fact that $\mathcal{T} : \mathbb{M} \to \mathbb{M}$ as well as the existence of maximizers of $v \in \mathbb{M}$ have been shown in Theorem 4.3.1. Since the bounding function is continuous, the set $I\!M$ is a closed subspace of $I\!B_b$. Theorem 7.3.6 implies d). □

Example 9.1.2 (Power Utility). We consider now a financial market with one bond and one stock and choose both utility functions as power utility functions, i.e. $U_c(x) = \frac{1}{\gamma}x^\gamma = U_p(x)$ with $0 < \gamma < 1$. It is assumed that $\beta p \alpha_b < 1$. Then we get $\mathcal{T} : I\!M \to I\!M$ with

$$I\!M := \{v \in I\!B_b \mid v(x) = dx^\gamma \text{ for } d > 0\}.$$

This follows as in Section 4.3. Hence we obtain

$$\lim_{n\to\infty} J_n(x) = J_\infty(x) = d_\infty x^\gamma$$

for some $d_\infty > 0$. Let us now compute d_∞ and the optimal strategy. As in Section 4.3 we consider the problem

$$\begin{cases} \mathbb{E}\left(1 + \alpha \cdot R_1\right)^\gamma \to \max \\ \alpha \in \tilde{A} \end{cases} \tag{9.2}$$

and denote the value of this problem by v_∞. By induction we obtain $J_n(x) = d_n x^\gamma$ and the sequence (d_n) satisfies the recursion

$$d_{n+1}^\delta = \gamma^{-\delta} + \Big(\frac{\beta(1-p)v_\infty}{\gamma} + \beta p v_\infty d_n\Big)^\delta$$

with $\delta := (1-\gamma)^{-1}$ and $d_0 = 0$. Theorem 9.1.1 implies that $d_\infty = \lim_{n\to\infty} d_n$ and d_∞ is the unique solution d of the following equation

$$d^\delta = \gamma^{-\delta} + \Big(\frac{\beta(1-p)v_\infty}{\gamma} + \beta p v_\infty d\Big)^\delta.$$

Then a maximizer f^* of J_∞ can be computed explicitly. Since

$$\mathcal{T} J_\infty(x) = x^\gamma \cdot \sup_{0\le\zeta\le 1} \Big\{\frac{1}{\gamma}\zeta^\gamma + (1-\zeta)^\gamma\big(\beta(1-p)v_\infty\frac{1}{\gamma} + \beta p v_\infty d_\infty\big)\Big\},$$

the optimal investment and consumption decisions separate. The amount invested in the assets is given by $a^*(x) = \alpha^* \cdot (x - c^*(x))$ where α^* is the optimal solution of the problem (9.2), and the optimal consumption $c^*(x)$ is given by (cf. proof of Theorem 4.3.6)

$$c^*(x) := \frac{x}{1 + (\beta(1-p)v_\infty + \beta p \gamma v_\infty d_\infty)^\delta}.$$

Hence the strategy $(f^*, f^*, \ldots)$ with $f^*(x) := (a^*(x), c^*(x))$ is optimal. Note that $c^*(x)$ is the limit of the optimal consumptions $c_n(x)$ at time n

$$c_n^*(x) = \frac{x}{1 + (\beta(1-p)v_\infty + \beta p \gamma v_\infty d_n)^\delta}.$$

Thus $(f^*, f^*, \ldots)$ is also obtained by policy iteration. ♦

Example 9.1.3 (Power Utility without Consumption). Here we consider the same financial market as in Example 9.1.2, but now we maximize the expected discounted power utility of the terminal wealth at the random time point τ i.e. $U_c(x) \equiv 0, U_p(x) = \frac{1}{\gamma}x^\gamma$ with $0 < \gamma < 1$. There is no consumption. It is assumed that $\beta p \alpha_b < 1$. As in the previous example we consider the set

$$I\!M := \{v \in I\!B_b \mid v(x) = dx^\gamma \text{ for } d > 0\}$$

and obtain: $J_\infty(x) = d_\infty x^\gamma$ for some $d_\infty > 0$. In this case we can identify d_∞ as the unique solution d of the equation

$$d = \frac{\beta(1-p)v_\infty}{\gamma} + \beta p v_\infty d.$$

This equation can be solved and we obtain

$$d_\infty = \frac{\beta(1-p)v_\infty}{\gamma(1-\beta p v_\infty)}.$$

Note that $\beta p v_\infty < 1$, since $v_\infty \le \alpha_b$ and $\beta p \alpha_b < 1$. The stationary strategy $(f^*, f^*, \ldots)$ with $f^*(x) = \alpha^* x$ is optimal, where α^* is the optimal solution of (9.2). Thus when we compare Examples 9.1.2 and 9.1.3, we see that the optimal relative asset allocation is independent from the consumption. ♦

9.2 A Dividend Problem in Risk Theory

The dividend pay-out problem is a classical problem in risk theory. There are many different variants of it in discrete and continuous time. Here we consider a completely discrete setting which has the advantage that the structure of the optimal policy can be identified.

Problem and Model Formulation

Imagine we have an insurance company which earns some premia on the one hand but has to pay out possible claims on the other hand. We denote by Z_n the difference between premia and claim sizes in the n-th time interval and assume that $Z_1, Z_2, \ldots$ are independent and identically distributed with distribution $(q_k,\ k \in \mathbb{Z})$, i.e. $\mathbb{P}(Z_n = k) = q_k$ for $k \in \mathbb{Z}$. At the beginning of each time interval the insurer can decide upon paying a dividend. Of course this can only be done if the risk reserve at that time point is positive. Once the risk reserve becomes negative (this happens when the claims are larger than the reserve plus premia in that time interval) we say that the company is ruined and has to stop its business. The aim now is to maximize the expected discounted dividend pay out until ruin. In the economic literature this value is sometimes interpreted as the value of the company.

We formulate this problem as a stationary Markov Decision Problem with infinite horizon. The state space is $E := \mathbb{Z}$ where $x \in E$ is the current risk reserve. At the beginning of each period we have to decide upon a possible dividend pay out $a \in A := \mathbb{N}_0$. Of course we have the restriction that $a \in D(x) := \{0, 1, \dots, x\}$ when $x \geq 0$ and we set $D(x) := \{0\}$ if $x < 0$. If we denote by z the risk reserve change, then the transition function is given by

$$T(x, a, z) := \begin{cases} x - a + z & \text{if } x \geq 0 \\ x, & \text{if } x < 0. \end{cases} \tag{9.3}$$

The dividend pay-out is rewarded by $r(x, a) := a$ and the discount factor is $\beta \in (0, 1)$. We summarize the data of the stationary Markov Decision Problem as follows:

- $E := \mathbb{Z}$ where $x \in E$ denotes the risk reserve,
- $A := \mathbb{N}_0$ where $a \in A$ is the dividend pay-out,
- $D(x) := \{0, 1, \dots, x\}, x \geq 0$, and $D(x) := \{0\}, x < 0$,
- $\mathcal{Z} := \mathbb{Z}$ where $z \in \mathcal{Z}$ denotes the change of the risk reserve in (9.3),
- $T(x, a, z)$ is given in (9.3),
- $Q^Z(\{k\}|x, a) := q_k$ (independent of (x, a)),
- $r(x, a) := a$, $a \in D(x)$,
- $\beta \in (0, 1)$.

When we define the *ruin time* by

$$\tau := \inf\{n \in \mathbb{N}_0 \mid X_n < 0\}$$

then for a policy $\pi = (f_0, f_1, \dots) \in F^\infty$ we obtain

$$J_{\infty\pi}(x) = \mathbb{E}_x^\pi \left[\sum_{k=0}^{\tau-1} \beta^k f_k(X_k) \right].$$

Obviously $J_{\infty\pi}(x) = 0$ if $x < 0$. In order to have a well-defined and non-trivial model we assume that for $Z := Z_1$

$$\mathbb{P}(Z < 0) > 0 \quad \textit{and} \quad \mathbb{E}\, Z^+ < \infty.$$

Remark 9.2.1. Suppose we allow that $A = \mathbb{R}_+$ and $D(x) = [0, x]$ in the above setting, i.e. we allow for an arbitrary, not necessarily integer dividend pay out. In this case we would also have $E = \mathbb{R}$. However, it is not difficult to see (cf. Schmidli (2008), Lemma 1.9) that for initial risk reserve $x \in \mathbb{N}_0$ and an arbitrary policy $\pi \in F^\infty$ we can define a policy $\pi' \in F^\infty$ by setting $f_0'(x) := x - \lfloor x - f_0(x) \rfloor$ and $f_n'(x) := \lfloor x \rfloor - \lfloor x - f_n(x) \rfloor$ for $n \geq 1$ such that $J_{\infty\pi} \leq J_{\infty\pi'}$. Thus, we can restrict to the integer setting. ◊

Next we show that the Convergence Assumption (C) is satisfied for our infinite horizon Markov Decision Model.

Lemma 9.2.2. *a) The function* $b(x) := 1 + x$, $x \geq 0$ *and* $b(x) := 0$, $x < 0$ *is a bounding function with*

$$\mathcal{T}_\circ^n b \leq \beta^n \Big(b + n \mathbb{E}\, Z^+ \Big), \quad n \in \mathbb{N}.$$

b) For $x \geq 0$ *we obtain*

$$\delta(x) \leq x + \frac{\beta \mathbb{E}\, Z^+}{1 - \beta},$$

and hence $\delta \in \mathbb{B}_b$.

Proof. a) It is obvious that r is bounded by b and we get for $x \geq 0$

$$\int b(x') Q(dx'|x,a) = 1 + \sum_{k=a-x}^{\infty} (x - a + k) q_k \leq 1 + x + \mathbb{E}\, Z^+ \tag{9.4}$$

which implies that b is a bounding function (the statement for $x < 0$ is obvious). The second statement can by shown by induction on n. Note that the statement for $n = 1$ is obtained by (9.4).

b) Now we consider the same problem with Q replaced by $\tilde{Q}$ where $\tilde{q}_k := q_k$ for $k > 0$ and

$$\tilde{q}_0 := \sum_{k=-\infty}^{0} q_k,$$

i.e. $\tilde{Q}$ is concentrated on $\mathbb{N}_0$. Since $Q \leq_{st} \tilde{Q}$, all conditions of Theorem 2.4.14 are satisfied and it follows with Theorem 2.4.23 that

$$J_n(x) \leq \tilde{J}_n(x), \quad x \in E.$$

Since the Markov Decision Model is *positive* we obtain by taking limits

$$\delta(x) = J_\infty(x) \leq \tilde{J}_\infty(x).$$

For the problem with $\tilde{Q}$ the stationary policy $(f, f, \ldots)$ with $f(x) = x^+$, is optimal. Therefore,

$$\tilde{J}(x) = \tilde{J}_f(x) = x + \beta \mathbb{E}\, Z^+ + \beta^2 \mathbb{E}\, Z^+ + \ldots = x + \frac{\beta \mathbb{E}\, Z^+}{1 - \beta}$$

which implies b). □

In view of Lemma 9.2.2 the Integrability Assumption (A) and the Convergence Assumption (C) are satisfied since $\mathcal{T}_\circ^n b \to 0$ for $n \to \infty$, and $\mathbb{M} := \mathbb{B}_b$ fulfills (SA). Moreover, Theorem 7.1.8 yields that $\lim_{n\to\infty} J_n = J_\infty$ and

$$J_\infty(x) = \mathcal{T} J_\infty(x) = \max_{a\in\{0,1,\ldots,x\}} \left\{ a + \beta \sum_{k=a-x}^{\infty} J_\infty(x-a+k) q_k \right\}, \quad x \geq 0.$$

Obviously, $J_\infty(x) = 0$ for $x < 0$. Further, every maximizer of J_∞ (which obviously exists) defines an optimal stationary policy $(f^*, f^*, \ldots)$. In what follows, let f^* be the largest maximizer of J_∞. It will be convenient to define

$$G(x) := \beta \sum_{k=-x}^{\infty} J_\infty(x+k) q_k, \quad x \geq 0.$$

Thus we can write

$$J_\infty(x) = \max_{a\in\{0,1,\ldots,x\}} \big\{ a + G(x-a) \big\}, \tag{9.5}$$

$$J_\infty(x) = f^*(x) + G\big(x - f^*(x)\big), \quad x \geq 0. \tag{9.6}$$

The following theorem summarizes some basic properties of the value function J_∞ and the optimal policy $(f^*, f^*, \ldots)$. Let $q_+ := \mathbb{P}(Z \geq 0)$.

Theorem 9.2.3. *a) It holds for $x \geq 0$ that*

$$x + \frac{\beta\, \mathbb{E}\, Z^+}{1 - \beta q_+} \leq J_\infty(x) \leq x + \frac{\beta\, \mathbb{E}\, Z^+}{1-\beta}.$$

b) The value function $J_\infty(x)$ is increasing and

$$J_\infty(x) - J_\infty(y) \geq x - y, \quad x \geq y \geq 0.$$

c) It holds for $x \geq 0$ that $f^*\big(x - f^*(x)\big) = 0$ and

$$J_\infty(x) - f^*(x) = J_\infty\big(x - f^*(x)\big).$$

Proof. a) Since $J_\infty(x) = \delta(x)$, Lemma 9.2.2 implies the upper bound. For the lower bound consider the stationary policy f^∞ with $f(x) := x^+$. Then

$$J_f(x) = x + \beta\, \mathbb{E}\, Z^+ + \beta^2 q_+\, \mathbb{E}\, Z^+ + \ldots = x + \frac{\beta\, \mathbb{E}\, Z^+}{1-\beta q_+} \leq J_\infty(x).$$

b) Theorem 2.4.14 directly implies that $J_n(x)$ is increasing for all $n \in \mathbb{N}$ and hence also $J_\infty(x)$. Using the optimality equation (9.5) we obtain for $0 \le y < x$

$$\begin{aligned} J_\infty(x) &= \max_{a \in \{0,1,\dots,x\}} \{a + G(x-a)\} \\ &= \max\big\{G(x), \dots, x-y-1+G(y+1),\ x-y+J_\infty(y)\big\} \\ &\ge x - y + J_\infty(y). \end{aligned}$$

c) It follows from (9.5) that $J_\infty\big(x - f^*(x)\big) \ge G\big(x - f^*(x)\big)$ which together with (9.6) implies

$$J_\infty(x) - f^*(x) \le J_\infty\big(x - f^*(x)\big) \le J_\infty(x) - f^*(x).$$

The last inequality is obtained from part b) for $y := x - f^*(x) \ge 0$. Thus, we have shown the stated equation for J_∞. This equation implies now with (9.6) that

$$J_\infty\big(x - f^*(x)\big) = 0 + G\big(x - f^*(x)\big).$$

Comparing it with (9.5) gives $f^*\big(x - f^*(x)\big) = 0$. Note that in state $x - f^*(x)$ the action $a = 0$ is the unique maximum point of (9.5). □

From Theorem 9.2.3 a) we immediately obtain the following corollary

Corollary 9.2.4. *a) If* $\mathbb{P}(Z \le 0) = 1$, *then* $J_\infty(x) = x^+$ *and* $f^*(x) = x^+$.
b) If $\mathbb{P}(Z \ge 0) = 1$, *then* $J_\infty(x) = x + \frac{\beta \mathbb{E} Z}{1-\beta}$ *for* $x \ge 0$ *and* $f^*(x) = x^+$.

Both statements have an easy explanation. If the random variables Z_n are all non-positive with probability one, the company is certainly not profitable and the best we can do is pay-out all the money in the beginning and stop business. If the risk reserve changes Z_n cannot be negative, then there is no risk of getting ruined and due to discounting it is optimal to pay-out the money as fast as possible.

Structure of the Optimal Policy

In this subsection we show that the optimal stationary policy $(f^*, f^*, \dots)$ is a so-called *band-policy.*

Definition 9.2.5. a) A stationary policy f^∞ is called a *band-policy*, if there exist numbers $n \in \mathbb{N}_0$ and $c_0, \dots c_n, d_1, \dots d_n \in \mathbb{N}_0$ such that $d_k - c_{k-1} \ge 2$ for $k = 1, \dots, n$ and $0 \le c_0 < d_1 \le c_1 < d_2 \le \dots < d_n \le c_n$ and

$$f(x) = \begin{cases} 0, & \text{if } x \le c_0 \\ x - c_k, & \text{if } c_k < x < d_{k+1} \\ 0, & \text{if } d_k \le x \le c_k \\ x - c_n, & \text{if } x > c_n. \end{cases}$$

b) Suppose f^∞ is a band-policy. The sets $\{c_{k-1}, \dots, d_k\}$ are called *waves* and $d_k - c_{k-1}$ is the *length of wave k*.

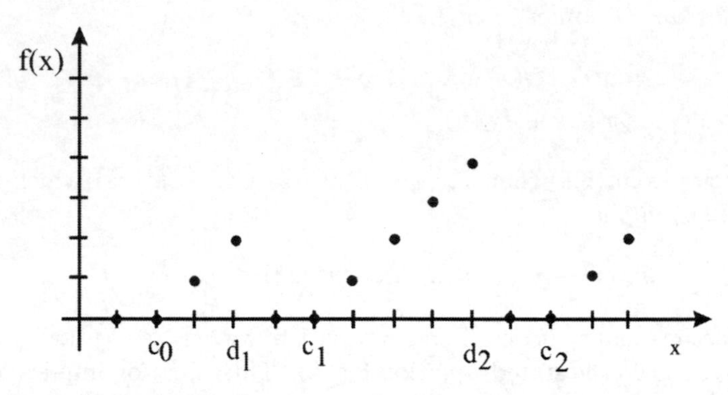

Fig. 9.1 Example of a band-policy.

A stationary policy f^∞ is called a *barrier-policy* if there exists a number $c \in \mathbb{N}_0$ such that

$$f(x) = \begin{cases} 0, & \text{if } x \le c \\ x - c, & \text{if } x > c. \end{cases}$$

Examples of a band-policy and a barrier-policy can be found in Figures 9.1 and 9.2.

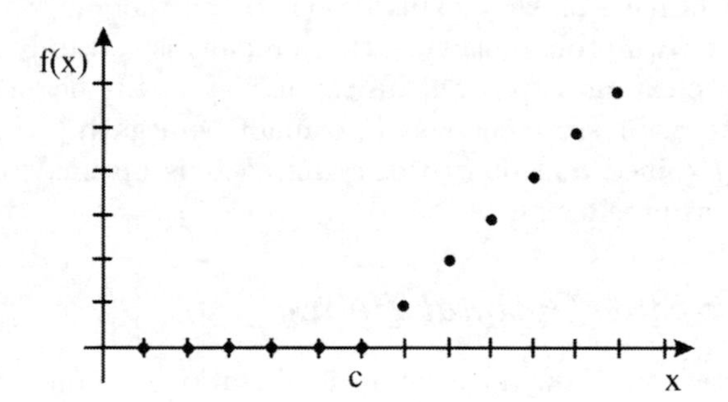

Fig. 9.2 Example of a barrier-policy.

A barrier-policy is a special band-policy with $c_n = c_0 = c$. In order to identify the structure of the optimal policy we need some preliminary results.

Proposition 9.2.6. *Let* $\xi := \sup\{x \in \mathbb{N}_0 \mid f^*(x) = 0\}$. *Then* $\xi < \infty$ *and*

$$f^*(x) = x - \xi \quad \text{for all } x \ge \xi.$$

Proof. For $x \geq 0$ with $f^*(x) = 0$ we obtain from Theorem 9.2.3

$$J_\infty(x) = G(x) = \beta \sum_{k=-x}^{\infty} q_k J_\infty(x+k) \leq \beta \sum_{k=-x}^{\infty} q_k \Big(x + k + \frac{\beta \mathbb{E} Z^+}{1-\beta}\Big)$$
$$\leq \beta x + \beta \mathbb{E} Z^+ + \frac{\beta^2 \mathbb{E} Z^+}{1-\beta} = \beta x + \frac{\beta \mathbb{E} Z^+}{1-\beta}$$

and

$$J_\infty(x) \geq x + \frac{\beta \mathbb{E} Z^+}{1-\beta q_+}.$$

Hence it holds

$$x \leq \frac{\beta^2 \mathbb{E} Z^+(1-q^+)}{(1-\beta)^2(1-\beta q^+)},$$

and ξ is finite. Moreover, $f^*(\xi) = 0$.
Now let $x \geq \xi$. From Theorem 9.2.3 we know that $f^*\big(x - f^*(x)\big) = 0$ which implies by the definition of ξ that $f^*(x) \geq x - \xi$. Since $x - f^*(x) \leq \xi \leq x$ it is admissible to pay out the dividend $f^*(x) - (x - \xi)$ in state ξ and we obtain by (9.6)

$$J_\infty(\xi) \geq f^*(x) - (x-\xi) + G\big(x - f^*(x)\big)$$
$$= J_\infty(x) - (x - \xi) \geq J_\infty(\xi).$$

Thus, we obtain $0 = f^*(\xi) \geq f^*(x) - (x - \xi)$. Together with the first inequality, we obtain $f^*(x) = x - \xi$ which concludes the proof. □

Remark 9.2.7. For $x \geq 0$ it holds $\mathbb{P}_x^{f^*}(\tau < \infty) = 1$, i.e. under the optimal stationary policy $(f^*, f^*, \ldots)$ ruin occurs almost surely. The proof of this statement is as follows. From Proposition 9.2.6 we know that the risk reserve after dividend payment is always less than or equal to ξ. The probability that there is a sequence of length $\xi + 1$ of negative incomes is positive:

$$\mathbb{P}_x^{f^*}(Z_1 < 0, \ldots, Z_{\xi+1} < 0) = (1 - q_+)^{\xi+1} > 0.$$

But this will lead to ruin and the probability is one that such a sequence occurs. ◊

Proposition 9.2.8. *Let $x_0 \geq 0$. If $f^*(x_0) = a_0$ and $f^*(x_0+1) > 0$, then $f^*(x_0 + 1) = a_0 + 1$.*

Proof. Since f^* is the largest maximizer of J_∞ it holds that

$$J_\infty(x_0) = a_0 + G(x_0 - a_0) \begin{cases} \geq a + G(x_0 - a) & \text{for } a = 0, \ldots, a_0 \\ > a + G(x_0 - a) & \text{for } a = a_0 + 1, \ldots, x_0. \end{cases} \tag{9.7}$$

Moreover we get

$$\begin{aligned} J_\infty(x_0+1) &= \max_{0\le a\le x_0+1} \{a+G(x_0+1-a)\} \\ &= \max\Big\{G(x_0+1)\,,\ \max_{0\le a\le x_0} \{a+1+G(x_0-a)\}\Big\}. \end{aligned}$$

The assumption $f^*(x_0+1)>0$ implies

$$G(x_0+1) \le \max_{0\le a\le x_0} \{a+1+G(x_0-a)\}. \tag{9.8}$$

From equation (9.7) we conclude

$$\begin{aligned} a+1+G(x_0-a) &\le a_0+1+G(x_0-a_0) \text{ for } a=0,\dots,a_0 \\ a+1+G(x_0-a) &< a_0+1+G(x_0-a_0) \text{ for } a=a_0+1,\dots,x_0 \end{aligned}$$

and by shifting the index

$$\begin{aligned} a+G(x_0+1-a) &\le a_0+1+G(x_0-a_0) \text{ for } a=0,\dots,a_0+1 \qquad (9.9)\\ a+G(x_0+1-a) &< a_0+1+G(x_0-a_0) \text{ for } a=a_0+2,\dots,x_0+1. \end{aligned}$$

Note that the first inequality (9.9) also holds for $a=0$ since by (9.7) and (9.8):

$$G(x_0+1) \le 1+\max_{0\le a\le x_0} \{a+G(x_0-a)\} = 1+a_0+G(x_0-a_0).$$

Now the last inequalities together with (9.6) imply $f^*(x_0+1)=a_0+1$, since f^* is the largest maximizer. □

Theorem 9.2.9. *The stationary policy $(f^*, f^*, \dots)$ is optimal and a band-policy.*

Proof. By Proposition 9.2.6 we have $f^*(x)=x-\xi$ for all $x\ge\xi$. For $x<\xi$ we have to distinguish different cases. If $f^*(x)=0$ for all $x=0,\dots,\xi$, then f^* is a barrier-policy. If there exists an $x_0<\xi$ such that $f^*(x)=0$ for $x=0,\dots,x_0-1$ and $f^*(x_0)>0$, then by Proposition 9.2.8 $f^*(x_0)=1$. If further $f^*(x_0+m)>0$ for $m=1,\dots,\xi-x_0-1$ then by induction

$$f^*(x_0+m) = f^*(x_0+m-1)+1 = \ldots = f^*(x_0)+m = m+1.$$

If $f^*(x_0+1)=0$ we either have $f^*(x)=0$ for $x=x_0+1,\dots,\xi$ or there exists an x_1 with $x_0<x_1<\xi$ and $f^*(x_1)>0$. Now we proceed in the same way as with x_0. After a finite number of steps we reach ξ. In any case the constructed policy is a band-policy. □

Theorem 9.2.10. *a) If* $\mathbb{P}(Z \geq -z_0) = 1$ *for some* $z_0 \in \mathbb{N}$*, then the length of the waves of* f^* *is bounded by* z_0*, i.e.*

$$c_{k+1} - d_k \leq z_0.$$

b) If $\mathbb{P}(Z \geq -1) = 1$ *then the stationary policy* $(f^*, f^*, \ldots)$ *is a barrier-policy.*

Proof. a) Suppose that f^∞ is a band-policy with parameters

$$0 \leq c_0 < d_1 \leq c_1 < d_2 \leq \ldots < d_n \leq c_n$$

and suppose there exists a $k_0 \leq n-1$ such that $d_{k_0+1} - c_{k_0} > z_0$. We consider now a history-dependent policy $\sigma = (g_0, g_1, \ldots)$ which is defined as follows

$$g_n(x_0, \ldots, x_n) := \begin{cases} 0 & \text{if } f(x_k + 1) = 0 \text{ for all } k \leq n \\ f(x_n) & \text{otherwise.} \end{cases}$$

After some cumbersome calculations it is possible to show that

$$J_f(d_{k_0+1}) < J_\sigma(d_{k_0+1} - 1) + 1.$$

Since $J_\sigma \leq J_\infty$ we obtain

$$J_f(d_{k_0+1}) < J_\infty(d_{k_0+1} - 1) + 1$$

which in view of Theorem 9.2.3 b) implies that f^∞ cannot be optimal. Hence the length of the waves of the optimal band-policy $(f^*, f^*, \ldots)$ is bounded by z_0.

b) Since the length of a wave is by definition at least 2 we obtain with part a) that no wave can exist. Hence the optimal band-policy is a barrier-policy. □

Remark 9.2.11. The problem where $Z \in \{-1, 1\}$ was considered by de Finetti (1957). The value function in this case is given by

$$J_\infty(x) = x + \frac{\beta p}{1 - \beta p}, \quad x \geq 0$$

where $p := \mathbb{P}(Z = 1) \in (0, 1)$. From Theorem 9.2.10 we conclude that a barrier-policy is optimal in this case. ◊

9.3 Terminal Wealth Problems in a Pure Jump Market

In this section we will reconsider the financial market of Section 3.2. We investigate a portfolio problem in this market which leads to a controlled Piecewise Deterministic Process. Using the results of Chapter 8 it is possible to solve the arising optimization problem by means of a discrete-time Markov Decision Model with infinite horizon.
In what follows we will consider the classical problem of maximizing the expected utility of terminal wealth at time T in the continuous-time jump market of Section 3.2. Recall that the price processes are for $t \in [0,T]$ defined as follows:

- The price process (S_t^0) of the riskless bond is given by

$$S_t^0 := e^{\rho t},$$

  where $\rho \geq 0$ denotes the continuous interest rate.
- The price processes (S_t^k) of the risky assets $k = 1, \dots, d$ are given by

$$dS_t^k = S_{t-}^k \left(\mu_k dt + dC_t^k\right)$$

  where $\mu_k \in \mathbb{R}$ and (C_t^k) is the k-th component of the multivariate compound Poisson process $C_t := \sum_{n=1}^{N_t} Y_n$ (see Section 3.2). The initial prices S_0^k are assumed to be strictly positive.

By $(\mathcal{F}_t)$ we denote the filtration which is generated by (S_t) where $S_t := (S_t^1, \dots, S_t^k)$. Then it holds $\mathcal{F}_t = \mathcal{F}_t^C$. Recall that the support of the distribution of Y_n is given by $(-1, \infty)^d$. This implies that the stock prices stay positive. Only those portfolio strategies are admissible which guarantee that the wealth process is almost surely positive. As a consequence, in models with jumps short-sellings are often prohibited, and we will also assume this in our analysis. The amount of money $a_t := (a_t^1, \dots, a_t^d)$ invested in the stocks when the current wealth is X_t, has to satisfy

$$a_t \geq 0 \quad \text{and} \quad a_t \cdot e \leq X_t.$$

In this situation the wealth process will always stay positive and it obviously makes no difference whether a portfolio is given in terms of amounts which are invested in the assets or in terms of fractions. In what follows we will mainly focus on fractions, i.e. we define

$$\pi_t^k := \frac{a_t^k}{X_t}.$$

Thus, the set of admissible fractions of wealth which can be invested in the stocks is given by

$$\mathcal{U} := \{u \in \mathbb{R}^d \mid u \geq 0,\ u \cdot e \leq 1\}.$$

Note that this set is compact. A portfolio strategy is given by an $(\mathcal{F}_t)$-predictable stochastic process $\pi = (\pi_t)$ with values in $\mathcal{U}$ where $\pi_t := (\pi_t^1, \ldots, \pi_t^d)$ gives the fractions of wealth invested in the stocks at time t. The quantity $1 - \pi_t \cdot e$ is the fraction invested in the bond. The dynamics of the wealth process is then given by

$$dX_t = X_{t-}\Big(\big(\rho + \pi_t \cdot (\mu - \rho e)\big)dt + \pi_t dC_t\Big), \tag{9.10}$$

see also equation (3.7).
The aim of the investor is now to maximize her expected utility of terminal wealth (at time T). Thus, we denote by $U : [0, \infty) \to \mathbb{R}_+$ a strictly increasing and strictly concave utility function and define for a portfolio strategy $\pi = (\pi_t)$ and $(t, x) \in E := [0, T] \times \mathbb{R}_+$

$$V_\pi(t, x) := \mathbb{E}_{tx}^\pi[U(X_T)],$$

the expected utility of terminal wealth when $X_t = x$. The maximal expected utility is given by

$$V(t, x) := \sup_\pi V_\pi(t, x). \tag{9.11}$$

Obviously it holds

$$V_\pi(T, x) = U(x) = V(T, x).$$

Throughout this section we assume that $\mathbb{E}\,\|Y_n\| < \infty$. Note that there exists no arbitrage strategy in the class of admissible portfolio strategies.
Since (9.11) is a Piecewise Deterministic Markov Decision Problem we can use the results of Chapter 8. However, this time we have a problem with finite time horizon which implies that the time component will enter the state space. We will first reduce the problem to a discrete-time Markov Decision Process. Thus, let us define

$$A := \{\alpha : [0, T] \to \mathcal{U} \text{ measurable}\}. \tag{9.12}$$

For $\alpha \in A$ the movement of the wealth between jumps is then given by

$$\phi_t^\alpha(x) = x \exp\left(\int_0^t \big(\rho + \alpha_s \cdot (\mu - \rho e)\big)ds\right), \quad t \geq 0 \tag{9.13}$$

since by (9.10) the drift has the form $\mu(x, u) := x\big(\rho + u \cdot (\mu - \rho e)\big)$. The function $\phi_t^\alpha(x)$ gives the wealth t time units after the last jump, when the state directly after the jump was x. In what follows denote by (T_n) the jump time points of the wealth (stocks) and let (Z_n) be the post-jump wealths. As in Chapter 8 a *Markov portfolio strategy* $\pi = (\pi_t)$ is defined by a sequence of measurable functions (f_n) with $f_n : E \to A$ such that

$$\pi_t = f_n(T_n, Z_n)(t - T_n) \text{ for } t \in (T_n, T_{n+1}].$$

We write $\pi = (\pi_t) = (f_n)$ and it holds

$$X_t = \phi^{\pi}_{t-T_n}(Z_n) := \phi^{f_n(T_n,Z_n)}_{t-T_n}(Z_n) \quad \text{for } t \in [T_n, T_{n+1}).$$

Note that $Z_n = X_{T_n}$. Given a strategy π and an initial wealth $x > 0$ there is a probability space $(\Omega, \mathcal{F}, \mathbb{P}^{\pi}_x)$ on which the random variables T_n and Z_n are defined such that $X_0 = Z_0 = x$ and for all Borel sets $C \subset (0, \infty)$

$$\begin{aligned}
&\mathbb{P}^{\pi}_x\big(T_{n+1} - T_n \le t, Z_{n+1} \in C \mid T_0, Z_0, \dots, T_n, Z_n\big) \\
&= \lambda \int_0^t e^{-\lambda s}\Big[\int 1_C\Big(\phi^{\pi}_s(Z_n)\big(1 + f_n(T_n, Z_n)(s) \cdot y\big)\Big) Q_Y(dy)\Big] ds.
\end{aligned}$$

We introduce now the following *infinite horizon discrete-time Markov Decision Model* in the same spirit as in Section 8.2. Since the time horizon is finite, the state now contains a time component.

- The state space $E = [0, T] \times \mathbb{R}_+$ is endowed with the Borel σ-algebra. A state $(t, x) \in E$ gives the jump time point t and the wealth x of the process directly after the jump. The state process of the Markov Decision Model is denoted by (T'_n, Z'_n). It coincides with the embedded process (T_n, Z_n) as long as $T_n \le T$. Since we are only interested in the state process up to time T, we fix some external state $\Delta \notin E$ (cemetery state) and set $(T'_n, Z'_n) := \Delta$ whenever $T_n > T$.
- The action space is given by A defined in (9.12). Recall that A is a Borel space.
- For $(t, x) \in E$, $\alpha \in A$ and a Borel set $B \subset E$ the transition probability Q is given by

$$\begin{aligned}
Q(B|t, x, \alpha) &:= \lambda \int_0^{T-t} e^{-\lambda s}\Big[\int 1_B\Big(t + s, \phi^{\alpha}_s(x)\big(1 + \alpha_s \cdot y\big)\Big) Q_Y(dy)\Big] ds \\
Q(\{\Delta\}|t, x, \alpha) &:= e^{-\lambda(T-t)} = 1 - Q(E|t, x, \alpha) \\
Q(\{\Delta\}|\Delta, \alpha) &:= 1.
\end{aligned}$$

- The one-stage reward function is for $(t, x) \in E$ and $\alpha \in A$ defined by

$$\begin{aligned}
r(t, x, \alpha) &:= e^{-\lambda(T-t)} U\big(\phi^{\alpha}_{T-t}(x)\big), \\
r(\Delta, \alpha) &:= 0.
\end{aligned}$$

Note that the reward function is non-negative and unbounded. A policy (f_n) for the discrete-time model consists of a sequence of decision rules f_n which are measurable mappings $f_n : E \cup \{\Delta\} \to A$ (where we set $f(\Delta) := \alpha_0 \in A$, α_0 arbitrary) and we denote by F the set of all decision rules. The expected reward of a policy (f_n) is given by

$$J_{\infty(f_n)}(t,x) := \mathbb{E}_{tx}^{(f_n)}\left[\sum_{k=0}^{\infty} r(T'_k, Z'_k, f_k(T'_k, Z'_k))\right], \quad (t,x) \in E$$

and we define the value function of the discrete-time Markov Decision Model by

$$J_\infty(t,x) := \sup_{(f_n)\in F^\infty} J_{\infty(f_n)}(t,x), \quad (t,x) \in E.$$

The continuous-time optimization problem with finite horizon T can be solved by the infinite-stage Markov Decision Model above (cp. Chapter 8).

Theorem 9.3.1. *a) For a Markov portfolio strategy $\pi = (\pi_t) = (f_n)$ we have*

$$V_\pi(t,x) = J_{\infty(f_n)}(t,x), \quad (t,x) \in E.$$

b) It holds: $V = J_\infty$.

Proof. a) For a Markov portfolio strategy $\pi = (\pi_t) = (f_n)$ we have

$$X_t = \phi^\pi_{t-T_k}(Z_k) \text{ for } t \in [T_k, T_{k+1}),$$

and $(T_k, Z_k) = (T'_k, Z'_k)$ as long as $T_k \le T$. The sequence (f_n) defines a policy for the discrete-time Markov Decision Model. We consider here only the time point $t = 0$. Arbitrary time points t can be treated analogously. Using monotone convergence we obtain

$$\begin{aligned}
V_\pi(0,x) &= \mathbb{E}_x^\pi[U(X_T)] = \mathbb{E}_x^\pi\left[\sum_{k=0}^{\infty} 1_{[T_k \le T < T_{k+1}]} U(X_T)\right] \\
&= \sum_{k=0}^{\infty} \mathbb{E}_x^\pi\left[\mathbb{E}_x^\pi\left[1_{[T_k \le T < T_{k+1}]} U\big(\phi^\pi_{T-T_k}(Z_k)\big) \,\Big|\, T_k \le T, Z_k\right]\right] \\
&= \mathbb{E}_{0x}^{(f_n)}\left[\sum_{k=0}^{\infty} r\Big(T'_k, Z'_k, f_k(T'_k, Z'_k)\Big)\right] = J_{\infty(f_n)}(0,x)
\end{aligned}$$

since

$$\mathbb{E}_x^\pi\left[1_{[T_k \le T < T_{k+1}]} U\big(\phi^\pi_{T-T_k}(Z_k)\big) \,\Big|\, T_k \le T, Z_k\right] = r\big(T'_k, Z'_k, f_k(T'_k, Z'_k)\big)$$

and the transition kernel is given by

$$\mathbb{P}_{0x}^{(f_n)}\Big((T'_{k+1}, Z'_{k+1}) \in B \,\Big|\, T'_k, Z'_k\Big) = Q\Big(B \,\Big|\, T'_k, Z'_k, f_k(T'_k, Z'_k)\Big)$$

for all Borel sets $B \subset E$.

b) For any (history-dependent) policy π there exists a sequence (f_n) of measurable functions $f_n : E^{n+1} \to A$ with $\pi_t = f_n(T_0, Z_0, \ldots, T_n, Z_n)(t - T_n)$

for $t \in (T_n, T_{n+1}]$. Due to the Markovian structure of the state process, we obtain (see Bertsekas and Shreve (1978), p.216)

$$V(t,x) = \sup_{\pi \text{ Markov}} V_\pi(t,x) = J_\infty(t,x), \quad (t,x) \in E$$

where the last equation follows from part a).

□

Hence we have reduced the continuous-time terminal wealth problem to a discrete-time Markov Decision Model which will be solved with the theory developed in Chapter 7. In the following we write $\bar{\mu} := \max\{\mu_1, \dots, \mu_d, \rho\}$ and $\bar{y} := \max\{\mathbb{E}\, Y_1, \dots \mathbb{E}\, Y_d, 0\}$ for the maximal appreciation rate of the stocks and for the maximal expected relative jump height, respectively.

Proposition 9.3.2. *a) The function* $b(t,x) := e^{\gamma(T-t)}(1+x)$ *is a* bounding function *for the discrete-time Markov Decision Model for all* $\gamma \geq 0$.
b) We obtain

$$\alpha_b \leq \frac{\lambda(1+\bar{y})}{\gamma + \lambda - \bar{\mu}}\Big(1 - e^{-T(\gamma+\lambda-\bar{\mu})}\Big) =: \alpha_\gamma. \tag{9.14}$$

In particular for γ *large enough, we have* $\alpha_\gamma < 1$ *and the discrete-time model is contracting.*

Proof. Since $\mathcal{U}$ is compact there exists a $c_0 > 0$ such that

$$\alpha_s \cdot (\mu - \rho e) \leq c_0, \quad \text{for all } \alpha \in A, s \geq 0.$$

Moreover the concave utility function U can be bounded from above by an affine-linear function $c_1(1+x)$ for some $c_1 > 0$. Thus, we obtain

$$\begin{aligned} 0 \leq r\big(t,x,\alpha\big) &= e^{-\lambda(T-t)} U\big(\phi^\alpha_{T-t}(x)\big) \leq e^{-\lambda(T-t)} U\big(x e^{T(\rho+c_0)}\big) \\ &\leq e^{-\lambda(T-t)} c_1\big(1 + x e^{T(\rho+c_0)}\big) \leq c_r b(t,x) \end{aligned}$$

for c_r large enough. Moreover, we have

$$\begin{aligned} \alpha_b &= \sup_{(t,x,\alpha)\in D} \frac{\int b(s,y) Q(ds,dy|t,x,\alpha)}{b(t,x)} \\ &= \sup_{(t,x,\alpha)\in D} \frac{\lambda \int_0^{T-t} e^{-\lambda s + \gamma(T-s-t)}\Big(1 + \phi^\alpha_s(x)\big(1 + \alpha_s \cdot \int y\, \nu(dy)\big)\Big) ds}{e^{\gamma(T-t)}(1+x)} \end{aligned}$$

$$\leq \sup_{(t,x)\in E} \frac{\lambda \int_0^{T-t} e^{-s(\lambda+\gamma)}\Big(1 + xe^{\bar{\mu}s}(1+\bar{y})\Big)ds}{1+x}$$

$$= \frac{\lambda(1+\bar{y})}{\gamma+\lambda-\bar{\mu}}\Big(1 - e^{-T(\gamma+\lambda-\bar{\mu})}\Big) = \alpha_\gamma.$$

This implies that b is a bounding function. Also part b) is shown. □

The operator $\mathcal{T}$ of the Markov Decision Model has the form

$$(\mathcal{T}v)(t,x) = \sup_{\alpha\in A}\Big\{e^{-\lambda(T-t)}U\big(\phi^\alpha_{T-t}(x)\big) \tag{9.15}$$
$$+ \lambda \int_0^{T-t} e^{-\lambda s}\Big[\int v\Big(t+s, \phi^\alpha_s(x)\big(1+\alpha_s\cdot y\big)\Big)Q_Y(dy)\Big]ds\Big\}.$$

Proposition 9.3.2 implies that $\mathcal{T}: I\!B_b \to I\!B_b$ is contracting with module $\alpha_b \leq \alpha_\gamma < 1$ if γ is large enough which we assume from now on. Later we will see that the value function $V(t,x)$ is the unique fixed point of $\mathcal{T}$ in a certain subset of $I\!B_b$. As in Chapter 8 we define for relaxed controls $\alpha \in \mathcal{R} := \{\alpha : [0,T] \to I\!P(\mathcal{U}) \text{ measurable}\}$

$$\bar{\alpha}_s := \int_{\mathcal{U}} u\alpha_s(du), \quad s\in[0,T].$$

Note that $\bar{\alpha} \in A$ since $\mathcal{U}$ is convex, and $\phi^\alpha_t(x) = \phi^{\bar{\alpha}}_t(x)$ since the drift $\mu(x,u)$ is linear in u. Moreover, we obtain $r(t,x,\alpha) = r(t,x,\bar{\alpha})$. The transition kernel Q of the Markov Decision Model is extended for $\alpha \in \mathcal{R}$ by

$$\int v(s,y)Q\big(ds,dy \mid t,x,\alpha\big)$$
$$= \lambda\int_0^{T-t} e^{-\lambda s}\int\Big[\int v\Big(t+s,\phi^\alpha_s(x)\big(1+u\cdot y\big)\Big)\alpha_s(du)\Big]Q_Y(dy)ds$$

for all $v \in I\!B_b$. Finally, we consider the set

$$I\!M_{cv} := \{v \in I\!B_b \mid v \text{ is continuous, } v(t,x) \text{ is concave and increasing in } x \text{ and decreasing in } t, v \geq U\}.$$

Proposition 9.3.3. *The sets $I\!M_{cv}$ and $\Delta := \{f : E \to A \text{ measurable}\}$ satisfy the assumptions of Theorem 7.3.5.*

Proof. First we remark that for $v \in I\!M_{cv}$ we have

$$\sup_{\alpha\in\mathcal{R}} Lv(t,x,\alpha) = \sup_{\alpha\in A} Lv(t,x,\alpha), \quad (t,x)\in E. \tag{9.16}$$

This follows since Jensen's inequality implies

$$Lv(t, x, \alpha) \leq Lv(t, x, \bar{\alpha}), \quad \alpha \in \mathcal{R}.$$

We have to check the conditions (i)–(iii) of Theorem 7.3.5. It is clear that $U \in I\!M_{cv}$. Here we can choose U instead of $v = 0$ since the Markov Decision Model is contracting. If $v \in I\!M_{cv}$ then $\mathcal{T}v \geq U$ and $\mathcal{T}v(t, x)$ is increasing in x and decreasing in t. This can be seen from the definition of $\mathcal{T}$.
Let us prove that $x \mapsto \mathcal{T}v(t, x)$ is concave. So far we have worked with portfolios in terms of fractions of invested wealth. Since our model guarantees that the wealth process never falls to zero (given $x > 0$) we can equivalently work with invested amounts a_t. More precisely, the fraction $\alpha \in A$ gives the same wealth as the amount $a_t := \alpha_t \phi_t^\alpha(x)$. Under $a = (a_t)$, the deterministic evolution between jumps is given by

$$\phi_t^a(x) := e^{\rho t}\left(x + \int_0^t e^{-\rho s} a_s \cdot (\mu - \rho e) ds\right).$$

The advantage is now that $(x, a) \mapsto \phi_t^a(x)$ is linear. We show first that $(x, a) \mapsto Lv(t, x, a)$ is concave. The concavity of $x \mapsto \mathcal{T}v(t, x)$ then follows as in Theorem 2.4.19. Fix $t \in [0, T]$, wealths $x_1, x_2 > 0$, controls a_1, a_2 and $\kappa \in (0, 1)$. Let $\hat{x} := \kappa x_1 + (1 - \kappa)x_2$ and $\hat{a} := \kappa a_1 + (1 - \kappa)a_2$. Note that $\hat{a}$ is again admissible and that

$$\phi_s^{\hat{a}}(\hat{x}) = \kappa \phi_s^{a_1}(x_1) + (1 - \kappa)\phi_s^{a_2}(x_2).$$

Then we obtain

$$\begin{aligned} Lv(t, \hat{x}, \hat{a}) &= e^{-\lambda(T-t)} U\big(\phi_{T-t}^{\hat{a}}(\hat{x})\big) \\ &\quad + \lambda \int_0^{T-t} e^{-\lambda s} \int v\Big(t + s, \phi_s^{\hat{a}}(\hat{x}) + \hat{a}_s \cdot y\Big) Q_Y(dy) ds \\ &\geq \kappa\, Lv(t, x_1, a_1) + (1 - \kappa)\, Lv(t, x_2, a_2) \end{aligned}$$

which is the desired statement.
Next we use Proposition 2.4.8 and Proposition 9.3.2 to show that $\mathcal{T}v$ is continuous for $v \in I\!M_{cv}$. For this task we use relaxed controls and the representation of $\mathcal{T}$ by (9.16). Due to Remark 8.2.3 we know that $\mathcal{R}$ is compact with respect to the Young topology. It remains to show that the functions

$$E \times \mathcal{R} \ni (t, x, \alpha) \mapsto r(t, x, \alpha) \tag{9.17}$$

$$E \times \mathcal{R} \ni (t, x, \alpha) \mapsto \int v(s, y) Q(ds, dy | t, x, \alpha) \tag{9.18}$$

are both continuous for all $v \in I\!M_{cv}$. First we show that the function

$$(t, x, \alpha) \mapsto \phi_t^\alpha(x)$$

is continuous. To this end, let $(t_n, x_n) \subset E$ and $(\alpha_n) \subset \mathcal{R}$ be sequences with $(t_n, x_n) \to (t, x) \in E$ and $\alpha_n \to \alpha \in \mathcal{R}$. Since

$$\phi_t^\alpha(x) = xe^{\rho t} \exp\Big(\int_0^t \int_{\mathcal{U}} u \cdot (\mu - \rho e)\alpha_s(du)ds\Big)$$

and the exponential function is continuous, it suffices to show the continuity of the integral expression. To ease notation we define

$$\mu_s^n := \int_{\mathcal{U}} u \cdot (\mu - \rho e)\alpha_s^n(du) \quad \text{and} \quad \mu_s := \int_{\mathcal{U}} u \cdot (\mu - \rho e)\alpha_s(du).$$

Thus we look at

$$\left|\int_0^{t_n} \mu_s^n ds - \int_0^t \mu_s ds\right| \leq \left|\int_0^{t_n} \mu_s^n ds - \int_0^t \mu_s^n ds\right| + \left|\int_0^t \mu_s^n ds - \int_0^t \mu_s ds\right|.$$

Since $u \mapsto w(s,u) := u \cdot (\mu - \rho e)$ for $0 \leq s \leq t$ and $u \mapsto w(s,u) = 0$ for $t \leq s \leq T$ are continuous, it follows from Remark 8.2.3 that the second term tends to 0 for $n \to \infty$. Obviously, the first term is bounded by $\hat{c}|t - t_n|$ which also tends to 0 and the continuity of $(t, x, \alpha) \mapsto \phi_t^\alpha(x)$ is shown.
Since U is continuous it follows from the continuity of $\phi_t^\alpha(x)$ that the function (9.17) is continuous.
The continuity of the function in (9.18) follows from the following *auxiliary result* which can be shown similarly as Lemma 8.2.5:
Let $w : E \times \mathcal{U} \to \mathbb{R}_+$ be continuous with $w(t, x, u) \leq c_w b(t, x)$ for some $c_w \geq 0$. Then

$$(t, x, \alpha) \mapsto \int_0^{T-t} e^{-\lambda s} \int \Big[\int w\Big(t + s, \phi_s^\alpha(x)(1 + u \cdot y), u\Big)\alpha_s(du)\Big] Q_Y(dy)ds$$

is continuous on $E \times \mathcal{R}$. Note that $b(t, x)$ is continuous.
Altogether, we have shown condition (ii). Condition (iii) follows from Theorem 2.4.6. Last but not least $\mathbb{M}_{cv}$ is a closed subset of $\mathbb{B}_b$. □

The main results for the terminal wealth problem are summarized in the next theorem.

Theorem 9.3.4. *a) The value function $V(t, x)$ of the terminal wealth problem satisfies*

$$V = J_\infty = J \in \mathbb{M}_{cv}$$

where $J := \lim_{n\to\infty} \mathcal{T}^n U$.
b) $V(t, x)$ is the unique fixed point of $\mathcal{T}$ in $\mathbb{M}_{cv}$.
c) It holds for $g \in \mathbb{M}_{cv}$ that

$$\|V - \mathcal{T}^n g\|_b \leq \frac{\alpha_b^n}{1 - \alpha_b}\|\mathcal{T}g - g\|_b.$$

d) There exists an optimal Markov portfolio strategy $\pi^* = (\pi_t^*)$ *such that*

$$\pi_t^* = f^*(T_n, Z_n)(t - T_n), \quad t \in (T_n, T_{n+1}],$$

for a decision rule $f^* : E \to A$.

e) The policy iteration *holds.*

f) Howard's policy improvement algorithm holds.

Proof. Part a)–d) follow from Proposition 9.3.3 and Theorem 7.3.5. Part e) is deduced from Theorem 7.3.6 and the last statement follows from Theorem 7.5.1. □

The optimal portfolio strategy π^* is predictable and characterized by one decision rule f^*. Between two jumps π_t^* depends on the last jump time point T_n, X_{T_n} and $t - T_n$. For solving the terminal wealth problem we can use value iteration, policy iteration or Howard's policy improvement algorithm. More computational aspects are discussed in the following section.

Computational Aspects

Now we want to solve the fixed point equation in some special cases or approximately. To this end, let us introduce the following notation for $v \in \mathbb{B}_b$

$$\ell_v(t, x, u) := \lambda \int v\Big(t, x(1 + u \cdot y)\Big) Q_Y(dy), \quad (t, x) \in E, u \in \mathcal{U}.$$

Then we can write the fixed point equation $V = \mathcal{T}V$ (or Bellman equation) for $(t, x) \in E$ in the following form:

$$V(t, x) = \sup_{\alpha \in A} \Big\{ e^{-\lambda(T-t)} U(\phi_{T-t}^\alpha(x)) + \int_0^{T-t} e^{-\lambda s} \ell_V(t + s, \phi_s^\alpha(x), \alpha_s) ds \Big\}.$$

This is a *deterministic control problem* which can be solved explicitly for special utility functions.

Example 9.3.5 (Power Utility). Let $U(x) := x^\beta$ with $0 < \beta < 1$. In this case we obtain the explicit solution:

$$V(t, x) = x^\beta e^{\delta(T-t)}, \quad (t, x) \in E$$
$$\pi_t^* \equiv u^*, t \in [0, T]$$

where u^* is the maximum point of

$$u \mapsto \beta u \cdot (\mu - \rho e) + \lambda \int (1 + u \cdot y)^\beta Q_Y(dy)$$

on $\mathcal{U}$ and $\delta := \beta\rho - \lambda + \beta u^* \cdot (\mu - \rho e) + \lambda \int (1 + u^* \cdot y)^\beta Q_Y(dy)$. ♦

Uncontrolled Drift
Let us assume that $\rho = \mu_i$ for all $i = 1, \dots, d$ which means that the deterministic drift of all assets is equal to the drift of the bond (or $\mu(x,u) = x\rho$). In this case we obtain $\phi_t^\alpha(x) = xe^{\rho t}$, and the fixed point equation reduces to:

$$V(t,x) = e^{-\lambda(T-t)} U(xe^{\rho(T-t)}) + \int_0^{T-t} e^{-\lambda s} \sup_{u \in \mathcal{U}} \{\ell_V(t+s, xe^{\rho s}, u)\} ds.$$

Note that there exists an $f^* \in F$ such that

$$f^*(t,x)(s) := \text{argmax}_{u \in \mathcal{U}} \{\ell_V(t+s, xe^{\rho s}, u)\}, \quad (t,x) \in E.$$

The stationary policy $(f^*, f^*, \dots)$ defines an optimal portfolio strategy (see Theorem 9.3.4 d)). In this case the value iteration is rather easy to execute as the maximization problem boils down to a pointwise maximization.

Moreover, it is not difficult to obtain necessary and sufficient conditions for the policy *invest all the money in the bond* to be optimal. Denote this decision rule by $f_* \equiv 0$. The corresponding value function is then

$$J_{f_*}(t,x) = U(xe^{\rho(T-t)}).$$

We try a policy improvement for this strategy.

$$\mathcal{T} J_{f_*}(t,x) = e^{-\lambda(T-t)} U(xe^{\rho(T-t)}) \\ +\lambda \int_0^{T-t} e^{-\lambda s} \sup_{u \in \mathcal{U}} \Big\{ \int U\Big(xe^{\rho(T-t-s)}(1 + u \cdot y)\Big) Q_Y(dy) ds \Big\}.$$

The maximizer is again f_* if and only if for all time points $s \in [0, T-t]$ it holds that

$$\sup_{u \in \mathcal{U}} \mathbb{E}\, U\Big(xe^{\rho(T-t-s)}(1 + u \cdot Y)\Big) \le U\Big(xe^{\rho(T-t-s)}\Big).$$

This condition can be interpreted as a *one-step-look-ahead* rule. We compare the utility of investing all the money in the bond after the next jump with what we get under an arbitrary policy. Due to the concavity of U this condition is always fulfilled if $\mathbb{E} Y \le 0$, i.e. if the stocks are on average not better than the bond, there is no incentive to invest in the stock for a risk averse investor. Assuming moreover that the utility function is differentiable we obtain:

Theorem 9.3.6. *If U is continuously differentiable and $U'(x + u \cdot Y)Y$ is integrable for all $x > 0$ and $\|u\|$ small, then 'invest all the money in the bond' is optimal if and only if $\mathbb{E} Y \le 0$.*

Proof. Since the mapping

$$u \mapsto \mathbb{E}\, U\Big(xe^{\rho(T-t-s)}(1+u\cdot Y)\Big) =: g(u)$$

is concave, $u \equiv 0$ is a maximum point if and only if $g'(0) \le 0$. Differentiating g (the condition in the proposition is such that differentiation and integration can be interchanged) and noting that $U' \ge 0$, it follows that $g'(0) \le 0$ if and only if $\mathbb{E}\, Y \le 0$. □

Approximation of the Utility Function

Another possibility to approximate the value function is as follows. Suppose that instead of the utility function U we take $\tilde{U}$ which is somehow close to U and the terminal wealth problem with $\tilde{U}$ is simpler to solve. Then one would expect that also the corresponding value function and optimal policy are close to each other. In order to formalize this idea let $U^{(n)}$ be a utility function and define

$$\mathcal{T}^{(n)}v(t,x) := \sup_{\alpha\in\mathcal{R}} \Big\{ e^{-\lambda(T-t)}U^{(n)}\big(\phi^{\alpha}_{T-t}(x)\big) + \int v(s,y)Q(ds,dy|t,x,\alpha)\Big\},$$

where we replace U by $U^{(n)}$ in $\mathcal{T}$ for $n \in \mathbb{N}$. $L^{(n)}$ is defined in an obvious way. Moreover, denote by $V^{(n)}$ the corresponding value function and by

$$A^*_n(t,x) := \{\alpha \in \mathcal{R} \mid \mathcal{T}^{(n)}V^{(n)}(t,x) = L^{(n)}V^{(n)}(t,x,\alpha)\}$$

$$A^*(t,x) := \{\alpha \in \mathcal{R} \mid \mathcal{T}V(t,x) = LV(t,x,\alpha)\}$$

the set of maximum points of the operator $\mathcal{T}^{(n)}$ and $\mathcal{T}$ respectively. Then we are able to state the following approximation result.

Theorem 9.3.7. *a) If U and $\tilde{U}$ are two utility functions with corresponding value functions V and $\tilde{V}$, then*

$$\|V - \tilde{V}\|_b \le \|U - \tilde{U}\|_b \frac{e^{T\bar{\mu}}}{1-\alpha_b}.$$

b) Let $\big(U^{(n)}\big)$ be a sequence of utility functions with $\lim_{n\to\infty}\|U^{(n)} - U\|_b = 0$. Then it holds

$$\emptyset \ne Ls A^*_n(t,x) \subset A^*(t,x) \text{ for all } (t,x) \in E,$$

i.e. in particular, the limit f^* of a sequence (f^*_n) with $f^*_n(t,x) \in A^*_n(t,x)$ for all $(t,x) \in E$ defines an optimal stationary policy for the given model (with utility function U).

Proof. For part a) we obtain with Theorem 9.3.4

$$\begin{aligned}\|V-\tilde{V}\|_b &= \|\mathcal{T}V-\tilde{\mathcal{T}}\tilde{V}\|_b \\ &\le \|U-\tilde{U}\|_b \sup_{(t,x,\alpha)\in D} e^{-\lambda(T-t)}\frac{b(t,\phi^\alpha_{T-t}(x))}{b(t,x)}+\alpha_b\|V-\tilde{V}\|_b \\ &\le \|U-\tilde{U}\|_b e^{\bar{\mu}T}+\alpha_b\|V-\tilde{V}\|_b.\end{aligned}$$

Solving this for $\|V-\tilde{V}\|_b$ yields the stated result.
Part b) is shown with the help of Theorem A.1.5. The functions v_n appearing in this lemma are now $v_n(\alpha) := L^{(n)}V^{(n)}(t,x,\alpha)$ for a fixed state $(t,x)\in E$ and $\alpha\in\mathcal{R}$. Since $\mathcal{R}$ is compact with respect to the Young topology, and following the considerations in the proof of Proposition 9.3.3, the mapping $\alpha\mapsto L^{(n)}V^{(n)}(t,x,\alpha)$ is upper semicontinuous. It remains to show that (v_n) is weakly decreasing. Since $\lim_{n\to\infty}\|U^{(n)}-U\|_b=0$ by assumption and $\lim_{n\to\infty}\|V^{(n)}-V\|_b=0$ by part a), there exists a sequence $(\delta_n)\subset\mathbb{R}$ with $\lim_{n\to\infty}\delta_n=0$ such that for all $n\ge m$ and $(t,x)\in E$

$$\begin{aligned}U^{(n)}(x)&\le\delta_m b(t,x)+U^{(m)}(x)\\ V^{(n)}(t,x)&\le\delta_m b(t,x)+V^{(m)}(t,x).\end{aligned}$$

Using these inequalities we obtain

$$\begin{aligned}v_n(\alpha)&=L^{(n)}V^{(n)}(t,x,\alpha)\\ &\le L^{(m)}V^{(m)}(t,x,\alpha)+\delta_m\Big(b\Big(t,\phi^\alpha_{T-t}(x)\Big)+\mathcal{T}_\circ b(t,x)\Big)\\ &=v_m(\alpha)+\delta_m c(t,x)\end{aligned}$$

for some $c(t,x)<\infty$. Thus the statement follows from Theorem A.1.5. □

Numerical Examples

Here we present some numerical examples for the results of the preceding sections and compare the value iteration on a grid (cf. Section 7.5.3) with the so-called *approximating Markov chain method.* In the approximating Markov chain approach the state process is approximated by a discrete-time Markov chain (see Section 4.10).
Though our results hold for general utility functions we have chosen a power utility $U(x)=\frac{1}{\beta}x^\beta$ with $\beta=0.5$, since this allows us to compare the numerical value with the true value function (cp. Example 9.3.5). We take one stock, i.e. $d=1$ and $\mu=\rho=0$. The density of the relative jump distribution is given by

$$0.5e^{-y}1_{[y\ge0]}+0.5\cdot1_{[-1<y<0]}.$$

The time horizon is one year, i.e. $T=1$. A straightforward implementation of the approximating Markov chain approach with grid size $h>0$ for the time interval $[0,T]$ yields for $N:=\frac{T}{h}$, $\tilde{V}_N(x):=U(x)$ and for $n=N-1,\dots,0$ the recursion

$$\tilde{V}_n(x) := \sup_{u \in \mathcal{U}} \left\{ e^{-\lambda h} \tilde{V}_{n+1}(x) + (1 - e^{-\lambda h}) \int \tilde{V}_{n+1}\big(x(1+yu)\big)\nu(dy) \right\}.$$

Then $\tilde{V}_0(x)$ is an approximation for $V(0,x)$. Figure 9.3 shows $\tilde{V}_0$ (lower dotted line) and the result of one step of value iteration with the grid operator (upper dotted line) $J_1 = \mathcal{T}_G J_0$ for $t = 0$ and for $\lambda \in \{40, 70\}$, where we have started with $J_0 = U$. The upper solid line shows the true value function $V(0,x)$, the lower solid line the utility function U. For both values of λ and both algorithms we have chosen the mesh size $h = 0.01$. It is remarkable that one iteration step already yields such a good result! Obviously in both cases, the implementation of the grid operator outperforms the approximating Markov chain method using the same grid size. A key reason is that the approximating Markov chain approach is a crude approximation of the time integral which appears in the $\mathcal{T}$ operator. Moreover, a small mesh size h leads for small λ to a large probability $e^{-\lambda h}$ of staying in the state which in turn leads to weak contraction and slow convergence (see Fleming and Soner (1993) p. 374 for a discussion). This problem is circumvented in our approach.

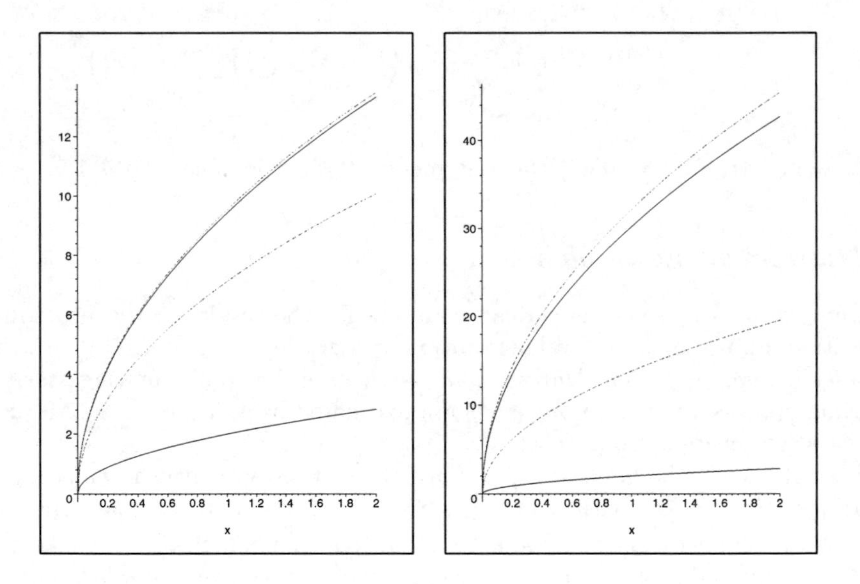

Fig. 9.3 Functions V, U and $J_1, \tilde{V}_0$ for $\lambda = 40$ (left) and $\lambda = 70$ (right).

Let us now start Howard's policy improvement algorithm with the decision rule $f_* \equiv 0$, *invest all the money in the bond.* The corresponding value function is given by

$$J_{f_*}(t,x) = U(xe^{\rho(T-t)}) = U(x).$$

In order to compute the first improvement we have to do exactly the same step as for the one-step value iteration above. The first improvement is given by $f^* :\equiv 27/40$. It is easily verified that this stationary policy $(f^*, f^*, \dots)$ defines an optimal portfolio strategy.

9.4 Trade Execution in Illiquid Markets

Suppose we have an agent who wants to sell a large amount of shares during a given time interval. Placing a large order in an order book will certainly lead to a price impact. Moreover in traditional markets, other participants may examine the order book and see the intention of the agent and then may try to trade against her. As a consequence it is recently possible to trade in *dark pools* where there is no order book and orders are matched electronically. This reduces the risk of adverse price manipulations but on the other hand may lead to lower liquidity since there is no market-maker.

We will set up a simple mathematical model to describe this situation. Suppose that the agent has initially $x_0 \in \mathbb{N}$ shares and is able to sell them in blocks only at the jump time points of a Poisson process to account for illiquidity. The execution horizon is T. All shares which have not been sold until time T will be placed at a traditional market and the order will be executed at once. The cost of selling a shares is given by $C(a)$ where $C : \mathbb{N}_0 \to \mathbb{R}_+$ is strictly increasing and strictly convex and satisfies $C(0) = 0$. Note that strictly convex means that

$$C(x) - C(x-1) < C(x+1) - C(x), \quad x \in \mathbb{N}. \tag{9.19}$$

The cost function C can be interpreted as a market depth function. Obviously this implies that it is better to sell small blocks, however if there are no trading epochs arriving anymore this will yield a large amount of shares which have to be liquidated at time T.

Let us now formalize this optimization problem: Suppose $N = (N_t)$ is a Poisson process with fixed intensity $\lambda > 0$. Denote by $0 = T_0 < T_1 < T_2 < \dots$ the jump time points of the Poisson process and by $(\mathcal{F}_t)$ the filtration generated by N. A control process $\pi = (\pi_t)$ has to be $(\mathcal{F}_t)$-predictable where π_t are the number of shares the agent would like to sell at time t, i.e. $\mathcal{U} = \mathbb{N}_0$. The order is only executed if t is also a jump time point of the Poisson process (which is not known to the agent at the time point of decision). The state process $X = (X_t)$ represents the number of shares which still have to be sold. Thus, if (π_t) is a control process we obtain

$$X_t = x_0 - \int_0^t \pi_s dN_s$$

where $\pi_t \le X_t$ has to be satisfied for all $t \in [0,T]$. A control process with this property is called admissible. The problem is now to minimize the function

$$V_\pi(t,x) := \mathbb{E}_{tx}^\pi \left[\int_t^T C(\pi_s) dN_s + C(X_T) \right],$$

i.e. to find

$$V(t,x) := \inf_\pi V_\pi(t,x), \quad (t,x) \in [0,T] \times \mathbb{N}_0 =: E$$

where the infimum is taken over all admissible policies.
Obviously this is a controlled Piecewise Deterministic Markov Decision problem with finite time horizon, where here $\phi_t^\alpha(x) = x$, i.e. the flow is uncontrolled. We will solve it by a discrete-time Markov Decision Problem along the lines of the last section. Let us denote by

$$A := \{\alpha : [0,T] \to \mathbb{N}_0 \text{ measurable } \} \tag{9.20}$$

and by $D(x) := \{\alpha \in A \mid \alpha_t \le x \text{ for all } t \in [0,T]\}$. As in the last section we consider Markov strategies $\pi = (\pi_t)$ which are given by a sequence of measurable functions (f_n) with $f_n : E \to A$ such that

$$\pi_t = f_n(T_n, Z_n)(t - T_n) \text{ for } t \in (T_n, T_{n-1}]$$

where $Z_n = X_{T_n}$ denotes the post-jump state (number of shares which still have to be sold). We obtain for a Markov strategy $\pi = (\pi_t) = (f_n)$ that

$$\begin{aligned} V_\pi(t,x) &= \mathbb{E}_{tx}^\pi \left[\int_t^T C(\pi_s) dN_s + C(X_T) \right] \\ &= \mathbb{E}_{tx}^\pi \left[\sum_{k=1}^\infty 1_{[T_k < T]} C(\pi_{T_k}) + 1_{[T_{k-1} \le T < T_k]} C(X_{T_{k-1}}) \right] \\ &= \sum_{k=1}^\infty \mathbb{E}_{tx}^\pi \left[c\Big(T_k, X_{T_k}, f_k(T_k, X_{T_k})\Big) \right], \end{aligned}$$

with

$$c(t,x,\alpha) := \int_0^{T-t} \lambda e^{-\lambda s} C(\alpha_s) ds + e^{-\lambda(T-t)} C(x). \tag{9.21}$$

Thus the optimization problem can be seen as a *continuous-time Markov Decision Chain* with finite time horizon (see Section 8.3) and with state space $\mathbb{N}_0$, action space A defined in (9.20), the set $D(x)$ of admissible actions, the reward rate $r(x,u) = -\lambda C(u)$, terminal reward function $g(x) = -C(x)$ and transition kernel $Q(\{y\}|x,u) = 1$ if $y = x - u$. Theorem 8.3.2 implies that $V(t,x)$ is also the value function of a discrete-time Markov Decision

Model. Moreover, the optimal policy for this discrete-time model determines an optimal control process for the trade execution problem.

Proposition 9.4.1. *The function* $b(t,x) := C(x)$ *is a* bounding function *for the discrete-time model and* $\alpha_b \le 1 - e^{-\lambda T} < 1$, *i.e. the discrete-time Markov Decision Model is contracting.*

Proof. First for $(t,x,\alpha) \in D$ we obtain since $\alpha_s \le x$ that

$$|c(t,x,\alpha)| \le \int_0^{T-t} \lambda e^{-\lambda s} C(x) ds + e^{-\lambda(T-t)} C(x) = C(x).$$

Moreover, we have

$$\begin{aligned}\alpha_b &= \sup_{(t,x,\alpha)\in D} \frac{\int b(s,y) Q(ds,dy|t,x,\alpha)}{b(t,x)} \\ &= \sup_{(t,x,\alpha)\in D} \frac{\int_0^{T-t} \lambda e^{-\lambda s} C(x-\alpha_s) ds}{C(x)} \le 1 - e^{-\lambda T} < 1.\end{aligned}$$

□

Properties of the value function V which can immediately be seen are

$$V(t,x) \le C(x), \; V(T,x) = C(x) \text{ and } V(t,0) = 0.$$

Now the dynamic programming operator $\mathcal{T}$ reads for $v \in \mathbb{B}_b$:

$$\begin{aligned}(\mathcal{T}v)(t,x) &= \inf_{\alpha \in D(x)} \left\{ \int_0^{T-t} \lambda e^{-\lambda s} \Big(C(\alpha_s) + v(t+s, x-\alpha_s) \Big) ds + e^{-\lambda(T-t)} C(x) \right\} \\ &= \int_0^{T-t} \lambda e^{-\lambda s} \min_{u\in\{0,\dots,x\}} \Big(C(u) + v(t+s, x-u) \Big) ds + e^{-\lambda(T-t)} C(x).\end{aligned}$$

Let us denote by

$$f^*(t,x) = \operatorname{argmin}_{u\in\{0,\dots,x\}} \Big(C(u) + v(t,x-u) \Big), \quad (t,x) \in E \tag{9.22}$$

the smallest minimizer of the right-hand side. Define

$$\begin{aligned}\mathbb{M}_{cx} := \Big\{ v \in \mathbb{B}_b \mid v(t,x) \le C(x), v(t,0) = 0, \; v \text{ is convex in } x, \\ v \text{ is continuous and increasing in } t,x \Big\}.\end{aligned}$$

Then we obtain the following statements for the trade execution problem:

Theorem 9.4.2. *a) The value function $V(t,x)$ is the unique fixed point of $\mathcal{T}$ in $I\!M_{cx}$.*
b) There exists an optimal Markov strategy $\pi^* = (\pi_t^*)$ such that $\pi_t^* = f^*(t, X_{t-})$ and f^* satisfies $f^*(t,x) \leq f^*(t,x+1) \leq f^*(t,x)+1$, and (X_t) is the corresponding number of share process.

Proof. We show that $\mathcal{T} : I\!M_{cx} \to I\!M_{cx}$ and that for $v \in I\!M_{cx}$, the minimizer f^* as defined in (9.22) has the properties $f^*(t,x) \leq f^*(t,x+1) \leq f^*(t,x)+1$. Let $v \in I\!M_{cx}$. Since $v(t,0) = 0$ we obtain $\mathcal{T}v \leq C$. That $\mathcal{T}v(t,0) = 0$ is obvious. The continuity of $(t,x) \mapsto \mathcal{T}v(t,x)$ follows immediately from the definition of $\mathcal{T}$. We next prove that $\mathcal{T}v$ is increasing in x, i.e. $\mathcal{T}v(t,x) \leq \mathcal{T}v(t,x+1)$, $x \in \mathbb{N}$. This can be seen since

$$C(a) + v(t+s, x-a) \leq C(a) + v(t+s, x+1-a), \text{ for } a = 0, \ldots, x$$
$$C(x) + v(t+s, 0) \leq C(x+1) + v(t+s, 0).$$

Next we show that $\mathcal{T}v$ is increasing in t. In what follows we write

$$G(t,x) := \min_{a \in \{0,\ldots,x\}} \Big(C(a) + v(t, x-a) \Big).$$

Let $t \geq t'$ and consider

$$\mathcal{T}v(t,x) - \mathcal{T}v(t',x) = \int_0^{T-t} \lambda e^{-\lambda s} \Big(G(t+s,x) - G(t'+s,x) \Big) ds$$
$$+ \int_{T-t}^{T-t'} \lambda e^{-\lambda s} \Big(C(x) - G(t'+s,x) \Big) ds.$$

Let $a^* = f^*(t+s,x)$ then we obtain

$$G(t+s,x) - G(t'+s,x) \geq C(a^*) + v(t+s, x-a^*) - C(a^*) - v(t'+s, x-a^*) \geq 0$$

and we obviously have

$$C(x) - G(t'+s,x) \geq C(x) - C(x) - v(t'+s,0) = 0$$

which implies that $\mathcal{T}v$ is increasing in t.
Next we show that $f^*(t,x+1) \leq f^*(t,x)+1$. If $f^*(t,x) = x$ the statement is clear, so suppose $a^* := f^*(t,x) \leq x-1$. Now suppose there exists an $a > a^*+1$ with

$$C(a) + v(t, x+1-a) < C(a^*+1) + v(t, x+1-(a^*+1)).$$

This implies

$$C(a-1) - C(a^*) < C(a) - C(a^*+1) < v(t, x-a^*) - v(t, x+1-a)$$

and hence

$$C(a-1) + v(t, x-(a-1)) < C(a^*) + v(t, x-a^*)$$

which contradicts the definition of a^*.
The remaining two statements $\mathcal{T}v(t, y+1) - \mathcal{T}v(t, y) \geq \mathcal{T}v(t, y) - \mathcal{T}v(t, y-1)$ and $f^*(t, y+1) \geq f^*(t, y)$ for $y \in \mathbb{N}$ are simultaneously shown by induction on y. For $y = 1$ we have

$$\mathcal{T}v(t, 2) - \mathcal{T}v(t, 1) \geq \mathcal{T}v(t, 1) - \mathcal{T}v(t, 0)$$

and $f^*(t, 2) = 1 = f^*(t, 1)$. Suppose the statement is true for $y = 1, \dots, x-1$. Let $a^* = f^*(t, x) \geq 1$. Suppose there exists an $0 < a < a^*$ (an easy argument gives us that $a = 0$ cannot be optimal) with

$$C(a) + v(t, x+1-a) \leq C(a^*) + v(t, x+1-a^*).$$

This implies that

$$C(a^*) - C(a) \geq v(t, x+1-a) - v(t, x+1-a^*) \geq v(t, x-a) - v(t, x-a^*)$$

where the last inequality follows from the induction hypothesis. Hence we conclude that

$$C(a^*) + v(t, x-a^*) \geq C(a) + v(t, x-a)$$

which is a contradiction to the definition of a^* and we obtain $f^*(t, x+1) \geq f^*(t, x)$. Now we have to show that $\mathcal{T}v(t, x+1) - \mathcal{T}v(t, x) \geq \mathcal{T}v(t, x) - \mathcal{T}v(t, x-1)$. Due to the convexity of C the statement is true when

$$G(t, x+1) - G(t, x) \geq G(t, x) - G(t, x-1).$$

Let us denote $f^*(t, x) =: a^* > 0$ and $b^* = f^*(t, x-1)$. Then $b^* \leq a^* \leq b^*+1$, i.e. $b^* \geq a^* - 1$. We discern the following cases:
Case 1: $f^*(t, x+1) = a^*$.
Thus we have

$$\begin{aligned} G(t, x) - G(t, x-1) &\leq v(t, x-b^*) - v(t, x-1-b^*) \\ &\leq v(t, x+1-a^*) - v(t, x-a^*) = G(t, x+1) - G(t, x). \end{aligned}$$

Case 2: $f^*(t, x+1) = a^* + 1$.
Here we have

$$\begin{aligned} G(t, x) - G(t, x-1) &\leq v(t, x-b^*) - v(t, x-1-b^*) \\ &\leq v(t, x-a^*) - v(t, x-a^*-1) \leq G(t, x+1) - G(t, x). \end{aligned}$$

Part a) follows from Theorem 7.3.5 since $\mathbb{M}_{cx}$ is a closed subset of $\mathbb{B}_b$. For part b) note that for $v = V$, the control function $f : E \to A$ defined by

$$f(t,x)(s) := f^*(t+s,x) \quad \text{for } s \geq 0$$

is a minimizer of V. Then by Theorem 7.3.5 the control process $\pi^* = (\pi_t^*)$ is optimal where for $t \in (T_n, T_{n+1}]$

$$\pi_t^* = f(T_n, Z_n)(t - T_n) = f^*(t, Z_n) = f^*(t, X_{t-}).$$

Altogether the statements are proven. □

Remark 9.4.3. In Bayraktar and Ludkovski (2011) it is also shown that $f^*(t,x)$ is increasing in t and jumps only by size one, i.e. there are thresholds $0 < t_1(x) < t_2(x) < \ldots < t_x(x)$ such that if we have $x \in \mathbb{N}$ shares, we try to sell k between time $t_{k-1}(x)$ and $t_k(x)$. ◊

9.5 Remarks and References

Section 9.2: The dividend payout problem was first considered in the case $Z \in \{-1, 1\}$ by de Finetti (1957) and Shubik and Thompson (1959). Miyasawa (1962) proved the existence of optimal band-policies under the assumption that the profit Z takes only a finite number of negative values. In this paper also a detailed proof of Theorem 9.2.10 can be found. The general case has been investigated in an informal way by Gerber (1969). Reinhard (1981) treats this problem with regime-switching. The distribution of Z depends here on an external Markov chain which represents the economic environment. Even in this case the optimality of a band-policy can be shown where now the parameters depend on the environment process. Waldmann (1988) treats (among others) the dividend payout problem under the constraint of an acceptable minimal expected lifetime. However, in this paper only barrier-policies are considered.
Other popular models in insurance consider the reinsurance and/or investment policies: Schäl (2004) studies reinsurance and investment policies and maximizes the expected exponential utility of the terminal surplus or minimizes the ruin probability. Schäl (2005) controls the probability of ruin by investment in a discrete-time financial market. For other applications of Markov Decision Process techniques in insurance see Venezia and Levy (1983), Martin-Löf (1994) and Schmidli (2008). Ruin problems can be found in Asmussen and Albrecher (2010).

Section 9.3: This section is essentially based on Bäuerle and Rieder (2009). For applications of controlled Piecewise Deterministic Markov Processes in insurance see in particular Schäl (1998) and the monograph Schmidli (2008). Kirch and Runggaldier (2005) used the discrete-time approach to solve a hedging problem in a continuous-time jump market. Their model is one with uncontrolled drift. Applications in queueing can be found in Bäuerle (2001) and Rieder and Winter (2009). The latter reference studied the optimal control of Markov jump processes with partial information and applications to a parallel queueing model. Portfolio problems with partial information and a piecewise deterministic process are investigated in Bäuerle and Rieder (2007). Jouini and Napp (2004) considered approximations of the utility function in a Black-Scholes-Merton market.

Section 9.4: The trade execution model is based on Bayraktar and Ludkovski (2011). In this paper also some extensions like order constraints, regime switching and partial observation are treated. Applications of continuous-time Markov Decision Chains in insurance can be found e.g. in Steffensen (2006) and Kraft and Steffensen (2008).

Part IV
Stopping Problems

Chapter 10
Theory of Optimal Stopping Problems

A very important subclass of the Markov Decision Problems considered so far are optimal stopping problems. There, a Markov process $(X_n)_{n\in\mathbb{N}}$ is given which cannot be influenced by the decision maker. However, this process has to be stopped at some time point n and a reward $g_n(X_n)$ is then obtained. Thus, the only decision at each time point is whether the process should be continued or stopped. Once it is stopped, no further decision is necessary. Sometimes costs have to be paid or an additional reward is obtained as long as the process is not stopped. Of course the aim is to find a stopping time such that the expected stopping reward is maximized.
In Section 10.1 we will first consider the simpler case of stopping problems with a *finite horizon*. We assume that the decision maker has to stop at the latest at time N. The problem is introduced using stopping times and it is then shown how it can be formulated as a Markov Decision Problem. Due to the simple action space it will turn out that the structure assumption in this case is always satisfied with the set of all measurable mappings and we immediately get a solution algorithm. In Section 10.2 we consider stopping problems with *infinite horizon*. They are much more delicate than the finite horizon case. We will treat them as limit cases of finite horizon Markov Decision Problems. In this case it is hard to identify optimal policies. Moreover, the optimal policy obtained by the Markov Decision Problem may lead to stopping times which are not admissible. However, in many applications we encounter the so-called *monotone case* which is considerably easier to solve and which is explained in the corresponding subsection. Finally in Section 10.3 some applications are given.

10.1 Stopping Problems with Finite Horizon

In this section we consider non-stationary stopping problems with finite horizon N. Note that we always have to stop, i.e. if we have not stopped before

N. Bäuerle and U. Rieder, *Markov Decision Processes with Applications to Finance*, Universitext, DOI 10.1007/978-3-642-18324-9_10,

time N we must stop at time N. We assume that the state process is given by a (non-stationary) Markov process (X_n) on a general state space E endowed with a σ-algebra $\mathfrak{E}$ and by a sequence of transition kernels (Q_n^X). As long as the process is not stopped, a measurable reward $c_n(X_n)$ is obtained which depends on the stage n. If the process is stopped at time n, a measurable reward $g_n(X_n)$ is received. We make no assumption on the sign of c_n and g_n so they may likewise stand for reward or cost. As in any decision problem we next have to define the information with respect to which the stopping time is chosen. We assume here that the decision may only depend on the observation of the process (X_n). Thus, we denote by $(\mathcal{F}_n)$ the filtration generated by the process (X_n), i.e. $\mathcal{F}_n := \sigma(X_0, \ldots, X_n)$.

Definition 10.1.1. A random time $\tau : \Omega \to \mathbb{N}_0 \cup \{\infty\}$ is an $(\mathcal{F}_n)$-*stopping time* if for all $n \in \mathbb{N}_0$

$$\{\tau \le n\} \in \mathcal{F}_n.$$

This condition means that upon observing the process (X_n) until time n we can decide whether or not τ has already occurred. In discrete time the condition is equivalent to $\{\tau = n\} \in \mathcal{F}_n$ for all $n \in \mathbb{N}_0$. Since the filtration will always be generated by (X_n) in our applications, we will not mention it explicitly. Thus, if we choose a stopping time τ with $\mathbb{P}_x(\tau \le N) = 1$ for all $x \in E$, we obtain the reward

$$R_\tau := \sum_{k=0}^{\tau-1} c_k(X_k) + g_\tau(X_\tau).$$

In order to have a well-defined problem we need the following general assumption (cp. with (A_N) of Chapter 2):

Assumption $(\mathbf{B}_N)$: *For $x \in E$*

$$\sup_{n \le \tau \le N} \mathbb{E}_{nx}\left[\sum_{k=n}^{\tau-1} c_k^+(X_k) + g_\tau^+(X_\tau)\right] < \infty, \quad 0 \le n \le N.$$

When we write $\mathbb{E}_{nx}$ (or $\mathbb{E}_x$ short for $\mathbb{E}_{0x}$), the expectation is taken with respect to the probability measure induced by the Markov process (X_n) given $X_n = x$. The problem is to find the value

$$V_N^*(x) := \sup_{\tau \le N} \mathbb{E}_x\left[R_\tau\right] \tag{10.1}$$

where the supremum is taken over all stopping times τ with $\mathbb{P}_x(\tau \le N) = 1$ for all $x \in E$. A stopping time $\tau^* \le N$ is *optimal* if for all $x \in E$

$$V_N^*(x) = \mathbb{E}_x\left[R_{\tau^*}\right].$$

Note that if we have a problem where it is admissible not to stop at all, then this problem can be embedded into our framework by choosing a time horizon of $N+1$ and setting $g_{N+1} := 0$.

We will now formulate the stopping problem as a (non-stationary) Markov Decision Model. Obviously the state space is E and the action space A consists only of two elements which represent the actions 'continue' and 'stop'. Since no decision has to be made after stopping, the problem is an absorbing Markov Decision Model: The decision process enters a 'cemetery' state after stopping (see Example 2.3.13). Following Example 2.3.13 we will not list the cemetery state explicitly but instead will consider the following *substochastic Markov Decision Model:*

- E is the state space of the (uncontrolled) Markov process,
- $A := \{0, 1\}$ where $a = 0$ means continue and $a = 1$ means stop,
- $D_n(x) := A$ for all $x \in E$,
- $Q_n(B|x, 0) := Q_n^X(B|x)$ and $Q_n(B|x, 1) \equiv 0$ for $x \in E$ and $B \in \mathfrak{E}$, i.e. the transition law in the case of continuation cannot be controlled,
- $r_n(x, 1) := g_n(x)$ and $r_n(x, 0) := c_n(x)$ for $x \in E$,
- g_N is the terminal reward function.

Following the theory of Chapter 2 – given all relevant assumptions are satisfied – the value functions of this Markov Decision Model can be computed recursively (cf. Theorem 2.3.8). In order to establish the connection to the stopping problem defined in (10.1), the following relation between stopping times and policies in the Markov Decision Model is crucial:
Suppose that $\pi = (f_0, \dots, f_{N-1})$ is an arbitrary policy for the Markov Decision Model and define

$$\tau_\pi := \inf\{n \in \mathbb{N}_0 \mid f_n(X_n) = 1\} \wedge N.$$

The random time τ_π is an $(\mathcal{F}_n)$-stopping time because

$$\{\tau_\pi = n\} = \{f_0(X_0) = 0, \dots, f_{n-1}(X_{n-1}) = 0, f_n(X_n) = 1\} \in \mathcal{F}_n.$$

By construction it is bounded by N. τ_π is a so-called *Markov stopping time* since $\pi = (f_n)$ is Markovian.
On the other hand suppose that τ is an $(\mathcal{F}_n)$-stopping time. Then by definition there exists for all $n \in \mathbb{N}_0$ a measurable function $f_n : E^{n+1} \to \{0, 1\}$ such that

$$1_{[\tau=n]} = f_n(X_0, X_1, \dots, X_n).$$

Thus, τ can be represented by a history-dependent policy $\pi = (f_0, \dots, f_{N-1})$. In particular it holds that

$$V_{0\pi}(x) = \mathbb{E}_x\left[\sum_{k=0}^{\tau_\pi - 1} c_k(X_k) + g_{\tau_\pi}(X_{\tau_\pi})\right] = \mathbb{E}_x[R_{\tau_\pi}] = \mathbb{E}_x[R_\tau].$$

From Theorem 2.2.3 we know already that in Markov Decision Models the optimal policy can be found among the Markov policies. Hence we can conclude here that the optimal stopping time, if it exists, can be found among the Markov stopping times. Thus it follows that $V_0 := \sup_\pi V_{0\pi}$ (the value of the Markov Decision model) is the same as V_N^* (the value of the stopping problem) and that an optimal policy π^* – if it exists – defines an optimal stopping time. Moreover, this discussion reveals that Assumption (B_N) is equivalent to the Integrability Assumption (A_N) for the corresponding Markov Decision Model. Altogether we have shown the following theorem.

Theorem 10.1.2. *It holds that*

a) $V_0(x) = V_N^*(x)$ *for all* $x \in E$.
b) Suppose $\pi^* = (f_0^*, \dots, f_{N-1}^*)$ *is an optimal policy for the Markov Decision Problem and define the stopping time*

$$\tau^* := \inf\{n \in \mathbb{N}_0 \mid f_n^*(X_n) = 1\} \wedge N,$$

then τ^* *is optimal for the stopping problem* (10.1).

Analogously it can be shown that V_n is the maximal expected reward for the stopping problem over the time period $[n, N]$, i.e.

$$V_n(x) = \sup_{n \le \tau \le N} \mathbb{E}_{nx}\Big[\sum_{k=n}^{\tau-1} c_k(X_k) + g_\tau(X_\tau)\Big]$$

where the supremum is taken over all stopping times τ which satisfy $\mathbb{P}_x(n \le \tau \le N) = 1$ for all $x \in E$. In view of assumption (B_N) we have $V_n(x) < \infty$. The theory of finite horizon Markov Decision Models now implies the next theorem.

Theorem 10.1.3. *Suppose a stopping problem with finite horizon N is given. Then it holds:*

a) $V_N = g_N$ *and* $V_n = \mathcal{T}_n V_{n+1}$ *for* $n = N-1, \dots, 0$ *where*

$$\mathcal{T}_n v(x) = \max\Big\{g_n(x),\ c_n(x) + \int v(x') Q_n^X(dx'|x)\Big\}, \quad x \in E.$$

b) Let $f_n^*(x) := 1$ *if* $V_n(x) = g_n(x)$ *and* $f_n^*(x) := 0$ *otherwise. Then* $(f_0^*, f_1^*, \dots, f_{N-1}^*)$ *is an optimal policy and the stopping time*

$$\tau^* := \min\{n \in \mathbb{N}_0 \mid V_n(X_n) = g_n(X_n)\}$$

is optimal for the stopping problem (10.1).

Proof. Part a) follows from our main Theorem 2.3.8 when we can show that the Structure Assumption (SA_N) is satisfied. However since the action space A consists of two elements only we obtain that (SA_N) is satisfied with

$$\mathbb{M}_n := \Big\{ v : E \to \mathbb{R} \,\Big|\, v(x) \le \sup_{n \le \tau \le N} \mathbb{E}_{nx} \Big[\sum_{k=n}^{\tau-1} c_k^+(X_k) + g_\tau^+(X_\tau) \Big],\ x \in E \Big\}$$

and $\Delta_n = F$ being the set of all possible decision rules.
Part b) follows again from Theorem 2.3.8 and Theorem 10.1.2 since f_n^* is a maximizer of V_{n+1}. Note that $V_N = g_N$ and thus $\mathbb{P}_x(\tau^* \le N) = 1$ for all $x \in E$. □

Remark 10.1.4 (Snell Envelope). If we define for $n = 0, \dots, N$ the random variables

$$Z_n := V_n(X_n),$$

then $Z_N = g_N(X_N)$ and

$$\begin{aligned} Z_n = V_n(X_n) &= \max\Big\{ g_n(X_n),\ c_n(X_n) + \int V_{n+1}(x') Q_n^X(dx'|X_n) \Big\} \\ &= \max\Big\{ g_n(X_n),\ c_n(X_n) + \mathbb{E}\big[Z_{n+1} \mid \mathcal{F}_n \big] \Big\}. \end{aligned}$$

The process (Z_n) is called the *Snell envelope.* It is the smallest c_n-supermartingale which dominates the process $(g_n(X_n))$, i.e. it is the smallest process which satisfies

$$\begin{aligned} Z_n &\ge c_n(X_n) + \mathbb{E}\big[Z_{n+1} \mid \mathcal{F}_n \big] \\ Z_n &\ge g_n(X_n). \end{aligned}$$

The value $Z_0 = V_0(X_0) = V_N^*(X_0)$ is the value of the stopping problem and the random time $\tau^* := \inf\{n \in \mathbb{N}_0 \mid Z_n = g_n(X_n)\}$ is an optimal stopping time. ◊

If the stopping problem is *stationary,* i.e. (X_n) is a stationary Markov process with transition kernel Q^X, the stopping reward is $g_n := \beta^n g$ and the intermediate reward is $c_n := \beta^n c$ for a discount factor $\beta \in (0, 1]$, then the recursive computation simplifies (cf. Section 2.5). Thus, we would like to solve the problem

$$J_N(x) := \sup_{\tau \le N} \mathbb{E}_x \left[\sum_{k=0}^{\tau-1} \beta^k c(X_k) + \beta^\tau g(X_\tau) \right], \quad x \in E$$

where the supremum is taken over all stopping times τ with $\mathbb{P}_x(\tau \le N) = 1$ for all $x \in E$. As before, this problem can also be formulated as a Markov Decision Problem (along the lines of the non-stationary case).

Theorem 10.1.5. *Suppose a stationary stopping problem with finite horizon N is given. Then it holds:*

a) $J_0 = g$ *and* $J_n = \mathcal{T} J_{n-1}$ *for* $n = 1, \dots, N$ *where*

$$\mathcal{T} v(x) = \max\Big\{ g(x),\ c(x) + \beta \int v(x') Q^X(dx'|x) \Big\}, \quad x \in E.$$

b) $g \le J_n \le J_{n+1}$ *for all* $n \in \mathbb{N}_0$.

c) Define $d_n(x) := g(x) - c(x) - \beta \int J_{n-1}(x') Q^X(dx'|x)$ *for* $n = 1, \dots, N$. *Then we obtain the following recursion:*

$$d_{n+1}(x) = d_1(x) - \beta \int d_n^-(x') Q^X(dx'|x).$$

d) Let $S_n^* := \{x \in E \mid J_n(x) = g(x)\}$ *and define* $f_n^* := 1_{S_n^*}$. *Then* $S_0^* = E$, $S_n^* = \{x \in E \mid d_n(x) \ge 0\}$ *and* $S_{n+1}^* \subset S_n^*$. *The policy* $(f_N^*, \dots, f_1^*)$ *is optimal and* $\tau^* := \min\{n \in \{0, 1, \dots, N\} \mid X_n \in S_{N-n}^*\}$ *is an optimal stopping time.*

The inclusion $S_{n+1}^* \subset S_n^*$ means that the tendency to stop is non-decreasing as time goes by.

Remark 10.1.6 (Threshold or Control Limit Policy). If E is a completely ordered space and $x \mapsto d_n(x)$ is non-decreasing, then either $S_n^* = \emptyset$, $S_n^* = E$ or there exists a state $x_n^* \in E$ such that

$$f_n^*(x) = \begin{cases} 1 & \text{if } x \ge x_n^* \\ 0 & \text{if } x < x_n^*. \end{cases}$$

Note that $x_n^* := \inf\{x \in E \mid d_n(x) \ge 0\}$. If this is satisfied for all n then such a policy is called *threshold policy* or *control limit policy*. ◊

Proof. a) Follows from Theorem 2.5.4 (stationary case) as in the non-stationary case.

b) Since by part a) J_n is the maximum of g and some other function, the statement $J_n \ge g$ is obvious. Since $g = J_0 \le J_1$ and due to the fact that the maximal reward operator is order preserving (Lemma 2.3.3 c)) it follows that $J_n \le J_{n+1}$.

c) First note that

$$d_1(x) = g(x) - c(x) - \beta \int g(x') Q^X(dx'|x).$$

By part a) and the definition of d_n we have

$$J_n(x) = g(x) + d_n^-(x). \tag{10.2}$$

Thus, we obtain

$$\begin{aligned} d_{n+1}(x) &= g(x) - c(x) - \beta \int J_n(x')Q^X(dx'|x) \\ &= g(x) - c(x) - \beta \int g(x')Q^X(dx'|x) - \beta \int d_n^-(x')Q^X(dx'|x) \end{aligned}$$

and the statement follows.

d) This part follows again from Theorem 2.5.4 and the monotonicity result in part b). □

10.2 Stopping Problems with Unbounded Horizon

Suppose a stationary stopping problem as in Section 10.1 is given. The reward under an unbounded stopping time τ is given by

$$R_\tau := \sum_{k=0}^{\tau-1} \beta^k c(X_k) + \beta^\tau g(X_\tau) \quad \text{for } \tau < \infty.$$

We do not define a reward for $\tau = \infty$ since we will restrict to stopping times with $\mathbb{P}_x(\tau < \infty) = 1$ for all $x \in E$, i.e. we have to stop with probability one in finite time. Throughout, we make the following general assumption:

Assumption (B): *For all $x \in E$ it holds:*

$$\sup_{\tau<\infty} \mathbb{E}_x \left[\sum_{k=0}^{\tau-1} \beta^k c^+(X_k) + \beta^\tau g^+(X_\tau) \right] < \infty \quad \textit{and}$$

$$\liminf_{n\to\infty} \mathbb{E}_x[R_{\tau\wedge n}] \geq E_x[R_\tau] \quad \textit{for all } \tau < \infty.$$

The task is to find the value of the stopping problem

$$V_\infty^*(x) := \sup_{\tau<\infty} \mathbb{E}_x[R_\tau] \tag{10.3}$$

where the supremum is taken over all stopping times τ with $\mathbb{P}_x(\tau < \infty) = 1$ for all $x \in E$. Obviously, Assumption (B) implies that $V_\infty^*(x) < \infty$ for all $x \in E$. A stopping time τ^* is *optimal* if $\mathbb{P}_x(\tau^* < \infty) = 1$ for all $x \in E$ and

$$V_\infty^*(x) = \mathbb{E}_x[R_{\tau^*}].$$

Remark 10.2.1. a) It follows from Fatou's Lemma that Assumption (B) is equivalent to: For all $x \in E$ it holds

$$\sup_{\tau<\infty} \mathbb{E}_x \left[\sum_{k=0}^{\tau-1} \beta^k c^+(X_k) + \beta^\tau g^+(X_\tau) \right] < \infty \quad \text{and}$$

$$\lim_{n\to\infty} \mathbb{E}_x[R_{\tau\wedge n}] = E_x[R_\tau] \quad \text{for all } \tau < \infty.$$

b) A sufficient condition for the second assumption in (B) is that

$$\liminf_{n\to\infty} \mathbb{E}_x[R_n 1_{[\tau>n]}] \geq 0 \quad \text{for all } \tau < \infty.$$

This condition in turn is satisfied if

$$\limsup_{n\to\infty} \mathbb{E}_x[R_n^- 1_{[\tau>n]}] = 0 \quad \text{for all } \tau < \infty.$$

Hence the second part of (B) certainly holds if $c \geq 0$ and $g \geq 0$. Condition (B) holds in particular when (R_n) is bounded. ◊

Solution via Finite Horizon Stopping Problems

We will solve problem (10.3) with the help of Markov Decision Theory as developed in Chapter 7. Here we consider the unbounded horizon model as a limit of finite horizon stopping problems. As in the previous subsection we have for $n \in \mathbb{N}$, a policy $\pi = (f_0, f_1, \ldots)$ and $x \in E$

$$J_{n\pi}(x) = \mathbb{E}_x \left[R_{\tau_\pi \wedge n} \right] \quad \text{and} \quad J_n(x) = \sup_\pi J_{n\pi}(x).$$

The limit of (J_n) for $n \to \infty$ exists, since the sequence is increasing by Theorem 10.1.5 b). We denote

$$J(x) := \lim_{n\to\infty} J_n(x), \quad x \in E.$$

Moreover, we define for $\pi = (f_0, f_1, \ldots)$

$$G_\pi(x) := \liminf_{n\to\infty} J_{n\pi}(x) \quad \text{and} \quad G(x) := \sup_\pi G_\pi(x), \quad x \in E.$$

We obtain the following main result which states that the functions V_∞^*, J and G which have been introduced before are indeed equal. Note that (B) is our standing assumption.

Theorem 10.2.2. *Suppose a stopping problem with unbounded horizon is given. Then it holds:*

a) $V_\infty^*(x) = G(x) = J(x)$ *for all* $x \in E$.

b) $J = \mathcal{T} J$, *i.e.* $J(x) = \max\left\{g(x),\ c(x) + \beta \int J(x')Q^X(dx'|x)\right\}$, $x \in E$.

c) J *is the smallest c-superharmonic function which majorizes* g, *i.e.* J *is the smallest function such that for all* $x \in E$

$$J(x) \geq c(x) + \beta \int J(x')Q^X(dx'|x) \quad \textit{and} \quad J(x) \geq g(x).$$

Proof. a) From the discussion in the previous subsection we know already that for all stopping times τ with $\mathbb{P}_x(\tau < \infty) = 1$ there exists a (possibly history dependent) policy π_τ such that for all n and $x \in E$:

$$J_{n\pi_\tau}(x) = \mathbb{E}_x[R_{\tau\wedge n}].$$

Thus, by the second part of Assumption (B) we have

$$G_{\pi_\tau}(x) = \liminf_{n\to\infty} J_{n\pi_\tau}(x) \geq \mathbb{E}_x[R_\tau].$$

This in turn implies

$$G(x) \geq \mathbb{E}_x[R_\tau].$$

Thus we obtain

$$G(x) \geq \sup_{\tau<\infty} \mathbb{E}_x[R_\tau] = V_\infty^*(x).$$

On the other hand, we obtain from Theorem 10.1.5 b) that

$$J(x) = \sup_n \sup_{\pi \ Markov} J_{n\pi}(x) = \sup_n \sup_\pi J_{n\pi}(x) = \sup_\pi \sup_n J_{n\pi}(x)$$

and hence for a (possibly history dependent) policy π

$$G_\pi(x) = \liminf_{n\to\infty} J_{n\pi}(x) \leq \sup_n J_{n\pi}(x) \leq J(x).$$

Altogether we have shown that

$$V_\infty^*(x) \leq G(x) \leq J(x), \quad x \in E.$$

Finally we get

$$J_n(x) = \sup_{\tau\leq n} \mathbb{E}_x[R_\tau] \leq \sup_{\tau<\infty} \mathbb{E}_x[R_\tau] = V_\infty^*(x)$$

which implies $J(x) \leq V_\infty^*(x)$ and part a) is shown.

b) This part follows from the monotonicity of (J_n) and Theorem A.1.6.
c) By part b) J is obviously c-superharmonic and majorizes g. Now let v be another c-superharmonic function which majorizes g. Then $v \geq \mathcal{T}v$ which yields by iteration $v \geq \mathcal{T}^n v \geq \mathcal{T}^n g = J_n$ for all $n \in \mathbb{N}$. Taking the limit $n \to \infty$ implies $v \geq J$. □

The last theorem characterizes the value function V_∞^*. The next step is to identify optimal stopping times. It seems to be reasonable that optimal stopping times are Markov and stationary, and we want to identify them using the Markov Decision Model. Thus, we are interested in stationary policies $(f, f, \ldots) \in F^\infty$ and induced stopping times

$$\tau_f := \inf\{n \in \mathbb{N}_0 \mid f(X_n) = 1\}.$$

Again note that τ_f might take the value ∞ with positive probability and thus may not be admissible for the given stopping problem. Recall the definition

$$G_f = \liminf_{n\to\infty} J_{nf} = \liminf_{n\to\infty} \mathcal{T}_f^n g \leq J$$

for $f \in F$ and the definition of d_n and S_n^* $n \in \mathbb{N}$ in Theorem 10.1.5. The following theorem gives conditions under which the maximizer of J defines an optimal stopping rule.

Theorem 10.2.3. *Suppose a stopping problem with unbounded horizon is given. Then it holds:*

a) The limit $d := \lim_{n\to\infty} d_n$ exists. Define $S^* := \{x \in E \,|\, d(x) \geq 0\}$, then $S^* = \cap_n S_n^* = \{x \in E \mid J(x) = g(x)\}$ and $f^* := 1_{S^*}$ is a maximizer of J.
b) If $G_{f^*} \geq \mathcal{T} G_{f^*}$ and $\mathbb{P}_x(\tau_{f^*} < \infty) = 1$ for all $x \in E$, then

$$\tau^* := \inf\{n \in \mathbb{N}_0 \mid X_n \in S^*\}$$

is an optimal stopping time and $G_{f^*}(x) = \mathbb{E}_x[R_{\tau^*}] = V_\infty^*(x)$, $x \in E$.

Proof. a) The existence of the limit $d(x)$ follows from the monotone convergence theorem. Note that $J(x) = g(x) + d^-(x)$ by (10.2) which implies the remaining statements.
b) Since $\mathbb{P}_x(\tau_{f^*} < \infty) = 1$ for all $x \in E$ it follows from Assumption (B) and Fatou's Lemma (see Remark 10.2.1) that

$$G_{f^*}(x) = \lim_{n\to\infty} J_{nf^*}(x) = \mathbb{E}_x\left[R_{\tau^*}\right] \leq V_\infty^*(x), \quad x \in E.$$

On the other hand since by assumption G_{f^*} is a c-superharmonic function which majorizes g it follows from Theorem 10.2.2 that $G_{f^*}(x) \geq J(x) = V_\infty^*(x)$ for all $x \in E$. Hence all statements are shown. □

Remark 10.2.4. It follows from Assumption (B) and Fatou's Lemma that

$$G_\pi(x) = \lim_{n\to\infty} J_{n\pi}(x) = \mathbb{E}_x[R_{\tau_\pi}], \quad x \in E$$

when $\mathbb{P}_x(\tau_\pi < \infty) = 1$ for all $x \in E$. If in addition $\pi = (f, f, \ldots)$ is stationary, we derive $G_f = \mathcal{T}_f G_f$ from dominated convergence. ◊

The following example shows that even though the reward is bounded and the stopping problem is well-defined, the maximizer of J may not define an optimal stopping rule, i.e. further conditions like in Theorem 10.2.3 are necessary.

Example 10.2.5. Let us consider the following Markov Decision Model (cf. Example 7.4.4): Suppose that the state space is $E := \mathbb{N}$ and the action space is $A := \{0, 1\}$ where $a = 0$ means 'continue' and $a = 1$ means 'stop'. Let $D(x) := A$. The transition probabilities of the uncontrolled Markov process are given by

$$q^X(x+1|x) = 1, \quad x \in \mathbb{N}.$$

For the reward function we assume that $c \equiv 0$ and $g(x) = 1 - \frac{1}{x+1}$ for $x \in \mathbb{N}$. The discount factor is assumed to be one. Since $c \equiv 0$ and g is bounded, Assumption (B) is certainly satisfied. It is not difficult to see that the value function of this stopping problem is given by $V_\infty^*(x) = 1$ for $x \in \mathbb{N}$. From Theorem 10.2.3 a) it follows that

$$S^* = \Big\{x \in \mathbb{N} \,\Big|\, V_\infty^*(x) = g(x)\Big\} = \emptyset$$

and $f^*(x) \equiv 0$ is a maximizer of V_∞^*. The stationary policy $(f^*, f^*, \ldots)$ corresponds to the stopping time $\tau_{f^*} = \infty$, i.e. we never stop. But such a stopping time is not admissible for the stopping problem. In this example no optimal stopping rule exists. ♦

This example shows that it is not enough to determine a maximizer of J. This maximizer also has to satisfy some extra conditions. The situation can be compared with the case of positive Markov Decision Models in Section 7.4.

Since J is characterized as the smallest c-superharmonic function which majorizes g, the condition $G_{f^*} \geq \mathcal{T} G_{f^*}$ is equivalent to $G_{f^*} = J$. The next corollary gives some sufficient conditions which imply that $G_{f^*} = J$ or $\mathbb{P}_x(\tau_{f^*} < \infty) = 1$ for all $x \in E$.

Corollary 10.2.6. *Suppose a stopping problem with unbounded horizon is given and f^* is a maximizer of J. Then it holds:*

a) If the corresponding Markov Decision Problem has a bounding function b with $\beta\alpha_b < 1$, then $G_{f^*} = J$. Moreover, if $\mathbb{P}_x(\tau_{f^*} < \infty) = 1$ for all $x \in E$ then τ_{f^*} is an optimal stopping time.
b) If $\sup_{x\in E} c(x) < 0$, $\beta = 1$ and $G_{f^*} = J$, then $\mathbb{P}_x(\tau_{f^*} < \infty) = 1$ for all $x \in E$ and τ_{f^*} is an optimal stopping time.
c) If $J - g$ is bounded from above and $\mathbb{P}_x(\tau_{f^*} < \infty) = 1$ for all $x \in E$, then τ_{f^*} is an optimal stopping time and $G_{f^*} = J$.

Proof. a) Since f^* is a maximizer of J we obtain $\mathcal{T}_{f^*} J = \mathcal{T} J = J$ and

$$\begin{aligned} J(x) &= \mathcal{T}^n_{f^*} g(x) + \beta^n \, \mathbb{E}_x \left[(J-g)(X_n) 1_{[\tau_{f^*} > n]} \right] \\ &\leq \mathcal{T}^n_{f^*} g(x) + (\beta\alpha_b)^n \|J - g\|_b b(x). \end{aligned}$$

Taking the limit $n \to \infty$ we obtain

$$J(x) \leq \liminf_{n\to\infty} \mathcal{T}^n_{f^*} g(x) + 0 = G_{f^*}(x) \leq G(x).$$

Since by Theorem 10.2.2 $G(x) = J(x)$ we obtain the statement with Theorem 10.2.3 b).

b) Suppose that $\mathbb{P}_x(\tau_{f^*} = \infty) > 0$ for some $x \in E$. Then

$$G_{f^*}(x) = \liminf_{n\to\infty} \mathbb{E}_x \left[R_{\tau_{f^*} \wedge n} \right] = -\infty.$$

Since $G_{f^*}(x) = J(x) \geq g(x) > -\infty$, this is a contradiction and the statement follows again from Theorem 10.2.3.

c) By assumption there exists a constant $d > 0$ such that $J(x) - g(x) \leq d$ for all $x \in E$. From part a) we know that

$$\begin{aligned} J(x) &= \mathcal{T}^n_{f^*} g(x) + \beta^n \, \mathbb{E}_x \left[(J-g)(X_n) 1_{[\tau_{f^*} > n]} \right] \\ &\leq \mathcal{T}^n_{f^*} g(x) + d \, \mathbb{P}_x(\tau_{f^*} > n). \end{aligned}$$

Since by assumption $\lim_{n\to\infty} \mathbb{P}_x(\tau_{f^*} > n) = 0$ we obtain the statement as in part a). □

The Monotone Case

Let us consider the following special class of stopping problems which is known as the *monotone case*. Here the optimal stopping time can often be identified explicitly.

Theorem 10.2.7. *Suppose a stopping problem is given. Let*

$$S_0 := \left\{ x \in E \;\middle|\; g(x) \geq c(x) + \beta \int g(x')Q^X(dx'|x) \right\}$$

be closed, i.e. $Q^X(S_0|x) = 1$ *for all* $x \in S_0$. *Then it holds:*

a) The decision rule $f^* = 1_{S_0}$ *is a maximizer of* J_n, $n \in \mathbb{N}$ *and also of* J.
b) Define $\tau^* := \inf\{n \in \mathbb{N}_0 \mid X_n \in S_0\}$. *Then* $\tau^* \wedge N$ *is optimal for the* N*-stage stopping problem. If* $\mathbb{P}_x(\tau^* < \infty) = 1$ *for all* $x \in E$, *then* τ^* *is optimal for the unbounded stopping problem.*

The decision rule defined in the last theorem is called the *One-Step-Look-Ahead Rule*: It compares the reward which is obtained when we stop immediately with the reward when we stop one step ahead.

Proof. a) We show by induction on $n \in \mathbb{N}$ that

$$J_n(x) = \begin{cases} g(x) & , x \in S_0, \\ c(x) + \beta \int J_{n-1}(x')Q^X(dx'|x) & , x \notin S_0. \end{cases}$$

For $n = 1$ the statement follows directly from the value iteration (Theorem 10.1.3 part a)). Now suppose it is true for $k = 1, 2, \ldots, n$:
For $x \notin S_0$ we obtain

$$g(x) < c(x) + \beta \int g(x')Q^X(dx'|x) \leq c(x) + \beta \int J_n(x')Q^X(dx'|x).$$

Hence $J_{n+1}(x) = c(x) + \beta \int J_n(x')Q^X(dx'|x)$.
If $x \in S_0$ we obtain since S_0 is closed and $J_n(x) = g(x)$:

$$\begin{aligned} g(x) &\geq c(x) + \beta \int g(x')Q^X(dx'|x) \\ &= c(x) + \beta \int g(x')1_{S_0}(x')Q^X(dx'|x) + \beta \int g(x')1_{S_0^c}(x')Q^X(dx'|x) \\ &= c(x) + \beta \int J_n(x')Q^X(dx'|x). \end{aligned}$$

Thus, $J_{n+1}(x) = g(x)$ for $x \in S_0$. In particular we have $S_n^* = S_0$ for all $n \in \mathbb{N}$ and $S^* = S_0$. Hence $f^* = 1_{S_0}$ is a maximizer of J_n and also of J.
b) The optimality of $\tau^* \wedge N$ follows from part a). Theorem 10.2.3 implies the last statement, since $T_{f^*}^n g = J_n$ and hence $\lim_{n\to\infty} T_{f^*}^n g = G_{f^*} = J$. □

Corollary 10.2.8. *Suppose* E *is a completely ordered space. Assume that we are in the monotone case, i.e. the assumptions of Theorem 10.2.7 are satisfied and that*

(i) $x \mapsto g(x)$ *is increasing,*
(ii) $x \mapsto c(x)$ *is decreasing,*
(iii) $x \mapsto \int g(x')Q^X(dx'|x)$ *is decreasing.*

Let τ^* *be defined as in Theorem 10.2.7. If* $\mathbb{P}_x(\tau^* < \infty) = 1$ *for all* $x \in E$*, then the optimal stopping time is of threshold type.*

Proof. From Theorem 10.2.7 it remains to show that the mapping

$$x \mapsto g(x) - c(x) - \beta \int g(x')Q^X(dx'|x)$$

is increasing. But this follows directly from our assumptions. □

10.3 Applications and Examples

In this section we consider a number of typical applications. Among them, stopping of a sequence of independent and identically distributed random variables, the quiz show problem, the secretary problem and some Bayesian stopping problems.

10.3.1 A House Selling Problem

Imagine a person who wants to sell her house. At the beginning of each week she receives an offer which is randomly distributed over the interval $[m, M]$ with $0 < m < M$. The offers are independent and identically distributed with distribution Q. The house seller has to decide immediately whether to accept or reject this offer. If she rejects, the offer is lost and she has maintenance cost of $c > 0$. After N weeks the house has to be sold. Which offer should she accept in order to maximize her expected reward?
This is a stationary stopping problem with the following data.

- $E := [m, M]$ where x denotes the current offer,
- $A := \{0, 1\}$ where $a = 0$ means reject the offer and $a = 1$ means accept.
- $Q^X(\cdot|x) := Q(\cdot)$ distribution of an offer (independent of x),
- $c(x) \equiv -c$ and $g(x) := x$,
- $\beta \in (0, 1]$.

Since the state space is compact, the stopping reward function g is bounded. Together with the fact that $c \leq 0$, Assumption (B_N) is satisfied. We can use Theorem 10.1.3 to solve this problem. We have the following recursion: $J_0(x) = x$ and for $n = 1, \dots, N$

$$J_n(x) = \max\Big\{x, -c + \beta \int J_{n-1}(x')Q(dx')\Big\}, \quad x \in E.$$

Let us define $c_n^* := -c+\beta \int J_{n-1}(x')Q(dx')$ which is independent of x. Thus, the optimal stopping set is given by

$$S_n^* = \{x \in E \mid x \geq c_n^*\}.$$

The thresholds $x_n^* := c_n^*$ can be computed recursively.

A different version of this problem is obtained when we use a utility function U to evaluate the reward, i.e. we stop the process (X_n) with $X_n = U(Y_n - nc)$ where Y_n is the random variable which gives the offer at time n and the Y_n are independent and identically distributed. Obviously the problem can be solved as an non-stationary stopping problem.

Now suppose the house seller has no fixed date by which the house has to be sold, i.e. we have a stopping problem with unbounded horizon. In this case we assume $\beta \in (0,1)$. Then Assumption (B) is satisfied.

Theorem 10.3.1. *In the unbounded horizon house selling problem it is optimal to accept the first offer which exceeds the threshold x^*, where x^* is the maximum point of the function*

$$x \mapsto \frac{-c\,Q\big([m,x)\big) + \int_x^\infty x'Q(dx')}{1 - \beta Q\big([m,x)\big)}$$

on the interval $E = [m,M]$ and $x^* < M$.

Proof. According to Theorem 10.2.2 we obtain

$$J(x) = \max\Big\{x, -c + \beta\, \mathbb{E}\, J(X)\Big\}, \quad x \in E.$$

Since the expression $x^* := -c + \beta\, \mathbb{E}\, J(X)$ is independent of x, a maximizer of J is given by $f^* = 1_{S^*}$ with $S^* := \{x \in E \mid x \geq x^*\}$ and if we define

$$\tau^* = \inf\{n \in \mathbb{N}_0 \mid X_n \geq x^*\}$$

then obviously τ^* has a geometric distribution and $\mathbb{P}_x(\tau^* < \infty) = 1$ for all $x \in E$. Moreover, since $c > 0$ and Q has bounded support we obtain that $J - g$ is bounded from above. Thus it follows with Corollary 10.2.6 a) that it is optimal to accept the first offer which exceeds x^*. Moreover, it is possible to compute the corresponding maximal expected reward explicitly: Let $f := 1_S$ with $S = \{x \in S | x \geq x_0\}$ be a threshold policy with threshold $x_0 < M$. Since $\mathbb{P}_x(\tau_f < \infty) = 1$ for all x we obtain $G_f = \mathcal{T}_f G_f$ and conclude:

$$
\begin{aligned}
\mathbb{E}\, G_f(X) &= \int_m^{x_0} G_f(x)Q(dx) + \int_{x_0}^M G_f(x)Q(dx)\\
&= \int_m^{x_0}\Big(-c+\beta\int G_f(y)Q(dy)\Big)Q(dx) + \int_{x_0}^M xQ(dx)\\
&= Q\big([m,x_0)\big)\Big(-c+\beta\,\mathbb{E}\, G_f(X)\Big) + \int_{x_0}^M xQ(dx).
\end{aligned}
$$

Hence the expected reward is given by

$$
\mathbb{E}\, G_f(X) = \frac{-c\,Q\big([m,x_0)\big) + \int_{x_0}^M xQ(dx)}{1-\beta\,Q\big([m,x_0)\big)}.
$$

Maximizing the expression on the right-hand side over $x_0 \in [m, M]$ yields the optimal threshold x^* and also the maximal expected reward of the house selling problem. Moreover it follows that $x^* < M$. □

10.3.2 Quiz Show

A contestant in a quiz show has to answer questions. For each correct answer she wins one Euro and she has the option of either leaving with her accumulated fortune or continuing with the next question. However, with a wrong answer she forfeits her complete fortune and has to quit the show. When should she stop in order to maximize her expected reward?
We suppose that the contestant answers each question independently with probability $p \in (0,1)$ correctly. Thus, the problem is stationary and has an unbounded horizon. Besides the contestant's own decision to stop, there is an external event (wrong answer) which leads to an absorbing state. Thus, the state space is $\mathbb{N}_0 \cup \{\infty\}$ where $x_n \in \mathbb{N}_0$ denotes the current fortune of the contestant and $x_n = \infty$ indicates that she has answered incorrectly. The transition probabilities are given by

$$
q^X(x+1|x) = p, \quad q^X(\infty|x) = 1-p,\ x \in \mathbb{N}_0
$$

and $q^X(\infty|\infty) = 1$. The reward is $g(x) = x$ if $x \in \mathbb{N}_0$ and $g(\infty) = 0$. There is no intermediate reward or cost, i.e. $c \equiv 0$ and there is no discounting. We summarize the data of the stopping problem:

- $E := \mathbb{N}_0 \cup \{\infty\}$ where $x \in \mathbb{N}_0$ denotes the current fortune and $x = \infty$ indicates that she has answered incorrectly,
- $A := \{0,1\}$ where $a = 0$ means continue and $a = 1$ means quit the game,
- $q^X(x+1|x) := p$, $q^X(\infty|x) := 1-p$ for $x \in \mathbb{N}_0$ and $q^X(\infty|\infty) := 1$,
- $c \equiv 0$ and $g(x) := x$ for $x \in \mathbb{N}_0$, $g(\infty) := 0$,
- $\beta := 1$.

Assumption (B) is satisfied which can be seen as follows. First note that $g \geq 0$, thus the second part of (B) is satisfied (see Remark 10.2.1). Suppose that $\hat{\tau}$ is the stopping time which gives the first time the contestant answers incorrectly. Then $\hat{\tau}$ has a geometric distribution and

$$\sup_{\tau<\infty} \mathbb{E}_x[R_\tau] \leq x + \mathbb{E}[\hat{\tau}] = x + \frac{1}{1-p} < \infty, \quad x \in \mathbb{N}_0.$$

Moreover, the conditions of Theorem 10.2.7 are fulfilled, i.e. we are in the monotone case: It is easy to see that the set

$$\begin{aligned} S_0 &:= \Big\{ x \in E \;\Big|\; g(x) \geq \sum_y q^X(y|x) g(y) \Big\} \\ &= \Big\{ x \in \mathbb{N}_0 \cup \{\infty\} \;\Big|\; x \geq x^* \Big\} \quad \text{with } x^* = \left\lceil \frac{p}{1-p} \right\rceil \end{aligned}$$

is closed (by $\lceil x \rceil$ we denote the smallest integer greater than or equal to x). Moreover, when we define

$$\tau^* := \inf\{n \in \mathbb{N}_0 \mid X_n \in S_0\}$$

then $\mathbb{P}_x(\tau^* < \infty) = 1$ for all $x \in E$. Indeed, it even holds for all $x \in E$ that $\mathbb{P}_x\left(\tau^* \leq x^*\right) = 1$.
Thus if the contestant starts with zero fortune, it is optimal for her to stop after x^* questions if she comes so far and her maximal expected reward is given by $V_\infty^*(0) = x^* p^{x^*}$.

10.3.3 The Secretary Problem

The secretary problem is a classical stopping problem which can be found in many textbooks. However in most books the problem is solved by martingale methods. Here we use the solution technique of Markov Decision Models: Imagine an executive who has to hire a new secretary. She has selected $N > 2$ applicants and interviews them one at a time. After an interview she directly has to decide whether to accept or reject that particular candidate (no recall is possible). The order in which the candidates appear is completely random. However we suppose that after each interview the executive is able to rank the candidate compared with her predecessors but of course no comparison with the remaining candidates can be made. The objective of the executive is to find a stopping time such that she maximizes the probability that the best candidate is chosen.
In what follows we denote by Z_n the *absolute rank* of candidate number n. We assume that N is the best rank and 1 the poorest and that $(Z_1, \dots, Z_N)$ has a uniform distribution over all permutations of the set $\{1, 2, \dots, N\}$. In order

to formulate the problem as a Markov Decision Model we need a Markov process which has to be stopped: It is not difficult to see that it can only be optimal to accept candidate number $n > 1$ if her relative rank so far is maximal (in this case we say that the candidate is leading), because if the relative rank is not maximal we can be sure that the candidate is not the best one and it is better to wait for the next one. Thus, we consider the process (X_n) with state space $E := \{1, 2, \dots, N, N+1\}$ and the interpretation:

X_n is the time point at which for the $(n+1)$-th time a candidate is leading.

More formally we have $X_0 = 1$ and set $X_N := N + 1$. Let us denote by R_n the *relative rank* of candidate n, i.e. $R_n \in \{1, \dots, n\}$. Then we define for $n = 1, \dots, N-1$

$$X_n = \inf \Big\{ k > X_{n-1} \mid R_k = k \Big\}$$

where $\inf \emptyset := N + 1$. If for example $N = 10$ and the candidates appear according to the rank permutation $(3, 2, 1, 7, 6, 9, 10, 4, 8, 5)$ we have

$$X_0 = 1,\ X_1 = 4,\ X_2 = 6,\ X_3 = 7,\ X_4 = 11 = X_5 = \ldots = X_{10}.$$

For an illustration see Figure 10.1.

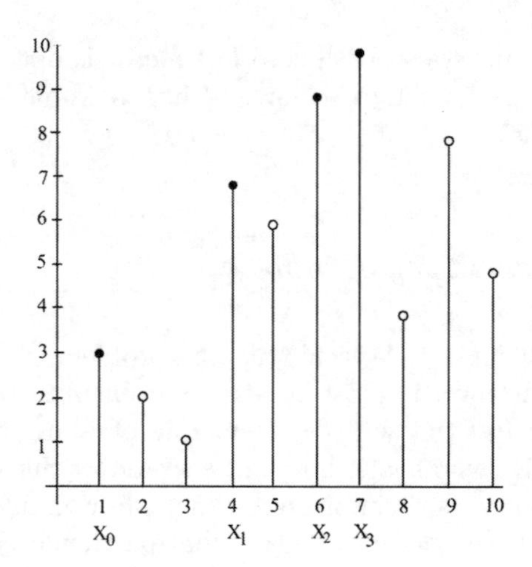

Fig. 10.1 Absolute ranks and the process (X_n) for the example.

It can now be shown that (X_n) is a Markov chain with transition probabilities (the calculation is indeed quite cumbersome; we refer to Schäl (1990) Section 1.3 or Suhov and Kelbert (2008) Section 1.11 for details):

$$
\begin{aligned}
q^X(y|x) &= \frac{x}{y(y-1)}, \quad 1 \le x < y \le N, x, y \in E, \\
q^X(N+1|x) &= \frac{x}{N}, \quad 1 \le x \le N, \qquad (10.4)\\
q^X(N+1|N+1) &= 1.
\end{aligned}
$$

All other probabilities are zero. Obviously $N+1$ is an absorbing state of the Markov chain. The aim is to maximize the *probability that the best candidate is chosen*. For a fixed stopping time τ on $\{0, 1, \dots, N-1\}$ this probability is given by

$$
\begin{aligned}
\mathbb{P}\left(Z_{X_\tau} = N\right) &= \mathbb{P}\left(X_\tau \le N, X_{\tau+1} = N+1\right) \\
&= \sum_{t=0}^{N-1} \sum_{y=1}^{N} \mathbb{P}\left(X_t = y, \tau = t\right) \mathbb{P}\left(X_{t+1} = N+1 | X_t = y, \tau = t\right) \\
&= \sum_{t=0}^{N-1} \sum_{y=1}^{N} \mathbb{P}\left(X_t = y, \tau = t\right) q^X(N+1|y) = \mathbb{E}\, g(X_\tau)
\end{aligned}
$$

where $g(N+1) := 0$ and $g(x) := \frac{x}{N}$ for $x = 1, \dots, N$. Thus, we have to solve the problem

$$\sup_{\tau \le N-1} \mathbb{E}\, g(X_\tau).$$

We summarize the data of the stopping problem:

- $E := \{1, 2, \dots, N, N+1\}$ where x denotes the time point a candidate is leading,
- $A = \{0, 1\}$ where $a = 0$ means reject the candidate and $a = 1$ means accept the candidate,
- q^X is given by equation (10.4),
- $c(x) \equiv 0$ and $g(N+1) := 0$, $g(x) := \frac{x}{N}$ for $x = 1, \dots, N$,
- $\beta := 1$.

Since the problem is bounded, (B_N) is satisfied. Though the problem is stationary we choose the non-stationary formulation because at every time point only a subset of the states are relevant. In particular, X_n takes only values in the set $\{n+1, \dots, N+1\}$ with positive probability for $n \in \{1, 2, \dots, N\}$. Thus, we obtain with Theorem 10.1.3 the recursion (note that we disregard the state $N+1$ since it is absorbing, cf. Example 2.3.13):

$$
\begin{aligned}
V_{N-1}(x) &= \frac{x}{N}, \\
V_n(x) &= \max\left\{\frac{x}{N}, \sum_{y=x+1}^{N} \frac{x}{y(y-1)} V_{n+1}(y)\right\}, \quad x = 1, \dots, N.
\end{aligned}
$$

Note that $V_0(1)$ is the maximal probability for choosing the best candidate. This problem can now be solved explicitly. For this instance, define the function

$$h(x) := \frac{1}{x} + \frac{1}{x+1} + \ldots + \frac{1}{N-1}, \quad x = 1, \ldots, N-1.$$

It is not difficult to show that the function h is decreasing and satisfies the inequality $h(1) > 1 > h(N-1)$. Denote by

$$k^* := \inf \big\{k \in \{1, \ldots, N-2\} \mid h(k) > 1 \geq h(k+1)\big\}. \tag{10.5}$$

Proposition 10.3.2. *We claim now that for $n = 0, \ldots, N-1$:*

$$V_n(x) = \begin{cases} 1, & x = N \\ \frac{x}{N}, & x = k^* + 1, \ldots, N-1 \\ \frac{k^*}{N} h(k^*), & x = n, \ldots, k^*. \end{cases}$$

Proof. We prove the proposition by induction. For $n = N-1$ we obtain obviously the assertion. Now suppose the statement is true for $n+1$. We will show that it is also true for n. Let $x \in \{n+1, \ldots, N\}$ be fixed. Then we obtain:

$$\begin{aligned} V_n(x) &= \max\Big\{\frac{x}{N}, \sum_{y=x+1}^{N} \frac{x}{y(y-1)} V_{n+1}(y)\Big\} \\ &= \frac{x}{N} \max\Big\{1, \sum_{y=x+1}^{N} \frac{N}{y(y-1)} V_{n+1}(y)\Big\}. \end{aligned}$$

Now if $x \geq k^* + 1$ this yields due to the induction hypothesis:

$$V_n(x) = \frac{x}{N} \max\Big\{1, \sum_{y=x+1}^{N} \frac{N}{y(y-1)} \frac{y}{N}\Big\} = \frac{x}{N} \max\{1, h(x)\} = \frac{x}{N}.$$

In case $x < k^* + 1$ we obtain

$$\begin{aligned} V_n(x) &= \frac{x}{N} \max\Big\{1, \sum_{y=k^*+1}^{N} \frac{1}{(y-1)} + \sum_{y=x+1}^{k^*} \frac{k^*}{y(y-1)} h(k^*)\Big\} \\ &= \frac{x}{N} \max\Big\{1, h(k^*) + \sum_{y=x+1}^{k^*} \frac{k^*}{y(y-1)} h(k^*)\Big\}. \end{aligned}$$

Now note that

$$h(k^*)k^*\Big(\frac{1}{k^*} + \sum_{y=x+1}^{k^*} \frac{1}{y(y-1)}\Big) = h(k^*)k^*\frac{1}{x} > 1$$

which implies the statement. □

Proposition 10.3.2 now directly implies the solution of the secretary problem:

Theorem 10.3.3. *The optimal stopping time for the secretary problem is as follows: make interviews with the first k^* candidates and reject them all (where k^* is given by equation* (10.5)*). Afterwards, take the first candidate who is better than her predecessors. The probability for choosing the best candidate is then given by $\frac{k^*}{N}h(k^*)$.*

Note that $k^* = k^*(N)$ depends on N. It holds that

$$\lim_{N\to\infty} \frac{k^*(N)}{N} = \frac{1}{e}.$$

Thus, if the number of candidates is large, approximately the first 37% will be rejected and the next one is accepted who is better than her predecessors.

10.3.4 A Bayesian Stopping Problem

In this section we consider the general problem of stopping a sequence of independent and identically distributed random variables. A special case has been solved in Section 10.3.1 (house selling problem). But this time we assume that the distribution of the random variables (offers) $Q(\cdot|\theta)$ depends on an unknown parameter $\theta \in \Theta \subset \mathbb{R}$. It is assumed that $Q(\cdot|\theta)$ has a density $q(z|\theta)$. Thus, we have to use the theory of Bayesian Markov Decision Models developed in Section 5.4 to solve the problem. In what follows we formulate the problem by a filtered Markov Decision Model and derive some general statements. We restrict to the N-stage problem and use a substochastic formulation. The data is given as follows (for the definition of $\hat{\Phi}$ and $\hat{\mu}$ we refer the reader to Section 5.4):

- $E := \mathbb{R} \times I$ where a state (x, i) gives the current offer x and the relevant information i about the unknown parameter,
- $A := \{0, 1\}$ where $a = 0$ means continue and $a = 1$ means stop the process,
- $\mathcal{Z} := \mathbb{R}$ where z denotes the offer,
- $\hat{T}\big((x,i), a, z\big) := \big(z, \hat{\Phi}(i, z)\big)$ is the transition function,
- $\hat{Q}^Z(\cdot|x, i, a) := \int Q(\cdot|\theta)\hat{\mu}(d\theta|i)$ (independent of (x, a))
- Q_0 is the prior distribution of ϑ,

- $c(x,i) \equiv -c$ and $g(x,i) := x$,
- $\beta := 1$.

In order to satisfy Assumption (B_N) we suppose that

$$\sup_{\theta \in \Theta} \int z Q(dz|\theta) < \infty.$$

It follows immediately from Theorem 10.1.5 that the n-stage value functions satisfy the recursion

$$\begin{aligned} J_0(x,i) &= x \\ J_n(x,i) &= \max\Big\{x,\ -c + \int\int J_{n-1}\big(z, \hat{\Phi}(i,z)\big) Q(dz|\theta)\hat{\mu}(d\theta|i)\Big\} \\ &= \max\{x,\ c_n(i)\}, \end{aligned}$$

where

$$c_n(i) := -c + \int\int J_{n-1}\big(z, \hat{\Phi}(i,z)\big) Q(dz|\theta)\hat{\mu}(d\theta|i), \quad i \in I.$$

Thus, the optimal stopping sets are determined by

$$S_n^* := \Big\{(x,i) \in E \ \Big|\ x \geq c_n(i)\Big\}$$

and the policy $(f_N^*, \dots, f_1^*)$ with $f_n^* := 1_{S_n^*}$ is optimal for the N-stage Bayesian stopping problem. Theorem 10.1.5 implies immediately that $c_1(i) \leq \ldots \leq c_N(i)$ for all $i \in I$. The optimal policy can be interpreted as a *state-dependent threshold policy.* Under some assumption it is possible to prove the monotonicity of the threshold levels in the information state. For this instance we use the same order relation on I as in Section 5.4:

$$i \leq i' \quad :\Leftrightarrow \quad \hat{\mu}(\cdot|i) \leq_{lr} \hat{\mu}(\cdot|i')$$

where $\leq_{lr}$ is the likelihood ratio order. Then we consider the following relation on $E := \mathbb{R} \times I$

$$(x,i) \leq (x',i') \quad :\Leftrightarrow \quad x \leq x' \text{ and } i \leq i'$$

and conclude the next result.

Theorem 10.3.4. *If the density $q(z|\theta)$ is MTP_2 in z and θ, then the functions $(x,i) \mapsto J_n(x,i)$ and $i \mapsto c_n(i)$ are increasing for all n.*

Proof. The monotonicity of $J_n(x,i)$ follows directly from Theorem 5.4.10. Moreover in the proof of Theorem 5.4.10 we have shown that

$$i \mapsto \int\int v(z, \hat{\Phi}(i,z))q(z|\theta)dz\hat{\mu}(d\theta|i)$$

is increasing for all increasing $v : E \to \mathbb{R}$ for which the integral exists. Now the statement follows from the definition of c_n. □

Example 10.3.5. In this example we consider the special case $c = 0$ and exponentially distributed random variables (offers)

$$q(z|\theta) = \frac{1}{\theta}e^{-\frac{1}{\theta}z}, \quad z \geq 0,\ \theta \in \Theta := (0,\infty).$$

Then $t_n(x_0, a_0, z_1, x_1, \ldots, a_{n-1}, z_n, x_n) = \left(\sum_{\nu=1}^n z_\nu, n\right)$ is a sufficient statistic (cf. Example 5.4.4). Thus, we have $I := \mathbb{R}_+ \times \mathbb{N}_0$ and denote $i = (s,n) \in I$. Moreover, $\hat{\Phi}((s,n),z) = (s+z, n+1)$. It can be shown that the conditional distribution of ϑ has the form

$$\hat{\mu}(d\theta|s,n) \propto \left(\frac{1}{\theta}\right)^n e^{-\frac{s}{\theta}} Q_0(d\theta)$$

if the information (s,n) is given (cf. Example 5.4.4). With this representation it is not difficult to verify that (cf. Example B.3.8)

$$i = (s,n) \leq i' = (s',n') \quad \Leftrightarrow \quad s \leq s' \text{ and } n \geq n'.$$

Further, the family of densities $q(z|\theta)$ is MTP_2 in $z \geq 0$ and $\theta \geq 0$, thus the last theorem applies.
If we assume now a special prior distribution of ϑ then we can solve the problem quite explicitly. We assume that the prior distribution Q_0 is a so-called *Inverse Gamma distribution*, i.e. the density is given by

$$Q_0(d\theta) = \frac{b^a}{\Gamma(a)}\left(\frac{1}{\theta}\right)^{a+1} e^{-\frac{b}{\theta}} d\theta, \quad \theta > 0$$

where $a > 1$ and $b > 0$ are fixed. The name relates to the fact that $\frac{1}{\vartheta}$ has a Gamma distribution with parameters a and b. Then the distribution $\hat{Q}^Z$ is given by

$$\hat{Q}^Z(dz|s,n) = \int Q(dz|\theta)\hat{\mu}(d\theta|s,n) = (n+a)\frac{(s+b)^{n+a}}{(z+s+b)^{n+a+1}}dz.$$

Hence $\hat{Q}^Z$ is a special *Second Order Beta distribution* which is for $z > 0$ in general given by

$$Be(\alpha,\beta,\gamma)(dz) = \frac{\Gamma(\alpha+\beta)}{\Gamma(\alpha)\Gamma(\beta)} z^{\alpha-1}\frac{\gamma^\beta}{(\gamma+z)^{\alpha+\beta}}dz$$

where we denote $\Gamma(\alpha) = (\alpha - 1)!$ for $\alpha \in \mathbb{N}$. Hence we can also write $\hat{Q}^Z(\cdot|s,n) = Be(1, a+n, s+b)$. The expectation is given by

$$\int z\hat{Q}^Z(dz|s,n) = \frac{s+b}{a+n-1}.$$

Note that it is reasonable to start in the information state $i = (0,0)$. Then in particular at stage k we can only have information states $i = (s,k)$ and it suffices to consider the value function $J_{N-k}(x,(s,k))$. Thus only the values $c_{N-k}(s,k)$ are interesting.

Theorem 10.3.6. *a) The functions c_{N-k} separate in the variables. More precisely:*

$$c_{N-k}(s,k) = (b+s)\hat{c}_{N-k}, \quad k = 0, \dots, N-1$$

and the $\hat{c}_k$ satisfy the following recursion:

$$\hat{c}_1 = \frac{1}{N+a-2},$$
$$\hat{c}_{N-k+1} = \frac{1}{k+a-2}\left[(k+a-1)\hat{c}_{N-k} + \left((1-\hat{c}_{N-k})^+\right)^{k+a-1}\right].$$

Moreover, the $\hat{c}_k$ are increasing in k and we can define

$$n^* = n^*(N) := \max\left\{k \in \{1,\dots,N\} \mid \hat{c}_{N-k+1} \geq 1\right\}$$

where $\max\emptyset := 0$.

b) The $\hat{c}_k$ are decreasing in the time horizon N and n^* is increasing in N.

c) The optimal policy $(f_N^*, \dots, f_1^*)$ satisfies $f_{N-k}^* \equiv 0$ for $k = 0, \dots, n^*-1$.

d) The maximal expected reward of the Bayesian stopping problem is given by $J_N(0,(0,0)) = b\hat{c}_N$.

Proof. a) We show the separation and the recursion by induction. First we obtain

$$c_1(s,N-1) = -c + \int\int zQ(dz|\theta)\hat{\mu}(d\theta|(s,N-1))$$
$$= \int z\hat{Q}^Z(dz|(s,N-1)) = \frac{s+b}{N+a-2} = (b+s)\hat{c}_1.$$

By definition we obtain for $k = N-1, \dots, 1$

$$
\begin{aligned}
c_{N-k+1}(s,k-1) &= \int \max\big\{z, c_{N-k}(s+z,k)\big\}\hat{Q}^Z\big(dz|(s,k-1)\big)\\
&= \int \max\big\{z,(b+s+z)\hat{c}_{N-k}\big\}\hat{Q}^Z\big(dz|(s,k-1)\big)\\
&= \int_0^\infty (k+a-1)\frac{(s+b)^{k+a-1}}{(z+s+b)^{k+a}}\max\big\{z,(b+s+z)\hat{c}_{N-k}\big\}dz\\
&= \int_0^\infty \frac{(k+a-1)(s+b)^{k+a}}{(z+s+b)^{k+a}}\max\Big\{\frac{z}{b+s},\Big(1+\frac{z}{b+s}\Big)\hat{c}_{N-k}\Big\}dz.
\end{aligned}
$$

Now we change the variable $\tilde{z} := \frac{z}{b+s}$ to obtain

$$
\begin{aligned}
c_{N-k+1}(s,k-1) &= (s+b)\int_0^\infty \frac{k+a-1}{(1+\tilde{z})^{k+a}}\max\big\{\tilde{z},(1+\tilde{z})\hat{c}_{N-k}\big\}d\tilde{z}\\
&= (s+b)\int_0^\infty \max\big\{\tilde{z},(1+\tilde{z})\hat{c}_{N-k}\big\}Be(1,k+a-1,1)(d\tilde{z})\\
&= (s+b)\hat{c}_{N-k+1}. \qquad (10.6)
\end{aligned}
$$

In order to evaluate the integral we distinguish between the following two cases.

Case 1: $\hat{c}_{N-k} \geq 1$: Here it follows that $\tilde{z} \leq (1+\tilde{z})\hat{c}_{N-k}$ is true for all $\tilde{z} \geq 0$ and we obtain

$$
\begin{aligned}
\hat{c}_{N-k+1} &= \hat{c}_{N-k}\int_0^\infty (1+\tilde{z})Be(1,k+a-1,1)(d\tilde{z})\\
&= \hat{c}_{N-k}\frac{k+a-1}{k+a-2}.
\end{aligned}
$$

Case 2: $\hat{c}_{N-k} < 1$: Let us denote $d_k := \frac{\hat{c}_{N-k}}{1-\hat{c}_{N-k}}$. Here we obtain.

$$
\begin{aligned}
\hat{c}_{N-k+1} &= \hat{c}_{N-k}\int_0^{d_k} (1+\tilde{z})Be(1,k+a-1,1)(d\tilde{z})\\
&\quad + \int_{d_k}^\infty \tilde{z}Be(1,k+a-1,1)(d\tilde{z})\\
&= \hat{c}_{N-k}\frac{k+a-1}{k+a-2} + \frac{1}{k+a-2}(1-\hat{c}_{N-k})^{k+a-1}\\
&= \frac{1}{k+a-2}\left[(k+a-1)\hat{c}_{N-k} + (1-\hat{c}_{N-k})^{k+a-1}\right].
\end{aligned}
$$

Hence the recursion is shown.

The fact that $\hat{c}_k \leq \hat{c}_{k+1}$ follows directly from the preceding representation.

b) Suppose that $N \leq N'$. We have to show that $\hat{c}_{k,N} \geq \hat{c}_{k,N'}$. For $k=1$ the statement is equivalent to $c_1(s,N-1) \geq c_1(s,N'-1)$ for all $s \geq 0$. Now by Theorem 10.3.4 we know that $c_1(i)$ is increasing in i which implies the result when we take into account the specific order relation on I. For $k = 2, \dots, N$ the statement follows by induction from (10.6). The fact that n^* is increasing in N follows directly from its definition.

c) By definition of $\hat{\Phi}$ it is only possible to have $s \geq x$. Moreover, it holds for $i = (s, k) \in I$

$$f^*_{N-k}(x, (s, k)) = 0 \quad \text{if and only if} \quad x < c_{N-k}(s, k) = (b + s)\hat{c}_{N-k}.$$

Since $\hat{c}_N \geq \dots \geq \hat{c}_{N-n+1*} \geq 1$ and $b > 0$ the statement holds.

d) From a) we get that $J_{N-k}(x, (s, k)) = \max\{x, (b + s)\hat{c}_{N-k}\}$ and hence $J_N(0, (0, 0)) = b\hat{c}_N$. □

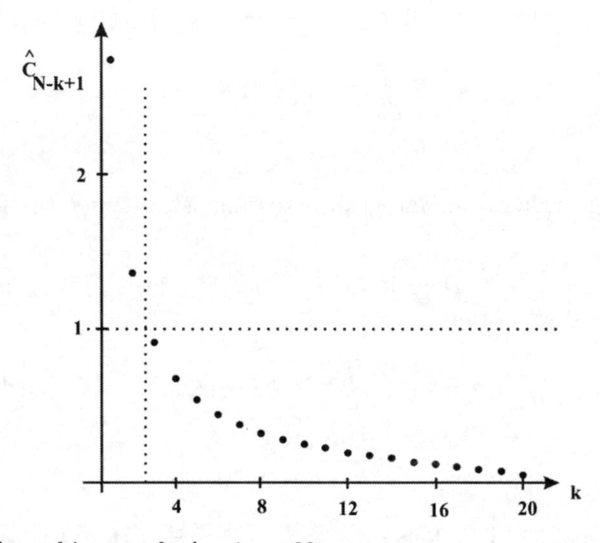

Fig. 10.2 Values of $\hat{c}_{N-k+1}$ for $k = 1, \dots, 20$.

In Figure 10.2 we see the values of $\hat{c}_k$ in the case $N = 20$ and for the parameters $a = 2, b = 1$. Theorem 10.3.6 c) implies that there is always a *training sample* of size n^* where every offer is rejected. In Table 10.1 some values for $n^*(N)$ are listed for different time horizons N.

N	5	10	20	50	100	1000
$n^*(N)$	1	2	2	3	4	6

Table 10.1 Training sample $n^*(N)$.

10.4 Exercises

Exercise 10.4.1 (Dice Game). You are allowed to throw a dice N times. After each throw you can decide to obtain the number of pips in Euro and quit the game or continue.

a) What is the optimal stopping time?
b) What is your maximal expected reward?
c) Would you play the game if $N = 3$ and you have to pay 4.50 Euro to enter the game?

Exercise 10.4.2 (Moving Particle). A particle moves on the set $E := \{0, 1, \dots, M\}$ like a symmetric random walk with absorbing barriers, i.e $q(x+1|x) = q(x-1|x) = \frac{1}{2}$ for $0 < x < M$ and $q(0|0) = q(M|M) = 1$. When you stop the particle at place x you receive the reward $g(x) \geq 0$.

a) Show that (B) is satisfied for this stopping problem.
b) Show that J is the smallest concave function v with $v \geq g$.
c) If $S^* := \{x \in E \mid J(x) = g(x)\}$ show that $\tau^* := \inf\{n \in \mathbb{N}_0 \mid X_n \in S^*\}$ is an optimal stopping time.
d) Compute J and S^* for the values $M = 7$, $g(0) = 0$, $g(1) = 4$, $g(2) = 2$, $g(3) = 1$, $g(4) = 2$, $g(5) = 5$, $g(6) = 7$, $g(7) = 0$.

Exercise 10.4.3 (Urn Game). An urn contains initially w_0 white balls and b_0 black balls. Balls are selected at random without replacement. For each selected white ball the player receives one Euro and for each selected black ball the player loses one Euro. When should the player quit the game in order to maximize her expected reward?

a) Set this up as a stopping problem and write down the optimality equation to solve the problem.
b) Suppose that a state is given by (w, b) where w and b are the number of white and black balls which are still in the urn and suppose that it is optimal to continue in this state. Prove or give a counterexample to the following claims:

   (i) it is optimal to play in state $(w+1, b)$,
   (ii) it is optimal to play in state $(w-1, b)$,
   (iii) it is optimal to play in state $(w, b+1)$,
   (iv) it is optimal to play in state $(w, b-1)$.

(This problem is taken from Ross (1983).)

Exercise 10.4.4 (Dice Game). A player is allowed to throw a dice arbitrarily often. The numbers are added and the player receives the sum in Euro as gain. However, when the player throws a 'one' the game ends and she receives nothing.

a) What is the optimal stopping time?

b) What is the maximal expected reward?

Exercise 10.4.5 (Burglar's Problem – Not for Imitation). A burglar loots some house each day. The daily gains are independent and identically distributed on $\mathbb{R}_+$. However with a certain probability $1 - p \in (0,1)$ she will be caught and loses her fortune. She uses an exponential utility $U(x) := 1 - e^{-\alpha x}, \alpha > 0$ to evaluate her fortune.

a) When should she stop?
b) What is her maximal expected reward?

Exercise 10.4.6 (Stopping Problem with Recall). Consider the house selling problem of Section 10.3.1. Now suppose that offers are not lost but a recall is possible. Show that the monotone case is on hand, i.e. Theorem 10.2.7 can be applied. What does the optimal stopping time look like?

Exercise 10.4.7 (Stock Selling). Suppose you own one stock which you would like to sell before time N. Let us assume the financial market of Chapter 3 with independent but not necessarily identically distributed relative price changes $\tilde{R}_1, \tilde{R}_2, \ldots$. The price is evaluated by a power utility function $U(x) = x^\gamma, \gamma \in (0,1)$. What is the optimal selling time? Show that $\mathbb{E}\,\tilde{R}_n^\gamma$ is an important quantity here.

10.5 Remarks and References

Most textbooks on optimal stopping problems use a martingale approach. Classical textbooks on optimal stopping problems are Chow et al. (1971) and Shiryaev (2008). The Markov Decision Process theory for non-stationary stopping problems can be found in Rieder (1975b). Optimal stopping problems have various applications. A prominent one is the sequential probability ratio test, introduced by Wald (see Exercise 5.6.4). Because of its importance some authors name the recursive equation in Theorem 10.1.3 the *Wald-Bellman* equation. A recent textbook which contains optimal stopping problems in discrete and continuous time is Peskir and Shiryaev (2006).
The examples which we have chosen here are quite classical and can be found in many textbooks. The Markov Decision Process formulation of the secretary problem also appears in Schäl (1990). Moreover, many extensions of the examples can be found in the literature. For example, an extension of the quiz show and burglar problem is given in Haigh and Roters (2000). Extensions of the secretary problem are in Freeman (1983) and Bearden and Murphy (2007). The house selling problem with a utility function is considered in Müller (2000). Partially observable stopping problems are considered in Monahan (1980, 1982a) and Nakai (1983, 1985). The Bayesian stopping problem treated in Section 10.3.4 is a modification of Tamaki (1984).

Chapter 11
Stopping Problems in Finance

Typical stopping problems in finance involve the *pricing of American options*. It can be shown by using no-arbitrage arguments that the price of an American option is the value of an optimal stopping problem under a risk neutral probability measure and the optimal stopping time is the optimal exercise time of the option. In order to have a complete financial market without arbitrage we restrict the first section on pricing American options to the binomial model. An algorithm is presented for pricing American options and the American put option is investigated in detail. In particular also perpetual American put options are studied. In Section 11.2 so-called *credit granting* problems are considered. Here the decision maker has to decide whether or not a credit is extended. In this context, a Bayesian Model is also presented.

11.1 Pricing of American Options

A classical application of optimal stopping problems in finance are American options: In order to find the fair price of an American option and its optimal exercise time, we have to solve an optimal stopping problem with finite horizon N. In contrast to a European option, the buyer of an American option can choose to exercise any time up to and including the expiration time N. In what follows we will consider the *binomial model* as underlying financial market (see Section 3.1) with the assumption $\boldsymbol{d} < 1 + i < \boldsymbol{u}$ which implies no arbitrage opportunities and the existence of a unique equivalent martingale measure $\mathbb{Q}$. This measure $\mathbb{Q}$ is used for pricing and is also called risk neutral probability measure. Under $\mathbb{Q}$, the probability for an up movement of the stock is given by

$$q = \frac{1 + i - \boldsymbol{d}}{\boldsymbol{u} - \boldsymbol{d}}. \tag{11.1}$$

We will first consider general American options with finite expiration date and concentrate on the case of path-independent options.

N. Bäuerle and U. Rieder, *Markov Decision Processes with Applications to Finance*, Universitext, DOI 10.1007/978-3-642-18324-9_11,

American Options

We concentrate our analysis on path-independent American options, i.e. the process (X_n) which has to be stopped is given by the stock price process $X_n = S_n$ itself. A path-independent American option yields the payoff $h(S_n)$ if it is exercised at time n, i.e. the payoff function depends only on the current stock price and not on its path. The expiration date is assumed to be N. However, the option may never be exercised in which case the payoff is zero. Thus, it is easy to see that it cannot be optimal to exercise when $h(S_n) < 0$ and we can equivalently choose $h^+(S_n)$ as a payoff. Let us denote $\beta := (1+i)^{-1}$. The price of this option at time zero is then computed as

$$\sup_{\tau \le N} \mathbb{E}_x^{\mathbb{Q}} \left[\beta^\tau h^+(S_\tau)\right]$$

where the supremum is taken over all stopping times τ with $\mathbb{P}(\tau \le N) = 1$. $S_0 = x$ is the stock price at time zero and the expectation is taken with respect to the risk neutral measure $\mathbb{Q}$. For example in the case of a European put option with strike price K, the payoff function h is given by $h(x) = K - x$. This stopping problem can be formulated as a stationary Markov Decision Problem (see Section 10.1). The data of the stopping problem is thus:

- $E := \mathbb{R}_+$, where x denotes the current stock price,
- $A := \{0, 1\}$ where $a = 0$ means continue and $a = 1$ means exercise,
- $Q^X(B|x) = q\delta_{x\boldsymbol{u}}(B) + (1-q)\delta_{x\boldsymbol{d}}(B), x \in E$ for Borel sets B where q is given by (11.1),
- $g(x) := h^+(x)$ and $c(x) \equiv 0$,
- $\beta := (1+i)^{-1} \in (0, 1]$ is the discount factor.

Note that when S_0 is the initial stock price, then at time n in the binomial model the only possible stock prices are given by

$$\{S_0 \boldsymbol{u}^k \boldsymbol{d}^{n-k} \mid k = 0, \dots, n\}.$$

However, it is sometimes convenient to choose a continuous state space. Assumption (B_N) is satisfied since

$$\sup_{n \le \tau \le N} \mathbb{E}_{nx}^{\mathbb{Q}} \left[\sum_{k=n}^{\tau-1} \beta^k c^+(X_k) + \beta^\tau h^+(X_\tau)\right] \le \mathbb{E}_{nx}^{\mathbb{Q}} \left[\sum_{k=n}^{N} h^+(X_k)\right] < \infty$$

because X_k can only take a finite number of possible values with positive probability for all $k = 1, \dots, N$. Moreover, the following value iteration holds for this problem (cf. Theorem 10.1.5).

Algorithm for pricing American options.
Suppose the payoff function is given by h.

1. Set $n := 0$ and define for $x \in \{S_0 \boldsymbol{d}^k \boldsymbol{u}^{N-k} \mid 0 \le k \le N\}$:
$$J_0(x) := h^+(x).$$
Set $f_0^*(x) := 1$ if $h(x) \ge 0$ and $f_0^*(x) := 0$ if $h(x) < 0$.
2. Set $n := n+1$ and compute for all $x \in \{S_0 \boldsymbol{d}^k \boldsymbol{u}^{N-n-k} \mid 0 \le k \le N-n\}$
$$J_n(x) = \max\Big\{h^+(x),\ \beta\big(qJ_{n-1}(x\boldsymbol{u}) + (1-q)J_{n-1}(x\boldsymbol{d})\big)\Big\}.$$
Set $f_n^*(x) := 1$ if $J_n(x) = h^+(x)$ and zero else.
3. If $n = N$, then the value function $J_N(S_0)$ is computed and an optimal policy π^* is given by $\pi^* = (f_N^*, \dots, f_1^*, f_0^*)$. Otherwise, go to step 2.

The price of the American option at time n is given by $\boldsymbol{\pi}_n(S_n) := J_{N-n}(S_n)$ and an optimal exercise time for the time period $[n, N]$ is
$$\tau_n^* := \inf\{k \in \{n, \dots, N\} \mid f_{N-k}^*(S_k) = 1\}.$$

Note that if we set $\inf \emptyset := N+1$, then $\tau_n^* = N+1$ means that the option is never exercised. The following iteration is equivalent and yields the same value.

$$\begin{aligned} J_0(x) &:= h^+(x) \\ J_n(x) &= \max\Big\{h(x),\ \beta\big(qJ_{n-1}(x\boldsymbol{u}) + (1-q)J_{n-1}(x\boldsymbol{d})\big)\Big\}. \end{aligned}$$

From a numerical point of view it is important that at every time point n only a subset of the stock prices in E can be attained and it is of course reasonable to compute the value function only for those prices. This is done in the algorithm. The immediate payoff $h(x)$ which is obtained when we exercise the option is called the *intrinsic value* of the option.

Example 11.1.1 (American Call Option). Let us consider the special case of an American call option with strike price K. The payoff when exercised is given by $h(x) = x - K$. It is well known that an optimal exercise strategy is to wait until the expiration date N and then exercise the option if the stock price is greater than K. Thus, the price is the same as for a European call option where the choice whether or not to exercise is only given at time N. Let us prove this statement in the framework of Markov Decision Models: We state that for $n = 0, 1, \dots, N-1$ and $x \in E$:

$$h^+(x) \le \beta\big(qJ_n(x\boldsymbol{u}) + (1-q)J_n(x\boldsymbol{d})\big)$$

which then implies that $f_1^*(x) = \ldots = f_N^*(x) = 0$ is an optimal exercise strategy, i.e. we do not exercise until time N. The inequality is true since

$$\begin{aligned}\beta\big(qJ_n(x\boldsymbol{u}) + (1-q)J_n(x\boldsymbol{d})\big) &\geq \beta\big(qJ_0(x\boldsymbol{u}) + (1-q)J_0(x\boldsymbol{d})\big)\\ &\geq qJ_0\big(\beta x\boldsymbol{u}\big) + (1-q)J_0\big(\beta x\boldsymbol{d}\big)\\ &\geq J_0\big(q\beta x\boldsymbol{u} + (1-q)\beta x\boldsymbol{d}\big)\\ &= J_0(x) \geq h^+(x)\end{aligned}$$

where we use the convexity of J_0 and the martingale property of the discounted stock price under $\mathbb{Q}$. ♦

Hedging strategy for an American option
Suppose we have an American option with payoff h and have computed the prices $\boldsymbol{\pi}_n$ according to the previous algorithm. The hedging strategy for this option is given as follows. Define for $n = 0, \ldots, N-1$:

$$\begin{aligned}a_n &:= \frac{\boldsymbol{\pi}_{n+1}(\boldsymbol{u}S_n) - \boldsymbol{\pi}_{n+1}(\boldsymbol{d}S_n)}{\boldsymbol{u}-\boldsymbol{d}}\\ c_n &:= \boldsymbol{\pi}_n(S_n) - \beta\Big(q\boldsymbol{\pi}_{n+1}(\boldsymbol{u}S_n) + (1-q)\boldsymbol{\pi}_{n+1}(\boldsymbol{d}S_n)\Big).\end{aligned}$$

As before, a_n is the amount of money invested in the stock at time n and c_n is the amount which is consumed at time n. Note that it follows from the algorithm that $c_n \geq 0$. We consider now the (self-financing) strategy (c_n, a_n) with initial wealth $\boldsymbol{\pi}_0(S_0)$. We obtain the following evolution of the wealth process (W_n) under the consumption and investment strategy (c_n, a_n):

$$\begin{aligned}W_0 &= \boldsymbol{\pi}_0(S_0)\\ W_{n+1} &= a_n\frac{S_{n+1}}{S_n} + (1+i)\Big(W_n - c_n - a_n\Big).\end{aligned}$$

We claim now that for all $n = 0, \ldots, N$:

$$W_n = \boldsymbol{\pi}_n(S_n). \tag{11.2}$$

The proof is as follows: For $n = 0$ the statement follows from the definition. Now suppose it is true for n. We obtain for $\tilde{R}_{n+1} := \frac{S_{n+1}}{S_n}$:

$$\begin{aligned}W_{n+1} &= a_n\tilde{R}_{n+1} + (1+i)\Big(W_n - c_n - a_n\Big)\\ &= a_n\tilde{R}_{n+1} + (1+i)\Big(\beta\big(q\boldsymbol{\pi}_{n+1}(\boldsymbol{u}S_n) + (1-q)\boldsymbol{\pi}_{n+1}(\boldsymbol{d}S_n)\big) - a_n\Big).\end{aligned}$$

Inserting the definition for a_n into this expression and also the definition of q we obtain

$$\begin{aligned} W_{n+1} &= \boldsymbol{\pi}_{n+1}(\boldsymbol{u}S_n)\Big(q + \frac{\tilde{R}_{n+1} - 1 - i}{\boldsymbol{u} - \boldsymbol{d}}\Big) \\ &\quad + \boldsymbol{\pi}_{n+1}(\boldsymbol{d}S_n)\Big(1 - q - \frac{\tilde{R}_{n+1} - 1 - i}{\boldsymbol{u} - \boldsymbol{d}}\Big) \\ &= \boldsymbol{\pi}_{n+1}(\boldsymbol{u}S_n)\Big(\frac{\tilde{R}_{n+1} - \boldsymbol{d}}{\boldsymbol{u} - \boldsymbol{d}}\Big) + \boldsymbol{\pi}_{n+1}(\boldsymbol{d}S_n)\Big(\frac{\boldsymbol{u} - \tilde{R}_{n+1}}{\boldsymbol{u} - \boldsymbol{d}}\Big). \end{aligned}$$

Now we discern the two cases:
Case 1: $\tilde{R}_{n+1} = \boldsymbol{u}$: In this case we obtain

$$W_{n+1} = \boldsymbol{\pi}_{n+1}(\boldsymbol{u}S_n).$$

Case 2: $\tilde{R}_{n+1} = \boldsymbol{d}$: In this case we obtain

$$W_{n+1} = \boldsymbol{\pi}_{n+1}(\boldsymbol{d}S_n).$$

Thus, in both cases we have $W_{n+1} = \boldsymbol{\pi}_{n+1}(S_{n+1})$ and the statement (11.2) is shown.
Since $W_n = \boldsymbol{\pi}_n(S_n) \geq h^+(S_n)$ this portfolio strategy hedges a short position in the American option and may further allow a consumption. However, this consumption is only positive if the option buyer does not exercise in an optimal way.

Path-dependent American options

Sometimes the payoff of an option depends on the history of the stock price evolution. For example in a Look-back option typically the maximum or the minimum of the stock price process plays a role: The payoff of a Look-back call with strike price K when exercised at time n is given by

$$\max_{1 \leq k \leq n} (S_k - K) = \max_{1 \leq k \leq n} S_k - K.$$

More generally, the payoff at time n is given by a function $h_n(S_0, \dots, S_n)$ and the task is to find

$$\sup_{\tau \leq N} \mathbb{E}_x^{\mathbb{Q}} \left[\beta^\tau h_\tau^+(S_0, \dots, S_\tau)\right]$$

where the supremum is taken over all stopping times with $\mathbb{P}_x(\tau \leq N) = 1$. In order to solve this stopping problem as a Markov Decision problem it is in general necessary to define the stock price history as the current state, i.e. a state at time n would be given by $(s_0, \dots, s_n) \in \mathbb{R}_+^n$. The pricing algorithm has then to be modified accordingly. However, sometimes partial information about the stock price history is sufficient. For example in the previous example of the Look-back call it is sufficient to take $X_n = (M_n, S_n)$ as the state process, where $M_n = \max_{1 \leq k \leq n} S_k$, since $X_{n+1} = \big(\max\{M_n, S_n R_{n+1}\}, S_n R_{n+1}\big)$ and thus (X_n) is a Markov chain.

American Put Options

In this section we consider the American put option in greater detail. In particular we establish some price properties and have a look at the so-called *perpetual American put option.*
Recall the pricing algorithm for an American put option with strike price K.

$$J_0(x) = (K - x)^+,$$
$$J_n(x) = \max\Big\{(K - x)^+,\ \beta\big(qJ_{n-1}(x\boldsymbol{u}) + (1 - q)J_{n-1}(x\boldsymbol{d})\big)\Big\}$$

where $\boldsymbol{\pi}_n(x) = J_{N-n}(x)$ is the price of the American put option at time n when the stock price is x at that time. Also recall that the equation

$$J_n(x) = \max\Big\{K - x,\ \beta\big(qJ_{n-1}(x\boldsymbol{u}) + (1 - q)J_{n-1}(x\boldsymbol{d})\big)\Big\}$$

is equivalent. The price of an American put option has the following properties.

Proposition 11.1.2. *The price $\boldsymbol{\pi}_n(x) := J_{N-n}(x)$ of an American put option has the following properties:*

a) $x \mapsto \boldsymbol{\pi}_n(x)$ is continuous.
b) $x \mapsto \boldsymbol{\pi}_n(x) + x$ is increasing.
c) $\boldsymbol{\pi}_n(x)$ is decreasing in n for all $x \in E$.
d) There exist real numbers $K =: x_N^* \geq x_{N-1}^* \geq \ldots \geq x_0^* \geq 0$ such that the optimal exercise time

$$\tau^* := \inf\{n \in \{0, \ldots, N\} \mid X_n \leq x_n^*\}$$

is of threshold type.

Proof. a) The statement follows easily by induction. Note that the maximum of continuous functions is continuous.
b) We prove the statement by induction. For $J_0 = \boldsymbol{\pi}_N$ the statement is obvious. Now suppose it is true for J_{n-1}. We will then show the property for n. To this end note that by definition of q we have $\beta q\boldsymbol{u} + \beta(1-q)\boldsymbol{d} = 1$ and thus:

$$J_n(x) + x = \max\Big\{K,\ \beta\big(q(J_{n-1}(x\boldsymbol{u}) + x\boldsymbol{u}) + (1 - q)(J_{n-1}(x\boldsymbol{d}) + x\boldsymbol{d})\big)\Big\}.$$

Now obviously the right-hand side is increasing in x by the induction hypothesis and the statement follows.
c) This follows from the general Theorem 10.1.5 b).
d) The existence of x_n^* follows from the monotonicity and continuity of J_n (cf. also Remark 10.1.6). More precisely we can define

$$x_n^* := \inf\Big\{x \in E \,|\, \beta\big(q(\boldsymbol{\pi}_{n+1}(x\boldsymbol{u}) + x\boldsymbol{u}) + (1-q)(\boldsymbol{\pi}_{n+1}(x\boldsymbol{d}) + x\boldsymbol{d})\big) \geq K\Big\}.$$

Note that $x_{N-1}^* \leq K$ since

$$\beta q(J_0(K\boldsymbol{u}) + K\boldsymbol{u}) + \beta(1-q)(J_0(K\boldsymbol{d}) + K\boldsymbol{d}) \geq K.$$

The fact that the x_n^* are decreasing follows from part c).

□

Proposition 11.1.2 implies in particular that the price of the put option is increasing in the expiration date and that it is optimal to exercise if the stock falls below a certain threshold which depends on the time to maturity and which is increasing when we approach the expiration date.

Next we consider *perpetual American put options.* The prefix 'perpetual' refers to the fact that the put has no expiration date, i.e. our stopping problem has an unbounded horizon. Options like this are not traded but serve as an approximation for large horizons which appear for example when the Black-Scholes-Merton model is approximated. The price of a perpetual American put option at time zero is given by

$$P(x) := \sup_{\tau \leq \infty} \mathbb{E}_x^{\mathbb{Q}}\left[\beta^\tau (K - S_\tau)\right]$$

where the stopping reward for $\tau = \infty$ is equal to zero.

Theorem 11.1.3. *a) The value $P(x)$ of the perpetual American put option with strike K and initial stock price $x > 0$ is given by $J(x) = \lim_{n\to\infty} J_n(x)$.*

b) P is a solution of the equation

$$P(x) = \max\Big\{(K-x)^+,\ \beta\big(qP(x\boldsymbol{u}) + (1-q)P(x\boldsymbol{d})\big)\Big\} =: \mathcal{T}P(x)$$

and $0 \leq P(x) \leq K$ for $x \in E$.

c) P is the smallest superharmonic function which majorizes $(K-x)^+$, i.e. P is the smallest solution of

$$P(x) \geq (K-x)^+, \quad P(x) \geq \beta\big(qP(x\boldsymbol{u}) + (1-q)P(x\boldsymbol{d})\big), \quad x \in E.$$

d) Let $E^* := \{x \in E \mid P(x) = (K-x)^+\}$ and $f^*(x) = 1_{E^*}(x)$. Moreover, let $J_{f^*} := \lim_{n\to\infty} \mathcal{T}_{f^*}^n 0$. If $J_{f^*} \geq \mathcal{T} J_{f^*}$ then $P(x) = J_{f^*}(x)$ for $x \in E$ and

$$\tau^* := \inf\{n \in \mathbb{N}_0 \mid X_n \in E^*\}$$

is an optimal exercise time.

e) There exists a constant $x^* \in [0, K]$ *such that*

$$E^* = \{x \in E \mid x \le x^*\}$$

i.e. it is optimal to exercise the perpetual put option the first time the stock falls below x^*.

Proof. a) Analogously to the proof of Theorem 10.1.2 it is possible to show that

$$P(x) = \sup_{\tau \le \infty} \mathbb{E}_x^{\mathbb{Q}} \left[\beta^\tau (K - S_\tau)^+\right] = J_\infty(x)$$

where J_∞ is the value function of the infinite-stage *positive* Markov Decision Model which is defined by the operator $\mathcal{T}$. From Theorem 7.4.3 part a) it follows with $h(x) = K - x$ that

$$J_\infty = \lim_{n\to\infty} \mathcal{T}^n 0 = \lim_{n\to\infty} \mathcal{T}^{n-1} h^+ = \lim_{n\to\infty} J_n = J.$$

b) This follows from part a) and Theorem 7.4.3 a). Since $0 \le J_n(x) \le K$, the same inequality holds for $P(x)$.

c) Again from Theorem 7.4.3 part b) we know that P is the smallest solution of $v \ge \mathcal{T} v$ which is equivalent to the statement.

d) Since P is the smallest solution of $v \ge \mathcal{T} v$ we obtain by our assumption that $J_{f^*} \ge P = J$. Since we always have $J_{f^*} \le J_\infty = J$ we obtain $J_{f^*} = J_\infty = P$, i.e. $(f^*, f^*, \ldots)$ is an optimal policy which is equivalent to saying that τ^* is an optimal exercise time.

e) Since $x \mapsto J_n(x) + x$ is increasing by Proposition 11.1.2, we obtain by taking the limit $n \to \infty$ that $x \mapsto P(x) + x$ is increasing and

$$P(x) + x = \max\Big\{K,\ \beta\big(q(P(x\boldsymbol{u}) + x\boldsymbol{u}) + (1-q)(P(x\boldsymbol{d}) + x\boldsymbol{d})\big)\Big\}.$$

Thus we have $E^* = \{x \in E \mid x \le x^*\}$ where

$$x^* := \inf\Big\{x \in E \mid \beta\big(q(P(x\boldsymbol{u}) + x\boldsymbol{u}) + (1-q)(P(x\boldsymbol{d}) + x\boldsymbol{d})\big) \ge K\Big\},$$

and the statement is shown. □

Remark 11.1.4. If the discount factor $\beta = (1+i)^{-1}$ is less than one, then $P = J_\infty = J = J_{f^*}$. In this case $P(x)$ is the unique bounded solution of $P = \mathcal{T} P$. Moreover, the exercise time τ^* is optimal, but note that τ^* is not finite in general. ◊

Example 11.1.5. Let us consider a perpetual American put option with the following specific data: $i = 0.25, \boldsymbol{u} = 2, \boldsymbol{d} = \frac{1}{2}$. Thus we obtain

$$\beta = (1+i)^{-1} = \frac{4}{5}, \quad q = \frac{1}{2}$$

and we are in the discounted case. The fixed point equation in this example

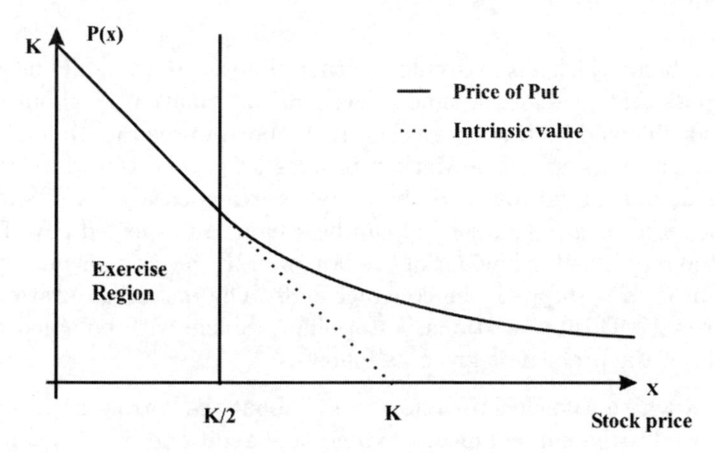

Fig. 11.1 Value and exercise region of a perpetual American option.

is

$$P(x) = \max\left\{K - x, \ \frac{2}{5}\big(P(2x) + P(\frac{1}{2}x)\big)\right\}.$$

It is possible to verify that

$$P(x) = \begin{cases} K - x \ , & \text{for } x \le \frac{K}{2} \\ \frac{K^2}{4x} \ , & \text{for } x > \frac{K}{2} \end{cases}$$

is the unique bounded solution of the fixed point equation (see Figure 11.1). Moreover, it is optimal to exercise when the stock falls below $x^* = \frac{K}{2}$. In this special case the stock price process is given by

$$S_n = S_0 2^{Z_n}, \quad n \in \mathbb{N}$$

where $Z_n = \sum_{j=1}^n Y_j$ and $\mathbb{Q}(Y_j = 1) = \mathbb{Q}(Y_j = -1) = \frac{1}{2}$, i.e. (Z_n) is a symmetric random walk. Now

$$S_n \le x^* \quad \Leftrightarrow \quad Z_n < \frac{\log x^* - \log S_0}{\log 2}.$$

Since the symmetric random walk is recurrent this will happen with probability one. Hence the exercise time

$$\tau^* = \inf\{n \in \mathbb{N}_0 \mid S_n \le x^*\}$$

satisfies $\mathbb{P}_x(\tau^* < \infty) = 1$ for all $x \in E$.

♦

11.2 Credit Granting

Imagine a bank which has to decide whether or not a credit should be granted. We suppose that the bank at time n has some information x_n about the borrower (which could be the rating class if the borrower is rated). This rating-class changes according to a Markov process (X_n) reflecting the changes in the solvency of the borrower. If the credit is extended a reward is obtained which depends on x_n. This reward can be seen as an expected payoff, anticipating the repayment behaviour of the borrower. If the credit is not extended, i.e. the process is stopped, the contract ends. The maximal duration of the contract is N. This is a stationary stopping problem with bounded horizon. The data of the problem is given as follows:

- $E := \mathbb{R}$ where x denotes the information about the borrower (rating class),
- $A := \{0, 1\}$ where $a = 0$ means extend the credit and $a = 1$ means cancel the contract,
- Q^X is the transition kernel of the information process (X_n),
- $c(x)$ is arbitrary and $g(x) \equiv 0$,
- $\beta \in (0, 1]$.

We suppose that an upper bounding function b exists for the corresponding Markov Decision Model, hence Assumption (B_N) is satisfied. Obviously Theorem 10.1.5 can be applied and we obtain the following recursion for the maximal expected discounted rewards.

$$J_0 \equiv 0$$

$$J_n(x) = \max\left\{0,\ c(x) + \beta \int J_{n-1}(y) Q^X(dy|x)\right\}, \quad x \in E.$$

To obtain some more structure we make the following assumptions:

(i) $x \mapsto c(x)$ *is increasing,*
(ii) Q^X *is stochastically monotone (for a definition see Section B.3).*

With these assumptions we obtain the following structural results.

Theorem 11.2.1. *For the credit granting model it holds:*

a) $J_n(x)$ *is increasing in* x *and* n.

b) There exist thresholds $x_N^* \le \ldots \le x_1^*$ such that the set of states in which the credit is cancelled is given by $S_n^* := \{x \in E \mid x < x_n^*\}$. The optimal credit policy $(f_N^*, \ldots, f_1^*)$ is defined by $f_n^* := 1_{S_n^*}$.

Proof. a) That $J_n(x)$ is increasing in x follows easily from Theorem 2.4.14. Note that by assumption $x \mapsto \int v(x')Q^X(dx'|x)$ is increasing for all increasing $v \in I\!B_b^+$. The monotonicity in n is implied by the general Theorem 10.1.5.

b) The existence of thresholds x_n^* follows since by part a) and our assumptions

$$x \mapsto c(x) + \beta \int J_{n-1}(x')Q^X(dx'|x)$$

is increasing (cf. Remark 10.1.6). The thresholds can be defined by

$$x_n^* := \inf\left\{x \in E \mid c(x) + \beta \int J_{n-1}(x')Q^X(dx'|x) \ge 0\right\}$$

where $\inf \emptyset = \infty$. The fact that $n \mapsto x_n^*$ is decreasing is again obtained from part a). □

Let us now consider the following *Bayesian version* of the problem: Suppose the borrower is not rated and the bank does not know the repayment probability p. It only has some prior information (distribution) μ_0 and receives a signal (either positive or negative) every period about the solvency of the borrower. Following the Bayesian approach in Section 5.4, the posterior distribution of the repayment probability after n signals has the form

$$\hat{\mu}(dp|s,n) \propto p^s(1-p)^{n-s}\mu_0(dp)$$

if $s \le n$ positive signals have been received (cf. Example 5.4.4). Thus, the expected repayment probability at that time is given by

$$q(s,n) := \frac{\int p^{s+1}(1-p)^{n-s}\mu_0(dp)}{\int p^s(1-p)^{n-s}\mu_0(dp)}, \quad s \le n.$$

The one-stage reward when the credit is granted is defined by

$$c(s,n) := K_1 q(s,n) + K_0\big(1 - q(s,n)\big) \tag{11.3}$$

where $K_1 > 0$ is the reward which is obtained if the borrower pays and $K_0 < 0$ the loss when she does not pay. Thus, we have a Markov Decision Model as in Section 5.4 with the following data:

- $E = \{(s,n) \in \mathbb{N}_0^2 \mid s \le n\}$ where (s,n) denotes number of positive signals and total number of signals received,

- $A := \{0, 1\}$ where $a = 0$ means extend the credit and $a = 1$ means cancel the contract,
- $q^X\Big((s+1, n+1)|(s, n)\Big) := q(s, n)$ and $q^X\Big((s, n+1)|(s, n)\Big) := 1 - q(s, n)$ are the transition probabilities of the information process,
- $c(s, n)$ is given as in (11.3) and $g \equiv 0$,
- $\beta \in (0, 1]$.

Obviously, Assumption (B_N) is satisfied and the value iteration is given by

$$\begin{aligned} J_0 &\equiv 0 \\ J_k(s, n) &= \max\Big\{0,\ c(s, n) + \beta q(s, n) J_{k-1}(s+1, n+1) \\ &\qquad + \beta(1 - q(s, n)) J_{k-1}(s, n+1)\Big\}, \quad (s, n) \in E. \end{aligned}$$

Again we can show the same properties of the value function and the optimal stopping time. But this time we have to introduce a relation on the state space first. We consider the following order relation on E

$$(s, n) \le (s', n') \quad :\Leftrightarrow \quad s \le s' \text{ and } n - s \ge n' - s'.$$

This means that an information state is larger if the number of positive signals is larger and the number of negative signals is smaller. In what follows a function $v : E \to \mathbb{R}$ is called increasing if $(s, n) \le (s', n')$ implies $v(s, n) \le v(s', n')$.
Note that it is reasonable to start in the state $(s, n) = (0, 0)$. Then at stage k we can only have states $(s, k) \in E$ and it suffices to consider the value function $J_{N-k}(s, k)$.

Theorem 11.2.2. *For the Bayesian credit granting model it holds:*

a) $J_k(s, n)$ is increasing in (s, n) and increasing in k.
b) There exist thresholds $t_N^* \le \ldots \le t_1^*$ such that the set of states in which the credit is cancelled is given by $S_k^* := \{(s, N-k) \in E \mid s < t_k^*\}$. The optimal credit policy $(f_N^*, \ldots, f_1^*)$ is defined by $f_k^* := 1_{S_k^*}$.

Proof. a) We can mimic the proof of the previous theorem. That J_k is increasing in k follows from the general stopping theory. For the other statement we have to show that

- $(s, n) \mapsto c(s, n) = (K_1 - K_0) q(s, n) + K_0$ is increasing,
- q^X is stochastically monotone.

Since $K_1 - K_0 \ge 0$, it is sufficient to show that $(s, n) \mapsto q(s, n)$ is increasing. But looking at the definition of $q(s, n)$ it is possible to see after some calculation that for $0 \le s \le n$

$$q(s,n+1) \le q(s,n) \le q(s+1,n+1)$$

which implies the first statement. The stochastic monotonicity of q^X is indeed simpler. According to Definition B.3.13 and Theorem B.3.3 we have to show that if $(s,n) \le (s',n')$ then $(s+1,n+1) \le (s'+1,n'+1)$ and $q(s,n) \le q(s',n')$. The first one is easy and the second one has just been shown. Finally Theorem 2.4.14 again implies the result.

b) The statement follows from part a) when we define

$$t_k^* := \min\left\{s \in \{0,1,\dots,N-k\} \mid \bar{c}_k(s,N-k) \ge 0\right\}$$

where $\min \emptyset := N-k+1$ and

$$\begin{aligned}\bar{c}_k(s,n) := c(s,n) &+ \beta q(s,n) J_{k-1}(s+1,n+1)\\ &+\beta(1-q(s,n))J_{k-1}(s,n+1).\end{aligned}$$

Then $\bar{c}_k(s,N-k) \le \bar{c}_{k+1}(s,N-k) \le \bar{c}_{k+1}(s,N-k-1)$ and hence $t_{k+1}^* \le t_k^*$. □

11.3 Remarks and References

The pricing algorithm in Section 11.1 for American options is very useful to get a numerical computation scheme. In Ben-Ameur et al. (2002) it has been used to price special Bermudan-American options with an Asian feature. In Ben-Ameur et al. (2007) the price of embedded call and put options in bonds are computed by a Markov Decision Problem. An application to installment options can be found in Ben-Ameur et al. (2006). Allaart and Monticino (2008) consider optimal buy and sell rules, i.e. multiple stopping problems. For more numerical aspects concerning the pricing algorithm for American options see Glasserman (2004), Section 8.

A discussion of the perpetual American option can be found in Shreve (2004a). Example 11.1.5 is worked out in Shreve (2004a), Section 5.4.

In Rogers (2002) a dual approach for pricing American options has been proposed. It is shown that the stopping problem is 'dual' to a problem where one has to minimize over a class of martingales. This dual problem provides in particular bounds on the prices.

The credit granting problem has been investigated by various authors. Waldmann (1998) treats the problem also with regime-switching and shows that the optimal credit policy is of threshold type.

Part V
Appendix

Appendix A
Tools from Analysis

A.1 Semicontinuous Functions

In order to prove existence of optimal policies, upper semicontinuous functions are important. For the following definition and properties we suppose that M is a metric space. We use the notation $\bar{\mathbb{R}} = \mathbb{R} \cup \{-\infty, \infty\}$.

Definition A.1.1. A function $v : M \to \bar{\mathbb{R}}$ is called *upper semicontinuous* if for all sequences $(x_n) \subset M$ with $\lim_{n\to\infty} x_n = x \in M$ it holds

$$\limsup_{n\to\infty} v(x_n) \le v(x).$$

A function $v : M \to \bar{\mathbb{R}}$ is called *lower semicontinuous* if $-v$ is upper semicontinuous.

A typical upper semicontinuous function is shown in Figure A.1.

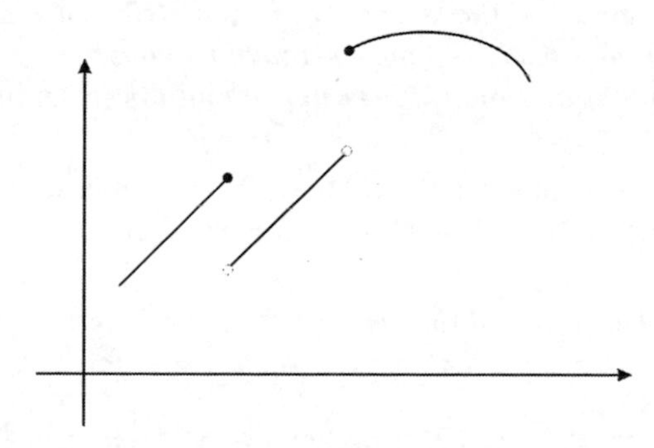

Fig. A.1 Graph of an upper semicontinuous function.

N. Bäuerle and U. Rieder, *Markov Decision Processes with Applications to Finance*, Universitext, DOI 10.1007/978-3-642-18324-9_12,

Theorem A.1.2. *Let M be compact. If $v : M \to \bar{\mathbb{R}}$ is upper semicontinuous then the function v attains its supremum.*

Proof. Let v be upper semicontinuous and denote $\alpha := \sup_{x\in M} v(x) \in \bar{\mathbb{R}}$. There exists a sequence $(x_n) \subset M$ with $\lim_{n\to\infty} v(x_n) = \alpha$. Since M is compact there exists a converging subsequence (x_{n_k}) of (x_n) with $\lim_{k\to\infty} x_{n_k} = b \in M$. Thus, we obtain

$$\alpha = \lim_{n\to\infty} v(x_n) = \lim_{k\to\infty} v(x_{n_k}) = \limsup_{k\to\infty} v(x_{n_k}) \le v(b).$$

Hence b is a maximum point of v. □

If $v : M \to \bar{\mathbb{R}}$ is upper semicontinuous and $v(x) < \infty$ for all $x \in M$, then v is bounded on every compact subset of M and attains its finite supremum. The next lemma summarizes some properties of semicontinuous functions (see e.g. Bertsekas and Shreve (1978), Puterman (1994)). Note that semicontinuous functions are also Baire functions and part a) is also called Baire's theorem on semicontinuous functions.

Lemma A.1.3. *Let $v : M \to \bar{\mathbb{R}}$ be a function.*

a) v is upper semicontinuous if and only if $\{x \in M \mid v(x) \ge \alpha\}$ is closed for all $\alpha \in \mathbb{R}$.
v is upper semicontinuous if and only if $\{x \in M \mid v(x) < \alpha\}$ is open for all $\alpha \in \mathbb{R}$.
b) v is upper semicontinuous and bounded from above if and only if there exists a sequence (v_n) of bounded and continuous functions such that $v_n \downarrow v$.
c) Let $v_i : M \to \bar{\mathbb{R}}$ be upper semicontinuous for all $i \in I$ (I arbitrary), then $\inf_{i\in I} v_i$ is upper semicontinuous.
d) If v is upper semicontinuous and M' is a metric space and $w : M' \to M$ is continuous then $v \circ w$ is upper semicontinuous.
e) v is continuous if and only if v is upper and lower semicontinuous.

In what follows we assume that $b : M \to \mathbb{R}_+$ is a measurable function and $I\!B_b^+ := \{v \in M(E) \mid v^+(x) \le cb(x) \text{ for some } c \in \mathbb{R}_+\}$.

Lemma A.1.4. *Let (v_n) and (δ_n) be sequences of functions with $v_n : M \to \mathbb{R}$ and $\delta_n : M \to \mathbb{R}_+$. Suppose that $\lim_{n\to\infty} \delta_n(x) = 0$ for all $x \in M$ and*

$$v_n(x) \le v_m(x) + \delta_m(x), \quad x \in M, n \ge m$$

i.e. (v_n) is weakly decreasing*. Then it holds:*

a) The limit $\lim_{n\to\infty} v_n =: v$ exists.

b) If $v_n \in I\!B_b^+$ *for all* $n \in \mathbb{N}$ *and* $\delta_0 \in I\!B_b^+$*, then* $v \in I\!B_b^+$.

c) If v_n *and* δ_n *are upper semicontinuous for all* $n \in \mathbb{N}$*, then* v *is upper semicontinuous.*

Proof. a) The assumptions imply that

$$\limsup_{n\to\infty} v_n(x) \le v_m(x) + \delta_m(x), \quad x \in M$$

for all $m \in \mathbb{N}$. Since $\lim_{m\to\infty} \delta_m(x) = 0$ we thus obtain

$$\limsup_{n\to\infty} v_n(x) \le \liminf_{m\to\infty} v_m(x)$$

and the limit exists.

b) Since $v \le v_0 + \delta_0 \in I\!B_b^+$ we conclude $v \in I\!B_b^+$.

c) The assumptions imply that $v \le v_m + \delta_m$ for all $m \in \mathbb{N}$. Thus, we obtain for a sequence $(x_n) \subset M$ with $x_n \to x \in M$ that for all $m \in \mathbb{N}$

$$\limsup_{n\to\infty} v(x_n) \le \limsup_{n\to\infty} \Big(v_m(x_n) + \delta_m(x_n)\Big) \le v_m(x) + \delta_m(x).$$

Taking the limit $m \to \infty$ we obtain

$$\limsup_{n\to\infty} v(x_n) \le v(x)$$

and thus v is upper semicontinuous. □

Now let (A_n) be a set sequence with $A_n \subset M$. Then we define by

$$LsA_n := \{a \in M \mid a \text{ is an accumulation point of a sequence } (a_n) \text{ with } a_n \in A_n \text{ for all } n \in \mathbb{N}\}$$

the so-called *upper limit of the set sequence* (A_n). The following theorem shows that under some continuity and compactness assumptions it is possible to interchange the limit and the supremum for a sequence of functions.

Theorem A.1.5. *Let* M *be compact and let* (v_n) *be a sequence of upper semicontinuous functions* $v_n : M \to \mathbb{R}$*. Moreover, there exists a sequence* $(\delta_n) \subset \mathbb{R}_+$ *with* $\lim_{n\to\infty} \delta_n = 0$ *and*

$$v_n(a) \le v_m(a) + \delta_m, \quad a \in M, n \ge m.$$

Then the limit $v_\infty := \lim v_n$ *exists and* v_∞ *is upper semicontinuous.*

a) Let $A_n := \{a \in M \mid v_n(a) = \sup_{x\in M} v_n(x)\}$ *for* $n \in \mathbb{N}$ *and* $n = \infty$*. Then*

$$\emptyset \ne LsA_n \subset A_\infty.$$

b) It holds:

$$\lim_{n\to\infty} \sup_{a\in M} v_n(a) = \sup_{a\in M} \lim_{n\to\infty} v_n(a) = \sup_{a\in M} v_\infty(a).$$

Proof. The first statements follow directly from Lemma A.1.4.
Since v_n is upper semicontinuous, we have $A_n \neq \emptyset$ and $\sup_{a\in M} v_n(a) < \infty$ for $n \in \mathbb{N} + \{\infty\}$. Obviously the assumption implies

$$\sup_{a\in M} v_n(a) \leq \sup_{a\in M} v_m(a) + \delta_m$$

for all $n \geq m$. Hence $\lim_{n\to\infty} \sup_{a\in M} v_n(a)$ exists. Since M is compact we have by definition that $LsA_n \neq \emptyset$. Now let $a_0 \in LsA_n$, i.e. a_0 is an accumulation point of a sequence (a_n) with $v_n(a_n) = \sup_{a\in M} v_n(a)$ for all n. Let us restrict to a subsequence (a_{n_k}) such that $\lim_{k\to\infty} a_{n_k} = a_0$. For $m \in \mathbb{N}$ it holds

$$\begin{aligned}\lim_{n\to\infty} \sup_{a\in M} v_n(a) &= \lim_{k\to\infty} \sup_{a\in M} v_{n_k}(a) = \lim_{k\to\infty} v_{n_k}(a_{n_k}) \\ &\leq \limsup_{k\to\infty} \left(v_m(a_{n_k}) + \delta_m\right) \leq v_m(a_0) + \delta_m.\end{aligned}$$

Since by assumption $v_\infty \leq v_n + \delta_n, n \in \mathbb{N}$ we further obtain together with the preceding inequality

$$\begin{aligned}\sup_{a\in M} v_\infty(a) \leq \lim_{n\to\infty} \sup_{a\in M} v_n(a) &\leq \lim_{m\to\infty} \left(v_m(a_0) + \delta_m\right) \\ &= v_\infty(a_0) \leq \sup_{a\in M} v_\infty(a).\end{aligned}$$

This implies now that $a_0 \in A_\infty$ and that

$$\sup_{a\in M} v_\infty(a) = \lim_{n\to\infty} \sup_{a\in M} v_n(a).$$

Thus, the statements in a) and b) follow. □

The interchange of supremum and limit is easier when the sequence of functions (v_n) is *weakly increasing.*

Theorem A.1.6. *Let* (v_n) *be a sequence of functions* $v_n : M \to \mathbb{R}$ *and* $(\delta_n) \subset \mathbb{R}_+$ *with* $\lim_{n\to\infty} \delta_n = 0$ *such that*

$$v_n(a) \geq v_m(a) - \delta_m, \quad a \in M, n \geq m.$$

Then the limit $v_\infty := \lim v_n$ *exists and*

$$\lim_{n\to\infty} \sup_{a\in M} v_n(a) = \sup_{a\in M} \lim_{n\to\infty} v_n(a) = \sup_{a\in M} v_\infty(a).$$

Proof. The existence of the limit follows similarly as in Lemma A.1.4. The inequality $v_n(a) \geq v_m(a) - \delta_m$ implies for all $a \in M$:

$$\lim_{n\to\infty} v_n(a) \geq v_m(a) - \delta_m.$$

Taking the supremum over all $a \in M$ and letting $n \to \infty$ yields

$$\sup_{a\in M} \lim_{n\to\infty} v_n(a) \geq \lim_{n\to\infty} \sup_{a\in M} v_n(a).$$

On the other hand we have for all $n \in \mathbb{N}$

$$\sup_{a\in M} v_n(a) \geq v_n(a), \quad a \in M.$$

Taking the limit $n \to \infty$ and then the supremum over all $a \in M$ yields

$$\lim_{n\to\infty} \sup_{a\in M} v_n(a) \geq \sup_{a\in M} \lim_{n\to\infty} v_n(a)$$

which implies the result. □

A.2 Set-Valued Mappings and a Selection Theorem

Here we suppose that E and A are Borel spaces. A set-valued mapping (also known as a multifunction or correspondence) $D(\cdot)$ from E to A is a function such that $D(x)$ is a non-empty subset of A for all $x \in E$. Here we consider only compact-valued mappings $x \mapsto D(x)$, i.e. $D(x)$ is compact for $x \in E$. In the sequel let $D := \{(x, a) \in E \times A \mid a \in D(x)\}$ be the graph of $D(\cdot)$.

Definition A.2.1. a) The set-valued mapping $x \mapsto D(x)$ is called *upper semicontinuous* if it has the following property for all $x \in E$: If $x_n \to x$ and $a_n \in D(x_n)$ for all $n \in \mathbb{N}$, then (a_n) has an accumulation point in $D(x)$.

b) The set-valued mapping $x \mapsto D(x)$ is called *lower semicontinuous* if it has the following property for all $x \in E$: If $x_n \to x$, then each point in $D(x)$ is an accumulation point of a sequence of points $a_n \in D(x_n)$ for all $n \in \mathbb{N}$.

c) The set-valued mapping $x \mapsto D(x)$ is called *continuous* if it is upper and lower semicontinuous.

Note that the definition of upper semicontinuity is slightly more restrictive than other definitions appearing in the literature (cp. Hinderer (1970), p.113). The next lemma provides some characterizations and examples for continuous or semicontinuous set-valued mappings.

Lemma A.2.2. *a) The set-valued mapping $x \mapsto D(x)$ is upper semicontinuous if and only if each sequence $(x_n, a_n) \subset D$ such that (x_n) converges in E, has an accumulation point in D. Then it follows:*

$$D \text{ is compact} \Rightarrow x \mapsto D(x) \text{ is upper semicontinuous} \Rightarrow D \text{ is closed.}$$

b) If A is compact, then $x \mapsto D(x)$ is upper semicontinuous if and only if D is closed.

c) If A is compact and $D(x) = A$ for all x, then $x \mapsto D(x)$ is continuous.

d) If $A = \mathbb{R}$ and $D(x) = [\underline{d}(x), \bar{d}(x)]$, then $x \mapsto D(x)$ is upper semicontinuous (continuous) if $\underline{d} : E \to \mathbb{R}$ is lower semicontinuous (continuous) and $\bar{d} : E \to \mathbb{R}$ is upper semicontinuous (continuous).

The following selection theorem of Kuratowski and Ryll-Nardzewski (1965) is basic for the existence of maximizers. For more selection theorems see Brown and Purves (1973), Himmelberg et al. (1976) and Rieder (1978).

Theorem A.2.3 (Selection Theorem). *Let $x \mapsto D(x)$ be a compact-valued mapping such that $D := \{(x, a) \in E \times A \mid a \in D(x)\}$ is a Borel subset of $E \times A$. Then there exists a Borel measurable selector f for D, i.e. there exists a Borel measurable function $f : E \to A$ such that $f(x) \in D(x)$ for all $x \in E$.*

From Theorem A.2.3 the existence of measurable maximizers can be derived.

Theorem A.2.4. *Let $x \mapsto D(x)$ be a compact-valued mapping such that $D := \{(x, a) \in E \times A \mid a \in D(x)\}$ is a Borel subset of $E \times A$. Let $w : D \to \mathbb{R}$ be Borel measurable. Then there exists a Borel measurable maximizer f of w, i.e. there exists a Borel measurable function $f : E \to A$ such that $f(x) \in D(x)$ for all $x \in E$ and*

$$w\big(x, f(x)\big) = \sup_{a \in D(x)} w(x, a) =: v(x), \quad x \in E.$$

Moreover, $v(x)$ is Borel measurable.

A.3 Miscellaneous

Supermodular functions are useful when monotonicity properties of maximizers are studied. They appear under different names in the literature. Sometimes they are called *L-superadditive* or functions with *increasing differences*. In what follows we denote for two vectors $x, y \in \mathbb{R}^d$

$$x \wedge y := (\min\{x_1, y_1\}, \dots, \min\{x_d, y_d\}),$$
$$x \vee y := (\max\{x_1, y_1\}, \dots, \max\{x_d, y_d\}).$$

Definition A.3.1. A function $f : \mathbb{R}^d \to \mathbb{R}$ is called *supermodular* if

$$f(x) + f(y) \le f(x \wedge y) + f(x \vee y), \quad x, y \in \mathbb{R}^d.$$

A function f is called *submodular* if $-f$ is supermodular. Note that f is supermodular if and only if all functions $(x_i, x_j) \to f(x)$ are supermodular for $i \neq j$. Thus, an alternative characterization of supermodular functions is given as follows: f is supermodular if and only if

$$\Delta_i^\varepsilon \Delta_j^\delta f(x) \ge 0, \quad x \in \mathbb{R}^d$$

for all $i, j = 1, \dots, d$ and $\varepsilon, \delta > 0$ where $\Delta_i^\varepsilon f(x) = f(x + \varepsilon e_i) - f(x)$ is the difference operator. Supermodular functions have the following useful properties (see e.g. Müller and Stoyan (2002) Theorem 3.9.3 or Bäuerle (1997)).

Lemma A.3.2. *a) If f is twice continuously differentiable, then f is supermodular if and only if for all $1 \le i, j \le n$*

$$\frac{\partial^2}{\partial x_i \partial x_j} f(x) \ge 0, \quad x \in \mathbb{R}^d.$$

b) If $g_1, \dots, g_d : \mathbb{R} \to \mathbb{R}$ are increasing and f is supermodular, then $f \circ (g_1, \dots, g_d)$ is supermodular.
c) If f, g are supermodular and $a, b \ge 0$, then $af + bg$ is supermodular.
d) If $f, g : \mathbb{R}^d \to \mathbb{R}_+$ are increasing and supermodular, then $f \cdot g$ is increasing and supermodular.

Next we introduce the concept of MTP_2 functions which is crucial when dependence properties of random vectors are discussed (for details see e.g. Müller and Stoyan (2002)).

Definition A.3.3. A function $f : \mathbb{R}^d \to \mathbb{R}_+$ is called MTP_2 (multivariate total positivity of order 2) if

$$f(x)f(y) \le f(x \wedge y) f(x \vee y), \quad x, y \in \mathbb{R}^d.$$

We obtain the following properties:

Lemma A.3.4. *a) A function $f : \mathbb{R}^d \to \mathbb{R}_+$ is MTP_2 if and only if $\log f$ is supermodular.*
b) If $f, g : \mathbb{R}^d \to \mathbb{R}_+$ are MTP_2, then also the product fg is MTP_2.

If f is a Lebesgue density of a random vector $X = (X_1, \dots, X_d)$, then f being MTP_2 implies a strong positive dependence between the random variables. In particular MTP_2 implies dependence concepts like *conditional increasing*, *association* and *positive orthant dependence*.

Finally we state Banach's fixed point theorem here since it will be important in the analysis of infinite horizon Markov Decision Models (see Section 7.3).

Theorem A.3.5 (Banach's Fixed Point Theorem). *Let M be a complete metric space with metric $d(x,y)$ and $\mathcal{T}$ an operator which satisfies*

(i) $\mathcal{T} : M \to M$.
(ii) *There exists a number $\beta \in (0,1)$ such that $d(\mathcal{T}v, \mathcal{T}w) \le \beta d(v,w)$ for all $v, w \in M$.*

Then it holds:

a) $\mathcal{T}$ has a unique fixed point v^* in M, i.e. $v^* = \mathcal{T}v^*$.
b) $\lim_{n\to\infty} \mathcal{T}^n v = v^*$ for all $v \in M$.
c) For $v \in M$ we obtain

$$d(v^*, \mathcal{T}^n v) \le \frac{\beta^n}{1-\beta} d(\mathcal{T}v, v), \quad n \in \mathbb{N}.$$

In our applications M is a closed subset of

$$\mathbb{B}_b := \{v \in \mathbb{M}(E) \,|\, \|v\|_b < \infty\}$$

with metric $d(v,w) := \|v - w\|_b$ for $v, w \in M$ and $\mathcal{T}$ is the maximal reward operator.

Appendix B
Tools from Probability

B.1 Probability Theory

In what follows we suppose that all random variables are defined on a complete probability space $(\Omega, \mathcal{F}, \mathbb{P})$. The following classical results about the interchange of expectation and limit can be found in every textbook on probability theory (see e.g. Billingsley (1995), Bauer (1996), Shiryaev (1996)).

- **(Monotone Convergence)** Suppose (X_n) is a sequence of random variables such that $X_n \uparrow X$, $X_n \geq Y$ $\mathbb{P}$-a.s. for all n and the random variable Y satisfies $\mathbb{E}\, Y > -\infty$. Then

$$\lim_{n\to\infty} \mathbb{E}\, X_n = \mathbb{E}\, X.$$

- **(Dominated Convergence)** Suppose (X_n) is a sequence of random variables such that $X_n \to X$, $|X_n| \leq Y$ $\mathbb{P}$-a.s. for all n and the random variable Y satisfies $\mathbb{E}\, Y < \infty$. Then

$$\lim_{n\to\infty} \mathbb{E}\, X_n = \mathbb{E}\, X.$$

- **(Fatou's Lemma)** Suppose (X_n) is a sequence of random variables such that $X_n \leq Y$ $\mathbb{P}$-a.s. for all n and the random variable Y satisfies $\mathbb{E}\, Y < \infty$. Then

$$\limsup_{n\to\infty} \mathbb{E}\, X_n \leq \mathbb{E}(\limsup_{n\to\infty} X_n).$$

For the infinite horizon Markov Decision Models we need the following result (see Hinderer (1970) Theorem A.3).

Theorem B.1.1. *Suppose (X_n) is a sequence of random variables with $\mathbb{E}[\sum_{k=1}^{\infty} X_k^+] < \infty$ or $\mathbb{E}[\sum_{k=1}^{\infty} X_k^-] < \infty$. Then*

N. Bäuerle and U. Rieder, *Markov Decision Processes with Applications to Finance*, Universitext, DOI 10.1007/978-3-642-18324-9_13,

$$\lim_{n\to\infty} \mathbb{E}\Big[\sum_{k=1}^{n} X_k\Big] = \mathbb{E}\Big[\sum_{k=1}^{\infty} X_k\Big] = \sum_{k=1}^{\infty} \mathbb{E}\, X_k.$$

The concept of weak convergence is needed when continuous-time processes have to be approximated by interpolations of discrete-time processes.

Definition B.1.2. Suppose that $X, X_1, X_2, \dots$ are random variables with values in a separable metric space M. Then (X_n) converges weakly against X if and only if

$$\lim_{n\to\infty} \mathbb{E}\, f(X_n) = \mathbb{E}\, f(X)$$

for all continuous and bounded functions $f : M \to \mathbb{R}$.

B.2 Stochastic Processes

In what follows we summarize definitions and facts from discrete-time Markov processes and martingales. For details on Markov processes we refer the reader to Meyn and Tweedie (2009) and for more information on continuous-time processes and predictability see e.g. Protter (2005).

Definition B.2.1. A family of random variables $(X_n)_{n\in\mathbb{N}_0}$ on a probability space $(\Omega, \mathcal{F}, \mathbb{P})$ with values in a measurable space $(E, \mathfrak{E})$ is called a *stochastic process (in discrete time)*.

Definition B.2.2. Let $(\Omega, \mathcal{F}, \mathbb{P})$ be a probability space.

a) A sequence of σ-algebras $(\mathcal{F}_n)$ is called a *filtration* if $\mathcal{F}_0 \subset \mathcal{F}_n \subset \mathcal{F}_{n+1} \subset \mathcal{F}$.

b) A stochastic process (X_n) on $(\Omega, \mathcal{F}, \mathbb{P})$ is called $(\mathcal{F}_n)$*-adapted* if X_n is $\mathcal{F}_n$-measurable for all n.

If $\mathcal{F}_n = \mathcal{F}_n^X := \sigma(X_0, \dots, X_n)$, i.e. $\mathcal{F}_n$ is the smallest σ-algebra such that the random variables $X_0, \dots, X_n$ are measurable with respect to $\mathcal{F}_n$, then $(\mathcal{F}_n)$ is called the *natural filtration* of (X_n). In this case (X_n) is trivially adapted to $(\mathcal{F}_n)$. Moreover, if a random variable Y is $\mathcal{F}_n^X$-measurable then there exists a measurable function $h : E^{n+1} \to \mathbb{R}$ such that $Y = h(X_0, \dots, X_n)$ $\mathbb{P}$-a.s.

Recall the following definition from Chapter 2.

Definition B.2.3. A mapping $Q : \mathfrak{E} \times E \to [0, 1]$ with the two properties

(i) $B \mapsto Q(B|x)$ is a probability measure for all $x \in E$,
(ii) $x \mapsto Q(B|x)$ is measurable for all $B \in \mathfrak{E}$,

is called a *stochastic (transition) kernel.*

The second property implies that whenever $v : E \times E \to \mathbb{R}$ is measurable, then

$$x \mapsto \int v(x, x')Q(dx'|x)$$

is again measurable whenever the integral exists.

Definition B.2.4. A stochastic process (X_n) is called a *(discrete-time) Markov process*, if there exists a sequence of stochastic kernels (Q_n) such that

$$\mathbb{P}(X_{n+1} \in B|\mathcal{F}_n^X) = \mathbb{P}(X_{n+1} \in B|X_n) = Q_n(B|X_n).$$

If (Q_n) does not depend on n the process is called a *stationary (or homogeneous) Markov process.* The first equality is called the *Markov property.*

A stochastic process (X_n) is a Markov process if and only if there exist independent random variables $Z_1, Z_2, \dots$ with values in a measurable space $(\mathcal{Z}, \mathfrak{Z})$ and measurable functions $T_n : E \times \mathcal{Z} \to E$, $n = 0, 1, 2, \dots$ such that X_0 is given and

$$X_{n+1} = T_n(X_n, Z_{n+1}),\ n = 0, 1, 2, \dots.$$

If the state space E of the stationary Markov process (X_n) is finite or countable, the transition kernel is represented by a stochastic matrix $P = (p_{ij})_{i,j \in E}$, i.e. $p_{ij} \geq 0$ and $\sum_j p_{ij} = 1$ for all $i \in E$ and it holds

$$\mathbb{P}(X_{n+1} = j|X_n = i) = p_{ij}.$$

In this case we also call (X_n) a *Markov chain.*

Proposition B.2.5 (Theorem of Ionescu-Tulcea). *Let ν be a probability measure on E and (Q_n) a sequence of stochastic kernels. Then there exists a unique probability measure $\mathbb{P}_\nu$ on E^∞ such that*

$$\mathbb{P}_\nu(B_0 \times \ldots \times B_N \times E \times \ldots) = \int_{B_0} \cdots \int_{B_N} Q_{N-1}(dx_N|x_{N-1}) \ldots Q_0(dx_1|x_0)\nu(dx_0)$$

for every measurable rectangle set $B_0 \times \ldots \times B_N \in \mathcal{E}^{N+1}$.

Definition B.2.6. A stochastic process (X_n) which is $(\mathcal{F}_n)$-adapted and satisfies $\mathbb{E}\,|X_n| < \infty$ for all n is called an $(\mathcal{F}_n)$*-martingale* if

$$\mathbb{E}[X_{n+1}|\mathcal{F}_n] = X_n \text{ for all } n \in \mathbb{N}.$$

The process (X_n) is called an $(\mathcal{F}_n)$-*supermartingale* if

$$\mathbb{E}[X_{n+1}|\mathcal{F}_n] \leq X_n \text{ for all } n \in \mathbb{N}.$$

The process (X_n) is called an $(\mathcal{F}_n)$-*submartingale* if

$$\mathbb{E}[X_{n+1}|\mathcal{F}_n] \geq X_n \text{ for all } n \in \mathbb{N}.$$

Note that the condition $\mathbb{E}[X_{n+1}|\mathcal{F}_n] = X_n$ for all n is equivalent to $\mathbb{E}[X_m|\mathcal{F}_n] = X_n$, for all $n < m$. Often the filtration is not explicitly mentioned in which case we assume that it is the natural filtration of the process.

Definition B.2.7. a) A continuous-time stochastic process $(N_t)_{t \geq 0}$ with values in $\mathbb{N}_0$ and $N_0 = 0$ is called a *homogeneous Poisson process* with intensity $\lambda > 0$ if it has independent increments and for $0 \leq s < t$ the increment $N_t - N_s$ is Poisson-distributed with parameter $\lambda(t-s)$.
b) A continuous-time stochastic process $(C_t)_{t \geq 0}$ with values in $\mathbb{R}^d$ and $C_0 = 0$ is called a *compound Poisson process* if it is given by

$$C_t := \sum_{k=1}^{N_t} Y_k,$$

where (N_t) is a Poisson process and $Y_1, Y_2, \dots$ is a sequence of independent and identically distributed random vectors with values in $\mathbb{R}^d$ which is independent of (N_t).

Definition B.2.8. A continuous-time stochastic process $(W_t)_{t \geq 0}$ with values in $\mathbb{R}$ and $W_0 = 0$ is called a *Wiener process* or *Brownian motion*, if it has almost surely continuous paths, independent increments and for $0 \leq s < t$ we have $W_t - W_s \sim \mathcal{N}(0, t-s)$.

Definition B.2.9. Denote by $\mathcal{P}$ the σ-algebra in $[0,T] \times \Omega$ generated by all adapted processes with left-continuous paths. A function $X : [0,T] \times \Omega \to \mathbb{R}^d$ which is measurable with respect to $\mathcal{P}$ is called a *predictable* process. For example any left-continuous process is predictable. ◊

B.3 Stochastic Orders

Stochastic orders are partial orders on a set of distributions. Here we consider the set of all distributions of real-valued random variables. The multivariate case is more delicate since there are different reasonable extensions of the univariate case. Stochastic orders are a valuable tool for obtaining bounds on performance measures or for deriving sensitivity results. General books on stochastic orders are Müller and Stoyan (2002) and Shaked and Shanthikumar (2007). The focus of the latter one is on applications. A particular view towards dependence issues can be found in Szekli (1995) and Nelsen (2006). In Kaas et al. (1998) one can find applications concerning the ordering of actuarial risks. Monotonicity properties of stochastic systems have been discussed extensively in Müller and Stoyan (2002), Chapters 5 and 6.

Definition B.3.1. A binary relation $\preceq$ on an arbitrary set S is called a *preorder* if

(i) $x \preceq x$ for all $x \in S$ (*Reflexivity*),
(ii) if $x \preceq y$ and $y \preceq z$, then $x \preceq z$ (*Transitivity*).

If $\preceq$ is also antisymmetric, i.e. if $x \preceq y$ and $y \preceq x$ imply $x = y$, then $\preceq$ is called a *(partial) order*. $(S, \preceq)$ is called *completely ordered* if for any $x, y \in S$ either $x \preceq y$ or $y \preceq x$.

The Usual Stochastic Order

Definition B.3.2. The random variable X is called smaller than the random variable Y with respect to the *stochastic order* (written $X \leq_{st} Y$), if $\mathbb{P}(X \leq t) \geq \mathbb{P}(Y \leq t)$ for all $t \in \mathbb{R}$.

In particular in the economic literature this order is often called *first order stochastic dominance* and the symbol $\leq_{FSD}$ is used. Note that $X \leq_{st} Y$ implies $\mathbb{P}(X > t) \leq \mathbb{P}(Y > t)$ for all $t \in \mathbb{R}$, i.e. Y takes larger values with larger probability. Since it follows from the definition that the order relation depends only on the distribution of X and Y it is common to write $P^X \leq_{st} P^Y$ also.
Useful characterizations of the stochastic order are given in the next theorem.

Theorem B.3.3. *Let X and Y be two random variables. The following statements are equivalent:*

(i) $X \leq_{st} Y$.
(ii) *For all increasing $f : \mathbb{R} \to \mathbb{R}$, it holds that $\mathbb{E} f(X) \leq \mathbb{E} f(Y)$, whenever the expectations exist.*
(iii) *There exists a probability space $(\Omega, \mathcal{F}, \mathbb{P})$ and random variables $\hat{X}$, $\hat{Y}$ on it such that $X \stackrel{d}{=} \hat{X}$, $Y \stackrel{d}{=} \hat{Y}$ and $\hat{X}(\omega) \leq \hat{Y}(\omega)$ for all $\omega \in \Omega$.*

Obviously $X \leq_{st} Y$ implies $\mathbb{E}X \leq \mathbb{E}Y$. Besides this, the stochastic order has a number of important properties. We mention only some of them. For examples see the next section.

Lemma B.3.4. *a) The stochastic order is* closed under convolution, *i.e. if X_1 and X_2 are independent random variables as well as Y_1 and Y_2 and $X_i \leq_{st} Y_i$ for $i = 1, 2$, then*

$$X_1 + X_2 \leq_{st} Y_1 + Y_2.$$

b) The stochastic order is closed under mixtures, *i.e. if X, Y and Z are random variables with $\mathbb{P}(X \leq t \mid Z = z) \geq \mathbb{P}(Y \leq t \mid Z = z)$ for all $t \in \mathbb{R}$ and $\mathbb{P}$-almost all z then $X \leq_{st} Y$.*

c) The stochastic order is closed with respect to weak convergence, *i.e. if $X_n \leq_{st} Y_n$ for all n and (X_n) converges weakly against X, (Y_n) converges weakly to Y, then $X \leq_{st} Y$.*

Likelihood Ratio Order

The likelihood ratio order is stronger than the stochastic order and important for comparison results in Bayesian models. Often it is easier to verify than the stochastic order.

Definition B.3.5. The random variable X is called smaller than the random variable Y with respect to the *likelihood ratio order* (written $X \leq_{lr} Y$), if X and Y have densities f_X and f_Y with respect to some dominating measure such that for all $s \leq t$:

$$f_X(t) f_Y(s) \leq f_X(s) f_Y(t).$$

Note that the definition is valid for continuous as well as discrete random variables or mixtures of both. In the multivariate case there are different possibilities to define a likelihood ratio order.

Theorem B.3.6. *If X and Y are random variables with $X \leq_{lr} Y$, then also $X \leq_{st} Y$.*

The following characterization is important for the understanding of the likelihood ratio order and particularly useful in Bayesian settings:

Theorem B.3.7. *Let X and Y be two random variables. The following statements are equivalent:*

(i) $X \leq_{lr} Y$.

(ii) *For all events A with $\mathbb{P}(X \in A) > 0$ and $\mathbb{P}(Y \in A) > 0$ we have*

$$\mathbb{P}(X \leq t \mid X \in A) \geq \mathbb{P}(Y \leq t \mid Y \in A), \quad t \in \mathbb{R}.$$

Example B.3.8. a) The discrete density of the Poisson distribution $Poi(\lambda)$ for $\lambda > 0$ is given by

$$p(k) = e^{-\lambda}\frac{\lambda^k}{k!}, \quad k \in \mathbb{N}_0.$$

It holds: $Poi(\lambda) \leq_{lr} Poi(\mu)$ if and only if $\lambda \leq \mu$.

b) The discrete density of the Binomial distribution $B(n,p)$ is for $n \in \mathbb{N}$, $p \in (0,1)$ given by

$$p(k) = \binom{n}{k} p^k (1-p)^{n-k}, \quad k = 0, \ldots, n.$$

It holds: $B(n,p) \leq_{lr} B(m,q)$ if $n \leq m$ and $p \leq q$.

c) The density of the Exponential distribution $Exp(\lambda)$ for $\lambda > 0$ is given by

$$f(x) = \lambda e^{-\lambda x} 1_{[x \geq 0]}.$$

It holds: $Exp(\lambda) \leq_{lr} Exp(\mu)$ if and only if $\lambda \geq \mu$.

d) The density of the Beta distribution $Be(\alpha, \beta)$ for $\alpha, \beta > 0$ is given by

$$f(x) \propto x^{\alpha-1}(1-x)^{\beta-1} 1_{[0 \leq x \leq 1]}.$$

It holds: $Be(\alpha,\beta) \leq_{lr} Be(\gamma,\delta)$ if and only if $\alpha \leq \gamma$ and $\beta \geq \delta$. ♦

Convex Orders

Convex orders compare the variability of random variables (instead of the size as done by the stochastic and likelihood ratio order). Remember that a function $f : \mathbb{R} \to \mathbb{R}$ is called *convex* if for all $\alpha \in (0,1)$ and for all $x, x' \in \mathbb{R}$:

$$f(\alpha x + (1-\alpha)x') \leq \alpha f(x) + (1-\alpha) f(x').$$

f is called *concave* if $-f$ is convex.

Definition B.3.9. Let X and Y be random variables with finite mean.

a) X is called smaller than Y with respect to the *convex order* (written $X \leq_{cx} Y$), if $\mathbb{E} f(X) \leq \mathbb{E} f(Y)$ for all convex functions f for which the expectations exist.
b) X is called smaller than Y with respect to the *increasing convex order* (written $X \leq_{icx} Y$), if $\mathbb{E} f(X) \leq \mathbb{E} f(Y)$ for all increasing, convex functions f for which the expectations exist.

c) X is called smaller than Y with respect to the *increasing concave order* (written $X \leq_{icv} Y$), if $\mathbb{E} f(X) \leq \mathbb{E} f(Y)$ for all increasing, concave functions f for which the expectations exist.

The $\leq_{icv}$ order is often called *second order stochastic dominance* in the economic literature and the symbol $\leq_{SSD}$ is used, whereas the $\leq_{icx}$ order is known as the *stop-loss order* $\leq_{sl}$ in the actuarial sciences. Moreover, it is not difficult to see that $X \leq_{icx} Y$ is equivalent to $-Y \leq_{icv} -X$, thus in what follows we restrict statements to the $\leq_{icx}$ case. Moreover, it follows immediately from the definition that $X \leq_{st} Y$ implies $X \leq_{icx} Y$.

Theorem B.3.10. *Let X and Y be two random variables. The following statements are equivalent:*

(i) $X \leq_{cx} Y$.
(ii) $X \leq_{icx} Y$ *and* $\mathbb{E} X = \mathbb{E} Y$.
(iii) *There exists a probability space $(\Omega, \mathcal{F}, \mathbb{P})$ and random variables $\hat{X}$, $\hat{Y}$ on it with $X \stackrel{d}{=} \hat{X}$, $Y \stackrel{d}{=} \hat{Y}$ such that $\mathbb{E}[\hat{Y} \mid \hat{X}] = \hat{X}$ $\mathbb{P}$-a.s. and that $\mathbb{P}(\hat{Y} \leq t \mid \hat{X} = x) \geq \mathbb{P}(\hat{Y} \leq t \mid \hat{X} = x')$ for all $t \in \mathbb{R}$ and $x < x'$.*

Theorem B.3.11. *Let X and Y be two random variables. The following statements are equivalent:*

(i) $X \leq_{icx} Y$.
(ii) $\mathbb{E}(X - t)_+ \leq \mathbb{E}(Y - t)_+$ *for all* $t \in \mathbb{R}$.
(iii) *There exists a probability space $(\Omega, \mathcal{F}, \mathbb{P})$ and random variables $\hat{X}$, $\hat{Y}$ on it with $X \stackrel{d}{=} \hat{X}$, $Y \stackrel{d}{=} \hat{Y}$ such that $\mathbb{E}[\hat{Y} \mid \hat{X}] \geq \hat{X}$ a.s. and that $\mathbb{P}(\hat{Y} \leq t \mid \hat{X} = x) \geq \mathbb{P}(\hat{Y} \leq t \mid \hat{X} = x')$ for all $t \in \mathbb{R}$ and $x < x'$.*

Since $\leq_{lr}$ implies $\leq_{icx}$ all examples of the previous section can also be used for the increasing convex order. Another interesting example is:

Example B.3.12. The density of the Normal distribution $\mathcal{N}(\mu, \sigma^2)$ for $\mu \in \mathbb{R}, \sigma > 0$ is given by

$$f(x) = \frac{1}{\sigma\sqrt{2\pi}} \exp\Big(- \frac{(x-\mu)^2}{2\sigma^2}\Big), \quad x \in \mathbb{R}.$$

We have $\mathcal{N}(\mu, \sigma^2) \leq_{cx} \mathcal{N}(\nu, \tau^2)$ if and only if $\mu = \nu$ and $\tau \geq \sigma$. ♦

Stochastic Monotonicity of Markov Processes

Suppose (X_n) is a stationary Markov process with state space $E \subset \mathbb{R}^d$ and stochastic transition kernel $Q(B|x)$.

Definition B.3.13. The stochastic kernel Q is said to be *stochastically monotone* if for all increasing $v : E \to \mathbb{R}$ the function

$$x \mapsto \int v(x')Q(dx'|x)$$

is increasing whenever the integral exists. If Q defines a Markov process (X_n) then we also say that (X_n) is *stochastically monotone.*

From the definition of the stochastic order it follows immediately that the kernel Q is stochastically monotone if and only if $x \le x', \ x, x' \in E$ implies $Q(\cdot|x) \le_{st} Q(\cdot|x')$. If the state space $E = \{1, \dots, m\}$ is finite and the transition probabilities are given by $P = (p_{ij})$ then the Markov chain (X_n) is stochastically monotone if and only if for all $i, j \in E$ with $i \le j$

$$\sum_{\nu=k}^{m} p_{i\nu} \le \sum_{\nu=k}^{m} p_{j\nu}, \quad k \in E.$$

If we denote by $\alpha \in \mathbb{R}^m$ the initial distribution, then αP^n is the distribution of the Markov chain at time $n \in \mathbb{N}$ and (X_n) is stochastically monotone if for any initial distributions α and α', $\alpha \le_{st} \alpha'$ implies $\alpha P^n \le_{st} \alpha' P^n$ for all $n \in \mathbb{N}$.

Example B.3.14. Suppose (X_n) is a discrete birth-and-death process with state space $E = \{0, 1, \dots, m\}$, i.e. $p_{i,i+1} = p$ for $i = 0, \dots, m-1$, $p_{i,i-1} = q$ for $i = 1, \dots, m$, $p_{0,0} = 1 - p$, $p_{mm} = 1 - q$ for $p + q = 1$ and $p \in (0, 1)$. Then it is not difficult to verify that this Markov chain is stochastically monotone by using the preceding criterion. ♦

Appendix C
Tools from Mathematical Finance

C.1 No Arbitrage Pricing Theory

In this section we summarize some facts from the fundamental no arbitrage pricing theory and shed some light on the role of martingales in option pricing. For details see Föllmer and Schied (2004) Chapter 5. In what follows suppose a filtered probability space $(\Omega, \mathcal{F}, (\mathcal{F}_n), \mathbb{P})$ is given where $\mathcal{F}_0 := \{\emptyset, \Omega\}$. On this space there exist $d+1$ assets and the price at time $n = 0, 1, \ldots, N$ of asset k is modelled by a random variable S_n^k (see Section 3.1 for a detailed description of the financial market). Asset S^0 is a riskless bond which is used as a numeraire.

Definition C.1.1. a) A probability measure $\mathbb{Q}$ on $(\Omega, \mathcal{F})$ is called *equivalent to* $\mathbb{P}$ if $\mathbb{Q}$ and $\mathbb{P}$ have the same null sets.
b) A probability measure $\mathbb{Q}$ on $(\Omega, \mathcal{F}_N)$ is called a *martingale measure* if the discounted stock price process $(\frac{S_n^k}{S_n^0})$ is an $(\mathcal{F}_n)$-martingale under $\mathbb{Q}$ for all $k = 1, \ldots, d$.

The following characterization of no arbitrage is crucial.

Theorem C.1.2. *The financial market is free of arbitrage if and only if there exists an equivalent martingale measure.*

Definition C.1.3. a) A *contingent claim* is a non-negative, $\mathcal{F}_N$-measurable random variable H.
b) A contingent claim is said to be *attainable* if there exists a self-financing portfolio strategy ϕ which replicates the payoff H, i.e.

$$H = X_N^\phi, \quad \mathbb{P}\text{-a.s.}$$

N. Bäuerle and U. Rieder, *Markov Decision Processes with Applications to Finance*, Universitext, DOI 10.1007/978-3-642-18324-9_14,

The portfolio strategy ϕ is called a *replicating strategy* or *hedging strategy.*

c) The financial market is called *complete* if every contingent claim is attainable.

Theorem C.1.4. *Suppose the financial market admits no arbitrage opportunities. Then the market is complete if and only if there exists a unique equivalent martingale measure.*

Definition C.1.5. Suppose there are no arbitrage opportunities and H is an attainable contingent claim. Its price $\boldsymbol{\pi}(H)$ is then defined as the initial amount which is necessary to replicate it, i.e. if ϕ is a hedging strategy then

$$\boldsymbol{\pi}(H) := \phi_0^0 + \phi_0 \cdot e.$$

Theorem C.1.6. *Suppose there are no arbitrage opportunities and $\mathbb{Q}$ is an equivalent martingale measure. Then the price of an attainable contingent claim H can be computed by*

$$\boldsymbol{\pi}(H) = \mathbb{E}_{\mathbb{Q}}\left[\frac{H}{S_N^0}\right].$$

If the contingent claim H is not attainable, then an interval of arbitrage-free prices can be computed by

$$\left(\inf_{\mathbb{Q}\in\mathcal{Q}} E_{\mathbb{Q}}\left[\frac{H}{S_N^0}\right], \sup_{\mathbb{Q}\in\mathcal{Q}} E_{\mathbb{Q}}\left[\frac{H}{S_N^0}\right]\right)$$

where $\mathcal{Q}$ is the set of all equivalent martingale measures.

Example C.1.7 (Binomial or Cox-Ross-Rubinstein Model). If the parameters in the binomial model satisfy

$$\boldsymbol{d} < 1 + i < \boldsymbol{u}$$

then the market admits no arbitrage opportunities (see Example 3.1.7) and is also complete. If we denote the sample space in this model by

$$\Omega := \Big\{(\omega_1, \dots, \omega_N) \mid \omega_i \in \{\boldsymbol{d}, \boldsymbol{u}\}\Big\}$$

and $\tilde{R}_n(\omega) = \omega_n$ then the unique equivalent martingale measure $\mathbb{Q}$ is determined by

$$\mathbb{Q}(\tilde{R}_n = \boldsymbol{u}) = q = \frac{1 + i - \boldsymbol{d}}{\boldsymbol{u} - \boldsymbol{d}}$$

and the fact that the random relative price changes $\tilde{R}_1, \ldots, \tilde{R}_N$ are independent under $\mathbb{Q}$. ♦

C.2 Risk Measures

In what follows suppose that we have a probability space $(\Omega, \mathcal{F}, \mathbb{P})$ and we denote by $L^1(\Omega, \mathcal{F}, \mathbb{P})$ the (equivalence classes of) integrable random variables. A random variable $X \in L^1(\Omega, \mathcal{F}, \mathbb{P})$ is interpreted as a risk with the convention that positive values are rewards and negative values are losses. A risk measure ρ maps a risk X on a real number with the interpretation that $\rho(X)$ is the amount of money which is necessary to make the risk acceptable. A theoretical foundation of the theory of risk measures can be found in Föllmer and Schied (2004) and Pflug and Römisch (2007). The latter book also contains a number of applications.

Definition C.2.1. A function $\rho : L^1(\Omega, \mathcal{F}, \mathbb{P}) \to \mathbb{R}$ is called a *risk measure.* Let $X_1, X_2 \in L^1(\Omega, \mathcal{F}, \mathbb{P})$.

a) ρ is called *monotone* if

$$X_1 \leq X_2 \quad \Longrightarrow \rho(X_1) \geq \rho(X_2).$$

b) ρ is called *cash invariant,* if for all $c \in \mathbb{R}$

$$\rho(X + c) = \rho(X) - c.$$

c) ρ is called *(positive) homogeneous,* if for all $\lambda \geq 0$

$$\rho(\lambda X) = \lambda \rho(X).$$

d) ρ is called *subadditive* if

$$\rho(X_1 + X_2) \leq \rho(X_1) + \rho(X_2).$$

Properties a)–d) are reasonable properties for a risk measure. If all are satisfied, the risk measure is called *coherent.* Note that subadditivity rewards diversification and that coherence implies $\rho(0) = 0$ and $\rho(X + \rho(X)) = 0$. A risk measure ρ is called *convex* if it is monotone, cash invariant and satisfies $\rho\big(\lambda X + (1 - \lambda)Y\big) \leq \lambda \rho(X) + (1 - \lambda)\rho(Y)$ for all $X, Y \in L^1$ and $\lambda \in (0, 1)$.

Example C.2.2. a) **Value-at-Risk**: For $X \in L^1(\Omega, \mathcal{F}, \mathbb{P})$ the Value-at-Risk of X at level $\gamma \in (0, 1)$ is defined by

$$VaR_\gamma(X) := \inf\{x \in \mathbb{R} \mid \mathbb{P}(x + X < 0) \le 1 - \gamma\}.$$

The Value-at-Risk is the smallest γ-quantile of $-X$. For applications γ has to be large, e.g. $\gamma = 0.995$ in the regulatory framework Solvency II. Note that Value-at-Risk is monotone, cash invariant and homogeneous, but in general not subadditive. Hence VaR is not *coherent*. The Value-at-Risk is increasing in γ.

b) **Average-Value-at-Risk**: For $X \in L^1(\Omega, \mathcal{F}, \mathbb{P})$ the Average-Value-at-Risk of X at level $\gamma \in (0,1)$ is given by

$$AVaR_\gamma(X) = \frac{1}{1-\gamma}\int_\gamma^1 VaR_u(X)du.$$

$AVaR_\gamma(X)$ is continuous and strictly increasing in γ. It holds that $\lim_{\gamma\to 0} AVaR_\gamma(X) = -\mathbb{E}X$ and $\lim_{\gamma\to 1} AVaR_\gamma(X)$ tends to the worst outcome. Moreover, it holds that

$$AVaR_\gamma(X) \ge VaR_\gamma(X), \quad \gamma \in (0,1)$$

and $AVaR_\gamma(X)$ is *coherent*. It is important to note that the Average-Value-at-Risk can be computed as the solution of the following optimization problem:

$$AVaR_\gamma(X) = \inf_{b\in\mathbb{R}} \left\{ b + \frac{1}{1-\gamma}\mathbb{E}[(X+b)^-] \right\}. \tag{C.1}$$

The minimum is attained at $b^* = VaR_\gamma(X)$. ♦

References

ABRAMS, R. A. and KARMARKAR, U. S. (1980) Optimal multiperiod investment-consumption policies. *Econometrica* **48**, 333–353.

ALBRIGHT, S. C. (1979) Structural results for partially observable Markov decision processes. *Oper. Res.* **27**, 1041–1053.

ALLAART, P. and MONTICINO, M. (2008) Optimal buy/sell rules for correlated random walks. *J. Appl. Probab.* **45**, 33–44.

ALMUDEVAR, A. (2001) A dynamic programming algorithm for the optimal control of piecewise deterministic Markov processes. *SIAM J. Control Optim.* **40**, 525–539.

ALTMAN, E. (1999) *Constrained Markov decision processes.* Chapman & Hall/CRC, Boca Raton, FL.

ALTMAN, E. and STIDHAM, JR., S. (1995) Optimality of monotonic policies for two-action Markovian decision processes, with applications to control of queues with delayed information. *Queueing Systems Theory Appl.* **21**, 267–291.

ASMUSSEN, S. and ALBRECHER, H. (2010) *Ruin probabilities.* World Scientific Publishing, River Edge, NJ.

AWANOU, G. (2007) Shortfall risk minimization in a discrete regime switching model. *Decis. Econ. Finance* **30**, 71–78.

BAIN, A. and CRISAN, D. (2009) *Fundamentals of stochastic filtering.* Springer, New York.

BANK, P. and FÖLLMER, H. (2003) American options, multi-armed bandits, and optimal consumption plans: a unifying view. In *Paris-Princeton Lectures on Mathematical Finance, 2002*, 1–42. Springer, Berlin.

BARONE, L. (2006) Bruno de Finetti, The problem of full-risk insurance. *Journal of Investment Management* **4**, 19–43.

BAUER, H. (1996) *Probability theory.* Walter de Gruyter, Berlin.

BAUER, H. (2004) Fluid approximation for controlled stochastic networks with delayed dynamics. Ph.D. thesis, Universität Ulm.

BAUER, H. and RIEDER, U. (2005) Stochastic control problems with delay. *Math. Methods Oper. Res.* **62**, 411–427.

N. Bäuerle and U. Rieder, *Markov Decision Processes with Applications to Finance*, Universitext, DOI 10.1007/978-3-642-18324-9,

Bäuerle, N. (1997) Inequalities for stochastic models via supermodular orderings. *Comm. Statist. Stochastic Models* **13**, 181–201.

Bäuerle, N. (2001) Discounted stochastic fluid programs. *Math. Oper. Res.* **26**, 401–420.

Bäuerle, N. (2005) Benchmark and mean-variance problems for insurers. *Math. Methods Oper. Res.* **62**, 159–165.

Bäuerle, N. and Mundt, A. (2009) Dynamic mean-risk optimization in a binomial model. *Math. Methods Oper. Res.* **70**, 219–239.

Bäuerle, N. and Rieder, U. (1997) Comparison results for Markov-modulated recursive models. *Probab. Engrg. Inform. Sci.* **11**, 203–217.

Bäuerle, N. and Rieder, U. (2004) Portfolio optimization with Markov-modulated stock prices and interest rates. *IEEE Trans. Automat. Control* **49**, 442–447.

Bäuerle, N. and Rieder, U. (2007) Portfolio optimization with jumps and unobservable intensity process. *Math. Finance* **17**, 205–224.

Bäuerle, N. and Rieder, U. (2009) MDP Algorithms for Portfolio Optimization Problems in Pure Jump Markets. *Finance Stoch.* **13**, 591–611.

Bäuerle, N. and Rieder, U. (2010) Optimal control of Piecewise Deterministic Markov Processes with finite time horizon. In Piunovskiy, A. (ed.), *Modern trends in controlled stochastic processes: theory and applications*, 123–143. Luniver Press.

Bäuerle, N. and Stidham, S. (2001) Conservation laws for single-server fluid networks. *Queueing Systems Theory Appl.* **38**, 185–194.

Bayraktar, E. and Ludkovski, M. (2011) Optimal trade execution in illiquid financial markets. *Math. Finance* .

Bearden, J. N. and Murphy, R. O. (2007) On generalized secretary problems. In *Uncertainty and risk*, 187–205. Springer, Berlin.

Beasley, J. E., Meade, N., and Chang, T.-J. (2003) An evolutionary heuristic for the index tracking problem. *European J. Oper. Res.* **148**, 621–643.

Bellman, R. (1954) The theory of dynamic programming. *Bull. Amer. Math. Soc.* **60**, 503–515.

Bellman, R. (1957) *Dynamic programming.* Princeton University Press, Princeton, NJ.

Bellman, R. (2003) *Dynamic programming.* Dover Publications, Mineola, NY.

Ben-Ameur, H., Breton, M., and François, P. (2006) A dynamic programming approach to price installment options. *European J. Oper. Res.* **169**, 667–676.

Ben-Ameur, H., Breton, M., Karoui, L., and L'Ecuyer, P. (2007) A dynamic programming approach for pricing options embedded in bonds. *J. Econom. Dynam. Control* **31**, 2212–2233.

Ben-Ameur, H., Breton, M., and L'Ecuyer, P. (2002) A dynamic programming procedure for pricing American-style Asian options. *Management Sci.* **48**, 625–643.

BENSOUSSAN, A. (1992) *Stochastic control of partially observable systems.* Cambridge University Press, Cambridge.

BENSOUSSAN, A. and ROBIN, M. (1982) On the convergence of the discrete time dynamic programming equation for general semigroups. *SIAM J. Control Optim.* **20**, 722–746.

BENZING, H., HINDERER, K., and KOLONKO, M. (1984) On the k-armed Bernoulli bandit: Monotonicity of the total reward under an arbitrary prior distribution. *Math. Operationsforschung Statistik, Ser. Optimization* **15**, 583–595.

BENZING, H. and KOLONKO, M. (1987) Structured policies for a sequential design problem with general distributions. *Math. Oper. Res.* **12**, 60–71.

BERGTHOLDT, P. (1998) Varianz-optimale Strategien für das Index-Tracking. Ph.D. thesis, Universität Ulm.

BERRY, D. A. and FRISTEDT, B. (1985) *Bandit problems.* Chapman & Hall, London.

BERTSEKAS, D. and TSITSIKLIS, J. (1996) *Neuro-dynamic programming.* Athena Scientific, Belmont, MA.

BERTSEKAS, D. P. (2001) *Dynamic programming and optimal control. Vol. II.* Athena Scientific, Belmont, MA, second edition.

BERTSEKAS, D. P. (2005) *Dynamic programming and optimal control. Vol. I.* Athena Scientific, Belmont, MA, third edition.

BERTSEKAS, D. P. and SHREVE, S. E. (1978) *Stochastic optimal control.* Academic Press, New York.

BERTSIMAS, D. and NIÑO MORA, J. (1996) Conservation laws, extended polymatroids and multiarmed bandit problems; a polyhedral approach to indexable systems. *Math. Oper. Res.* **21**, 257–306.

BIELECKI, T., HERNÁNDEZ-HERNÁNDEZ, D., and PLISKA, S. R. (1999) Risk sensitive control of finite state Markov chains in discrete time, with applications to portfolio management. *Math. Methods Oper. Res.* **50**, 167–188.

BIELECKI, T. and PLISKA, S. R. (1999) Risk-sensitive dynamic asset management. *Appl. Math. Optim.* **39**, 337–360.

BILLINGSLEY, P. (1995) *Probability and measure.* John Wiley & Sons, New York.

BINGHAM, N. H. and KIESEL, R. (2004) *Risk-neutral valuation.* Springer-Verlag, London.

BJÖRK, T. (2004) *Arbitrage theory in continuous time.* Oxford University Press, Oxford, UK.

BJÖRK, T., DAVIS, M. H., and LANDÉN, C. (2010) Optimal investment under partial information. *Math. Methods Oper. Res.* **71**, 371–399.

BLACKWELL, D. (1965) Discounted dynamic programming. *Ann. Math. Statist.* **36**, 226–235.

BOBRYK, R. V. and STETTNER, Ł. (1999) Discrete time portfolio selection with proportional transaction costs. *Probab. Math. Statist.* **19**, 235–248.

BODA, K., FILAR, J. A., LIN, Y., and SPANJERS, L. (2004) Stochastic target hitting time and the problem of early retirement. *IEEE Trans. Automat. Control* **49**, 409–419.

BODILY, S. E. and WHITE, C. C. (1982) Optimal consumption and portfolio strategies in a discrete-time model with summary-dependent preferences. *J. Financ. Quant. Anal.* **17**, 1–14.

BORKAR, V. S. (1991) *Topics in controlled Markov chains*, volume 240 of *Pitman Research Notes in Mathematics Series*. Longman Scientific & Technical, Harlow.

BOUAKIZ, M. and SOBEL, M. J. (1992) Inventory control with an exponential utility criterion. *Oper. Res.* **40**, 603–608.

BRENNAN, M. J. (1998) The role of learning in dynamic portfolio decisions. *Europ. Finance Rev.* **1**, 295–306.

BROWN, L. and PURVES, R. (1973) Measuarble selection of extrema. *Ann. Math. Statist.* **1**, 903–912.

ÇAKMAK, U. and ÖZEKICI, S. (2006) Portfolio optimization in stochastic markets. *Math. Methods Oper. Res.* **63**, 151–168.

ÇANAKOĞLU, E. and ÖZEKICI, S. (2009) Portfolio selection in stochastic markets with exponential utility functions. *Ann. Oper. Res.* **166**, 281–297.

ÇANAKOĞLU, E. and ÖZEKICI, S. (2010) Portfolio selection in stochastic markets with HARA utility functions. *European J. Oper. Res.* **201**, 520–536.

CARMONA, R. (ed.) (2009) *Indifference pricing.* Princeton University Press, Princeton, NJ.

CHANG, H. S., FU, M. C., HU, J., and MARCUS, S. I. (2007) *Simulation-based algorithms for Markov decision processes.* Springer-Verlag, London.

CHEN, R. W., SHEPP, L. A., YAO, Y.-C., and ZHANG, C.-H. (2005) On optimality of bold play for primitive casinos in the presence of inflation. *J. Appl. Probab.* **42**, 121–137.

CHEN, R. W., SHEPP, L. A., and ZAME, A. (2004) A bold strategy is not always optimal in the presence of inflation. *J. Appl. Probab.* **41**, 587–592.

CHEUNG, K. C. and YANG, H. (2004) Asset allocation with regime-switching: discrete-time case. *Astin Bull.* **34**, 99–111.

CHEUNG, K. C. and YANG, H. (2007) Optimal investment-consumption strategy in a discrete-time model with regime switching. *Discrete Contin. Dyn. Syst. Ser. B* **8**, 315–332.

CHOW, Y. S., ROBBINS, H., and SIEGMUND, D. (1971) *Great expectations: the theory of optimal stopping.* Houghton Mifflin, Boston, MA.

CONNELLY, R. (1974) Say red. *Pallbearers Review* **9**, 702.

CONSTANTINIDES, G. M. (1979) Multiperiod consumption and investment behavior with convex transactions costs. *Management Sci.* **25**, 1127–1137.

CONT, R. and TANKOV, P. (2004) *Financial modelling with jump processes.* Chapman & Hall/CRC, Boca Raton, FL.

CORSI, M., PHAM, H., and RUNGGALDIER, W. (2008) Numerical approximation by quantization of control problems in finance under partial observation. In *Mathematical modelling and numerical methods in finance*, 325–360. North Holland.

COSTA, O. L. V. and ARAUJO, M. V. (2008) A generalized multi-period mean-variance portfolio optimization with Markov switching parameters. *Automatica J. IFAC* **44**, 2487–2497.

COSTA, O. L. V. and NABHOLZ, R. B. (2007) Multiperiod mean-variance optimization with intertemporal restrictions. *J. Optim. Theory Appl.* **134**, 257–274.

COX, J. C., ROSS, S. A., and RUBINSTEIN, M. (1979) Options pricing: A simplified approach. *J. Fin. Econon.* **7**, 229–263.

COX, J. C. and RUBINSTEIN, M. (1985) *Option markets.* Prentice-Hall, Englewood Cliffs, NJ.

CREMER, L. (1998) Arbitrage und Preisgrenzen in endlichen Finanzmärkten. Ph.D. thesis, Universität Ulm.

DAVIS, M. H. A. (1984) Piecewise-deterministic Markov processes: a general class of nondiffusion stochastic models. *J. Roy. Statist. Soc. Ser. B* **46**, 353–388.

DAVIS, M. H. A. (1993) *Markov models and optimization.* Chapman & Hall, London.

DAYANIK, S., POWELL, W., and YAMAZAKI, K. (2008) Index policies for discounted bandit problems with availability constraints. *Adv. in Appl. Probab.* **40**, 377–400.

DEGROOT, M. (2004) *Optimal statistical decisions.* John Wiley & Sons, New York.

DEMPSTER, M. A. H. and YE, J. J. (1992) Necessary and sufficient optimality conditions for control of piecewise deterministic Markov processes. *Stochastics Stochastics Rep.* **40**, 125–145.

DERMAN, C. (1970) *Finite state Markovian decision processes.* Academic Press, New York.

DI MASI, G. B. and STETTNER, L. (1999) Risk sensitive control of discrete time partially observed Markov processes with infinite horizon. *Stochastics Stochastics Rep.* **67**, 309–322.

DUBINS, L. E. and SAVAGE, L. J. (1965) *How to gamble if you must. Inequalities for stochastic processes.* McGraw-Hill, New York.

DUFFIE, D. (1988) *Security markets: stochastic models.* Academic Press, Boston.

DUFFIE, D. (2001) *Dynamic asset pricing theory.* Princeton University Press, Princeton, NJ, third edition.

DUFFIE, D. and PROTTER, P. (1992) From discrete- to continuous-time finance: weak convergence of the financial gain process. *Math. Finance* **2**, 1–15.

DUMAS, B. and LUCIANO, E. (1991) An exact solution to a dynamic portfolio choice problem under transaction cost. *The Journal of Finance* **46**, 577–595.

DURST, J. (1991) Bayessche nichtparametrische Modelle zur Portfolioselektion. Ph.D. thesis, Universität Ulm.

DYNKIN, E. B. and YUSHKEVICH, A. A. (1979) *Controlled Markov processes.* Springer-Verlag, Berlin.

EDIRISINGHE, N. C. P. (2005) Multiperiod portfolio optimization with terminal liability: bounds for the convex case. *Comput. Optim. Appl.* **32**, 29–59.

EL KAROUI, N. and KARATZAS, I. (1994) Dynamic allocation problems in continuous time. *Ann. Appl. Probab.* **4**, 255–286.

ELLIOTT, R. J., AGGOUN, L., and MOORE, J. B. (1995) *Hidden Markov models.* Springer-Verlag, New York.

ELLIOTT, R. J. and KOPP, P. E. (2005) *Mathematics of financial markets.* Springer-Verlag, New York.

FAVERO, G. (2001) Shortfall risk minimization under model uncertainty in the binomial case: adaptive and robust approaches. *Math. Methods Oper. Res.* **53**, 493–503.

FAVERO, G. and VARGIOLU, T. (2006) Shortfall risk minimising strategies in the binomial model: characterisation and convergence. *Math. Methods Oper. Res.* **64**, 237–253.

FEINBERG, E. A. and SHWARTZ, A. (1994) Markov decision models with weighted discounted criteria. *Math. Oper. Res.* **19**, 152–168.

FEINBERG, E. A. and SHWARTZ, A. (eds.) (2002) *Handbook of Markov decision processes.* Kluwer Academic Publishers, Boston, MA.

FILAR, J. and VRIEZE, K. (1997) *Competitive Markov decision processes.* Springer-Verlag, New York.

DE FINETTI, B. (1940) Il probleme dei pieni. *Giornale dell'Istituto Italiano degl Attuari* **11**, 1–88.

DE FINETTI, B. (1957) Su unímpostazione alternativa della teoria collettiva del rischio. *Transactions of the XVth International Congress of Actuaries* **2**, 433–443.

FITZPATRICK, B. G. and FLEMING, W. H. (1991) Numerical methods for an optimal investment-consumption model. *Math. Oper. Res.* **16**, 823–841.

FLEMING, W. H. and SONER, H. M. (1993) *Controlled Markov processes and viscosity solutions.* Springer-Verlag, New York.

FÖLLMER, H. and SCHIED, A. (2004) *Stochastic finance.* Walter de Gruyter, Berlin.

FORWICK, L., SCHÄL, M., and SCHMITZ, M. (2004) Piecewise deterministic Markov control processes with feedback controls and unbounded costs. *Acta Appl. Math.* **82**, 239–267.

FREEMAN, P. R. (1983) The secretary problem and its extensions: a review. *Internat. Statist. Rev.* **51**, 189–206.

FRIIS, S.-H., RIEDER, U., and WEISHAUPT, J. (1993) Optimal control of single-server queueing networks. *Z. Oper. Res.* **37**, 187–205.

FRISTEDT, B., JAIN, N., and KRYLOV, N. (2007) *Filtering and prediction: a primer.* American Mathematical Society, Providence, RI.

Gennotte, G. and Jung, A. (1994) Investement strategies under transaction cost: The finite horizon case. *Management Sci.* **40**, 385–404.

Gerber, H. (1969) Entscheidungskriterien für den zusammengesetzten Poisson-Prozess. *Schweiz. Verein. Versicherungsmath. Mitt.* 185–228.

Gittins, J. C. (1979) Bandit processes and dynamic allocation indices. *J. Roy. Statist. Soc. Ser. B* **41**, 148–177.

Gittins, J. C. (1989) *Multi-armed bandit allocation indices.* John Wiley & Sons, Chichester.

Glasserman, P. (2004) *Monte Carlo methods in financial engineering.* Springer-Verlag, New York.

Glazebrook, K. D., Niño-Mora, J., and Ansell, P. S. (2002) Index policies for a class of discounted restless bandits. *Adv. in Appl. Probab.* **34**, 754–774.

Guo, X. and Hernández-Lerma, O. (2009) *Continuous-time Markov Decision Processes.* Springer-Verlag, New York.

Haigh, J. and Roters, M. (2000) Optimal strategy in a dice game. *J. Appl. Probab.* **37**, 1110–1116.

Hakansson, N. (1970) Optimal investment and consumption strategies under risk for a class of utility functions. *Econometrica* **38**, 587–607.

Hakansson, N. (1971a) Capital growth and the Mean-Variance approach to portfolio selection. *J. Finance Quant. Analysis* **6**, 517–557.

Hakansson, N. (1971b) Optimal entrepreneurial decisions in a completely stochastic environment. *Management Sci.* **17**, 427–449.

Hakansson, N. (1974) Convergence to isoelastic utility and policy in multiperiod portfolio choice. *J. Financ. Economics* **1**, 201–224.

He, H. (1991) Optimal consumption-portfolio policies: a convergence from discrete to continuous time models. *J. Econom. Theory* **55**, 340–363.

He, H. and Pearson, N. (1991a) Consumption and portfolio policies with incomplete markets and short-sale constraints: the finite-dimensional case. *Math. Finance* **1**, 1–10.

He, H. and Pearson, N. (1991b) Consumption and portfolio policies with incomplete markets and short-sale constraints: the infinite-dimensional case. *J. Econ. Theory* **54**, 259–304.

van Hee, K. M. (1978) *Bayesian control of Markov chains.* Mathematisch Centrum, Amsterdam.

Heilmann, W.-R. (1979) Solving a general discounted dynamic program by linear programming. *Z. Wahrsch. Verw. Gebiete* **48**, 339–346.

Hernández-Lerma, O. (1989) *Adaptive Markov control processes.* Springer-Verlag, New York.

Hernández-Lerma, O. and Lasserre, J. B. (1996) *Discrete-time Markov control processes.* Springer-Verlag, New York.

Hernández-Lerma, O. and Lasserre, J. B. (1999) *Further topics on discrete-time Markov control processes.* Springer-Verlag, New York.

HERNÁNDEZ-LERMA, O. and LASSERRE, J. B. (2002) The linear programming approach. In *Handbook of Markov decision processes*, 377–408. Kluwer Acad. Publ., Boston, MA.

HEYMAN, D. P. and SOBEL, M. J. (2004a) *Stochastic models in operations research. Vol. I.* Dover Publications, Mineola, NY.

HEYMAN, D. P. and SOBEL, M. J. (2004b) *Stochastic models in operations research. Vol. II.* Dover Publications, Mineola, NY.

HIMMELBERG, C. J., PARTHASARATHY, T., and VANVLECK, F. S. (1976) Optimal plans for dynamic programming problems. *Math. Oper. Res.* **1**, 390–394.

HINDERER, K. (1970) *Foundations of non-stationary dynamic programming with discrete time parameter.* Springer-Verlag, Berlin.

HINDERER, K. (1971) Instationäre dynamische Optimierung bei schwachen Voraussetzungen über die Gewinnfunktionen. *Abh. Math. Sem. Univ. Hamburg* **36**, 208–223.

HINDERER, K. (1985) On the structure of solutions of stochastic dynamic programs. In *Proceedings of the seventh conference on probability theory (Braşov, 1982)*, 173–182. VNU Sci. Press, Utrecht.

HINDERER, K. (2005) Lipschitz continuity of value functions in Markovian decision processes. *Math. Methods Oper. Res.* **62**, 3–22.

HINDERER, K. and WALDMANN, K.-H. (2001) Cash management in a randomly varying environment. *European J. Oper. Res.* **130**, 468–485.

VAN DER HOEK, J. and ELLIOTT, R. (2006) *Binomial models in finance.* Springer-Verlag, New York.

HORDIJK, A. (1974) *Dynamic programming and Markov potential theory.* Mathematisch Centrum, Amsterdam.

HORDIJK, A. and VAN DER DUYN SCHOUTEN, F. A. (1984) Discretization and weak convergence in Markov decision drift processes. *Math. Oper. Res.* **9**, 112–141.

HORDIJK, A. and VAN DER DUYN SCHOUTEN, F. A. (1985) Markov decision drift processes: Conditions for optimality obtained by discretization. *Math. Oper. Res.* **10**, 160–173.

HOWARD, R. A. (1960) *Dynamic programming and Markov processes.* The Technology Press of MIT, Cambridge, MA.

HUANG, C. and LITZENBERGER, R. (1988) *Foundations for financial economics.* North-Holland, New York.

HUANG, Y. and KALLENBERG, L. C. M. (1994) On finding optimal policies for Markov decision chains: a unifying framework for mean-variance-tradeoffs. *Math. Oper. Res.* **19**, 434–448.

IIDA, T. and MORI, M. (1996) Markov decision processes with random horizon. *J. Oper. Res. Soc. Japan* **39**, 592–603.

INGERSOLL, J. E. (1987) *Theory of financial decision making.* Rowman & Littlefield, Lanham, MD.

JACOBSEN, M. (2006) *Point process theory and applications.* Birkhäuser, Boston, MA.

JAQUETTE, S. C. (1973) Markov decision processes with a new optimality criterion: discrete time. *Ann. Statist.* **1**, 496–505.

JEANBLANC, M., YOR, M., and CHESNEY, M. (2009) *Mathematical methods for financial markets.* Springer-Verlag, London.

JOUINI, E. and NAPP, C. (2004) Convergence of utility functions and convergence of optimal strategies. *Finance Stoch.* **8**, 133–144.

KAAS, R., VAN HEERWAARDEN, A., and GOOVAERTS, M. (1998) *Ordering of Actuarial Risk.* Caire, Brussels.

KAHNEMAN, D. and TVERSKY, A. (1979) Prospect theory: An analysis of decision under risk. *Econometrica* **47**, 263–291.

KALLENBERG, L. C. M. (1983) *Linear programming and finite Markovian control problems.* Mathematisch Centrum, Amsterdam.

KAMIN, J. H. (1975) Optimal portfolio revision with a proportional transactions cost. *Management Sci.* **21**, 1263–1271.

KARATZAS, I. (1984) Gittins indices in the dynamic allocation problem for diffusion processes. *Ann. Probab.* **12**, 173–192.

KASPI, H. and MANDELBAUM, A. (1998) Multi-armed bandits in discrete and continuous time. *Ann. Appl. Probab.* **8**, 1270–1290.

KATEHAKIS, M. N. and VEINOTT, JR., A. F. (1987) The multi-armed bandit problem: decomposition and computation. *Math. Oper. Res.* **12**, 262–268.

KIRCH, M. and RUNGGALDIER, W. J. (2005) Efficient hedging when asset prices follow a geometric Poisson process with unknown intensities. *SIAM J. Control Optim.* **43**, 1174–1195.

KITAEV, M. and RYKOV, V. (1995) *Controlled queueing systems.* CRC Press, Boca Raton, FL.

KLEIN-HANEVELD, W. K. (1986) *Duality in stochastic linear and dynamic programming.* Springer-Verlag, Berlin.

KOLONKO, M. (1986) A note on a general stopping rule in dynamic programming with finite horizon. *Stat. Decisions* **4**, 379–387.

KOLONKO, M. and BENZING, H. (1985) On monotone optimal decision rules and the stay-on-a-winner rule for the two-armed bandit. *Metrika* **32**, 395–407.

KORN, R. (1997) *Optimal portfolios. Stochastic models for optimal investment and risk management in continuous time.* World Scientific, Singapore.

KORN, R. and SCHÄL, M. (1999) On value preserving and growth optimal portfolios. *Math. Methods Oper. Res.* **50**, 189–218.

KRAFT, H. and STEFFENSEN, M. (2008) Optimal consumption and insuarnce: A continuous-time Markov chain approach. *ASTIN Bulletin* **38**, 231–257.

KREPS, D. and PORTEUS, E. (1979) Dynamic choice theory and dynamic programming. *Econometrica* **47**, 91–100.

KRISHNAMURTHY, V. and WAHLBERG, B. (2009) Partially observed Markov decision process multiarmed bandits—structural results. *Math. Oper. Res.* **34**, 287–302.

KUMAR, P. and VARAIYA, P. (1986) *Stochastic Systems: Estimation, identification and adaptive control.* Prentice-Hall, Englewood Cliffs, NJ.

KURATOWSKI, K. and RYLL-NARDZEWSKI, C. (1965) A general theorem on selectors. *Bull. Acad. Polon. Sci.* **13**, 397–403.

KUSHNER, H. J. and DUPUIS, P. (2001) *Numerical methods for stochastic control problems in continuous time.* Springer-Verlag, New York.

LEVY, H. (2006) *Stochastic dominance.* Springer, New York.

LI, D. and NG, W.-L. (2000) Optimal dynamic portfolio selection: multiperiod mean-variance formulation. *Math. Finance* **10**, 387–406.

LI, P. and WANG, S.-Y. (2008) Optimal martingale measure maximizing the expected total utility of consumption with applications to derivative pricing. *Optimization* **57**, 691–703.

LOVEJOY, W. S. (1987) Some monotonicity results for partially observed Markov decision processes. *Oper. Res.* **35**, 736–743.

LOVEJOY, W. S. (1991a) Computationally feasible bounds for partially observed Markov decision processes. *Oper. Res.* **39**, 162–175.

LOVEJOY, W. S. (1991b) A survey of algorithmic methods for partially observed Markov decision processes. *Ann. Oper. Res.* **28**, 47–65.

LUENBERGER, D. G. (1998) *Investment Science.* Oxford University Press, New York, Oxford.

MARKOWITZ, H. (1952) Portfolio selection. *Journal of Finance* 77–91.

MARKOWITZ, H. (1987a) *Mean-Variance analysis in portfolio choice and capital markets.* Blackwell, Oxford.

MARKOWITZ, H. (1987b) *Portfolio selection.* John Wiley & Sons, New York.

MARTIN, J. (1967) *Bayesian decision problems and Markov chains.* John Wiley & Sons, New York.

MARTIN-LÖF, A. (1994) Lectures on the use of control theory in insurance. *Scand. Actuar. J.* 1–25.

MEYN, S. (2008) *Control techniques for complex networks.* Cambridge University Press, Cambridge.

MEYN, S. and TWEEDIE, R. L. (2009) *Markov chains and stochastic stability.* Cambridge University Press, Cambridge, second edition.

MIYASAWA, K. (1962) An economic survival game. *Operations Research Society of Japan* **4**, 95–113.

MONAHAN, G. E. (1980) Optimal stopping in a partially observable Markov process with costly information. *Oper. Res.* **28**, 1319–1334.

MONAHAN, G. E. (1982a) Optimal stopping in a partially observable binary-valued Markov chain with costly perfect information. *J. Appl. Probab.* **19**, 72–81.

MONAHAN, G. E. (1982b) A survey of partially observable Markov decision processes: theory, models, and algorithms. *Management Sci.* **28**, 1–16.

MOSSIN, J. (1968) Optimal multiperiod portfolio policies. *Journal of Business* 215–229.

MOTOCZYŃSKI, M. (2000) Multidimensional variance-optimal hedging in discrete-time model—a general approach. *Math. Finance* **10**, 243–257.

MÜLLER, A. (1997) How does the value function of a Markov decision process depend on the transition probabilities? *Math. Oper. Res.* **22**, 872–885.

MÜLLER, A. (2000) Expected utility maximization of optimal stopping problems. *European J. Oper. Res.* **122**, 101–114.

MÜLLER, A. and STOYAN, D. (2002) *Comparison Methods for Stochastic Models and Risks.* John Wiley & Sons, Chichester.

MUNDT, A. (2007) Dynamic risk management with Markov decision processes. Ph.D. thesis, Universität Karlsruhe (TH). URL http://digbib.ubka.uni-karlsruhe.de/volltexte/1000007340.

MUNK, C. (2003) The Markov chain approximation approach for numerical solution of stochastic control problems: experiences from Merton's problem. *Appl. Math. Comput.* **136**, 47–77.

MUSIELA, M. and RUTKOWSKI, M. (2005) *Martingale methods in financial modelling.* Springer-Verlag, Berlin.

MUSIELA, M. and ZARIPHOPOULOU, T. (2004) A valuation algorithm for indifference prices in incomplete markets. *Finance Stoch.* **8**, 399–414.

NAKAI, T. (1983) Optimal stopping problem in a finite state partially observable Markov chain. *J. Inform. Optim. Sci.* **4**, 159–176.

NAKAI, T. (1985) The problem of optimal stopping in a partially observable Markov chain. *J. Optim. Theory Appl.* **45**, 425–442.

NELSEN, R. B. (2006) *An introduction to copulas. 2nd ed.* Springer Series in Statistics. New York.

NELSON, D. B. and RAMASWAMY, K. (1990) Simple binomial processes as diffusion approximations in financial models. *Re. Fin. Stud.* **3**, 393–430.

VON NEUMANN, J. and MORGENSTERN, O. (1947) *Theory of games and economic behavior.* Princeton University Press, Princeton, NJ.

ØKSENDAL, B. and SULEM, A. (2005) *Applied stochastic control of jump diffusions.* Springer-Verlag, Berlin.

PESKIR, G. and SHIRYAEV, A. (2006) *Optimal stopping and free-boundary problems.* Birkhäuser Verlag, Basel.

PFLUG, G. and RÖMISCH, W. (2007) *Modeling, measuring and managing risk.* World Scientific, Singapore.

PFLUG, G. C. (2001) Scenario tree generation for multiperiod financial optimization by optimal discretization. *Math. Program.* **89**, 251–271.

PHAM, H. (2009) *Continuous-time stochastic control and optimization with financial applications.* Springer-Verlag, Berlin.

PHAM, H., RUNGGALDIER, W., and SELLAMI, A. (2005) Approximation by quantization of the filter processand applications to optimal stopping problems under partial observation. *Monte Carlo Methods Appl.* **11**, 57–81.

PIUNOVSKIY, A. B. (1997) *Optimal control of random sequences in problems with constraints.* Kluwer Academic Publishers, Dordrecht.

PLISKA, S. R. (2000) *Introduction to Mathematical Finance. Discrete time models.* Blackwell Publishers, Oxford.

POWELL, W. B. (2007) *Approximate dynamic programming.* Wiley-Interscience [John Wiley & Sons], Hoboken, NJ.

PRESMAN, E. and SONIN, I. (1990) *Sequential control with incomplete information.* Academic Press, San Diego, CA.

PRIGENT, J.-L. (2003) *Weak convergence of financial markets.* Springer-Verlag, Berlin.

PRIGENT, J.-L. (2007) *Portfolio optimization and performance analysis.* Chapman & Hall/CRC, Boca Raton, FL.

PROTTER, P. E. (2005) *Stochastic integration and differential equations.* Springer-Verlag, Berlin.

PUTERMAN, M. L. (1994) *Markov decision processes: discrete stochastic dynamic programming.* John Wiley & Sons, New York.

RÁSONYI, M. and STETTNER, L. (2005) On utility maximization in discrete-time financial market models. *Ann. Appl. Probab.* **15**, 1367–1395.

REINHARD, J.-M. (1981) A semi-Markovian game of economic survival. *Scand. Actuar. J.* 23–38.

RIEDER, U. (1975a) Bayesian dynamic programming. *Advances in Appl. Probability* **7**, 330–348.

RIEDER, U. (1975b) On stopped decision processes with discrete time parameter. *Stochastic Processes Appl.* **3**, 365–383.

RIEDER, U. (1976) On optimal policies and martingales in dynamic programming. *J. Appl. Probability* **13**, 507–518.

RIEDER, U. (1978) Measurable selection theorems for optimization problems. *Manuscripta Math.* **24**, 115–131.

RIEDER, U. (1991) Structural results for partially observed control models. *Z. Oper. Res.* **35**, 473–490.

RIEDER, U. and BÄUERLE, N. (2005) Portfolio optimization with unobservable Markov-modulated drift process. *J. Appl. Probab.* **42**, 362–378.

RIEDER, U. and WAGNER, H. (1991) Structured policies in the sequential design of experiments. *Ann. Oper. Res.* **32**, 165–188.

RIEDER, U. and WENTGES, P. (1991) On Bayesian group sequential sampling procedures. *Ann. Oper. Res.* **32**, 189–203.

RIEDER, U. and WINTER, J. (2009) Optimal control of Markovian jump processes with partial information and applications to a parallel queueing model. *Math. Methods Oper. Res.* **70**, 567–596.

RIEDER, U. and ZAGST, R. (1994) Monotonicity and bounds for convex stochastic control models. *Z. Oper. Res.* **39**, 187–207.

ROGERS, L. (2001) The relaxed investor and parameter uncertainty. *Finance Stoch.* **5**, 131–154.

ROGERS, L. (2002) Monte Carlo valuation of American options. *Math. Finance* **12**, 271–286.

ROGERS, L. C. G. (1994) Equivalent martingale measures and no-arbitrage. *Stochastics Stochastics Rep.* **51**, 41–49.

ROSS, S. (1983) *Introduction to stochastic dynamic programming.* Academic Press [Harcourt Brace Jovanovich Publishers], New York.

ROSS, S. M. (1970) *Applied probability models with optimization applications.* Holden-Day, San Francisco, CA.

RUBINSTEIN, M. (2006) Bruno de Finetti and mean-variance portfolio selection. *Journal of Investment Management* **4**, 3–4.

RUNGGALDIER, W. and STETTNER, L. (1994) *Approximations of discrete time partially observed control problems.* Applied Mathematics Monographs 6, Giardini Editori, Pisa.

RUNGGALDIER, W., TRIVELLATO, B., and VARGIOLU, T. (2002) A Bayesian adaptive control approach to risk management in a binomial model. In *Seminar on Stochastic Analysis, Random Fields and Applications, III (Ascona, 1999)*, 243–258. Birkhäuser, Basel.

SAMUELSON, P. A. (1969) Lifetime portfolio selection by dynamic stochastic programming. *The Review of Economics and Statistics* 239–246.

SASS, J. (2005) Portfolio optimization under transaction costs in the CRR model. *Math. Methods Oper. Res.* **61**, 239–259.

SASS, J. and HAUSSMANN, U. G. (2004) Optimizing the terminal wealth under partial information: the drift process as a continuous time Markov chain. *Finance Stoch.* **8**, 553–577.

SCHÄL, M. (1975) Conditions for optimality in dynamic programming and for the limit of n-stage optimal policies to be optimal. *Z. Wahrscheinlichkeitstheorie und Verw. Gebiete* **32**, 179–196.

SCHÄL, M. (1990) *Markoffsche Entscheidungsprozesse.* B. G. Teubner, Stuttgart.

SCHÄL, M. (1994) On quadratic cost criteria for option hedging. *Math. Oper. Res.* **19**, 121–131.

SCHÄL, M. (1998) On piecewise deterministic Markov control processes: control of jumps and of risk processes in insurance. *Insurance Math. Econom.* **22**, 75–91.

SCHÄL, M. (1999) Martingale measures and hedging for discrete-time financial markets. *Math. Oper. Res.* **24**, 509–528.

SCHÄL, M. (2000) Portfolio optimization and martingale measures. *Math. Finance* **10**, 289–303.

SCHÄL, M. (2002) Markov decision processes in finance and dynamic options. In *Handbook of Markov decision processes*, 461–487. Kluwer Acad. Publ., Boston, MA.

SCHÄL, M. (2004) On discrete-time dynamic programming in insurance: exponential utility and minimizing the ruin probability. *Scand. Actuar. J.* 189–210.

SCHÄL, M. (2005) Control of ruin probabilities by discrete-time investments. *Math. Methods Oper. Res.* **62**, 141–158.

SCHMIDLI, H. (2008) *Stochastic control in insurance.* Springer, London.

SCHWEIZER, M. (1995) Variance-optimal hedging in discrete time. *Math. Oper. Res.* **20**, 1–32.

SEIERSTAD, A. (2009) *Stochastic control in discrete and continuous time.* Springer, New York.

SENNOTT, L. I. (1999) *Stochastic dynamic programming and the control of queueing systems.* John Wiley & Sons, New York.

SHAKED, M. and SHANTHIKUMAR, J. G. (2007) *Stochastic orders.* Springer, New York.

SHAPLEY, L. S. (1953) Stochastic games. *Proc. Nat. Acad. Sci.* **39**, 1095–1100.

SHAPLEY, L. S. (2003) Stochastic games. In *Stochastic games and applications (Stony Brook, NY, 1999)*, 1–7. Kluwer Acad. Publ., Dordrecht.

SHIRYAEV, A. N. (1967) Some new results in the theory of controlled random processes. In *Trans. Fourth Prague Conf. on Information Theory, Statistical Decision Functions, Random Processes (Prague, 1965)*, 131–203. Academia, Prague.

SHIRYAEV, A. N. (1996) *Probability.* Springer-Verlag, New York.

SHIRYAEV, A. N. (1999) *Essentials of stochastic finance*, volume 3. World Scientific Publishing, River Edge, NJ.

SHIRYAEV, A. N. (2008) *Optimal stopping rules.* Springer-Verlag, Berlin.

SHREVE, S. E. (2004a) *Stochastic calculus for finance. I.* Springer-Verlag, New York.

SHREVE, S. E. (2004b) *Stochastic calculus for finance. II.* Springer-Verlag, New York.

SHUBIK, M. and THOMPSON, G. L. (1959) Games of economic survival. *Naval Res. Logist. Quart.* **6**, 111–123.

SMALLWOOD, R. and SONDIK, E. (1973) The optimal control of partially observable Markov decision processes over a finite horizon. *Oper. Res.* **21**, 1071–1088.

SMITH, J. E. and MCCARDLE, K. F. (2002) Structural properties of stochastic dynamic programs. *Oper. Res.* **50**, 796–809.

SONDIK, E. J. (1978) The optimal control of partially observable Markov processes over the infinite horizon: discounted costs. *Operations Res.* **26**, 282–304.

STEFFENSEN, M. (2006) Quadratic optimization of life and pension insurance payments. *ASTIN Bulletin* **36**, 245–267.

STEINBACH, M. C. (2001) Markowitz revisited: mean-variance models in financial portfolio analysis. *SIAM Rev.* **43**, 31–85.

STETTNER, L. (1999) Risk sensitive portfolio optimization. *Math. Methods Oper. Res.* **50**, 463–474.

STETTNER, L. (2004) Risk-sensitive portfolio optimization with completely and partially observed factors. *IEEE Trans. Automat. Control* **49**, 457–464.

STOKEY, N. L. and LUCAS, JR., R. E. (1989) *Recursive methods in economic dynamics.* Harvard University Press, Cambridge, MA.

SUDDERTH, W. D. (1971) On the Dubins and Savage characterization of optimal strategies. *Ann. Math. Statist.* **43**, 498–512.

SUHOV, Y. and KELBERT, M. (2008) *Probability and statistics by example. (II).* Cambridge University Press, Cambridge.

SUTTON, R. and BARTO, A. (1998) *Reinforcement learning: An introduction.* MIT Press, Cambridge, MA.

SZEKLI, R. (1995) *Stochastic ordering and dependence in applied probability.* Springer-Verlag, New York.

TAKSAR, M. and ZENG, X. (2007) Optimal terminal wealth under partial information: Both the drift and the volatility driven by a discrete-time Markov chain. *SIAM J. Control Optim.* **46**, 1461–1482.

TAMAKI, M. (1984) Optimal selection from a gamma distribution with unknown parameter. *Z. Oper. Res. Ser. A-B* **28**, 47–57.

TIJMS, H. (2003) *A first course in stochastic models.* John Wiley & Sons, Chichester.

TOBIN, J. (1958) Liquidity preference as behavior towards risk. *Review of Economic Studies* **25**, 65–86.

TOPKIS, D. M. (1978) Minimizing a submodular function on a lattice. *Operations Res.* **26**, 305–321.

TRIVELLATO, B. (2009) Replication and shortfall risk in a binomial model with transaction cost. *Math. Methods Oper. Res.* **69**, 1–26.

TSITSIKLIS, J. N. (1986) A lemma on the multiarmed bandit problem. *IEEE Trans. Automat. Control* **31**, 576–577.

VAN ROY, B. (2002) Neuro-dynamic programming: overview and recent trends. In *Handbook of Markov decision processes*, 431–459. Kluwer Acad. Publ., Boston, MA.

VARAIYA, P. P., WALRAND, J. C., and BUYUKKOC, C. (1985) Extensions of the multiarmed bandit problem: the discounted case. *IEEE Trans. Automat. Control* **30**, 426–439.

VENEZIA, I. and LEVY, H. (1983) Optimal multiperiod insurance contracts. *Insurance Math. Econom.* **2**, 199–208.

WALDMANN, K.-H. (1988) On optimal dividend payments and related problems. *Insurance Math. Econom.* **7**, 237–249.

WALDMANN, K.-H. (1998) On granting credit in a random environment. *Math. Methods Oper. Res.* **47**, 99–115.

WEBER, R. (1992) On the Gittins index for multiarmed bandits. *Ann. Appl. Probab.* **2**, 1024–1033.

WEBER, R. R. and WEISS, G. (1990) On an index policy for restless bandits. *J. Appl. Probab.* **27**, 637–648.

WHITE, D. J. (1988) Mean, variance, and probabilistic criteria in finite Markov decision processes: a review. *J. Optim. Theory Appl.* **56**, 1–29.

WHITE, D. J. (1993) *Markov decision processes.* John Wiley & Sons, Chichester.

WHITTLE, P. (1980) Multi-armed bandits and the Gittins index. *J. Roy. Statist. Soc. Ser. B* **42**, 143–149.

WHITTLE, P. (1982) *Optimization over time. Vol. I.* John Wiley & Sons, Chichester.

WHITTLE, P. (1983) *Optimization over time. Vol. II.* John Wiley & Sons, Chichester.

WHITTLE, P. (1988) Restless bandits: activity allocation in a changing world. *J. Appl. Probab.* 287–298.

WHITTLE, P. (1990) *Risk-sensitive optimal control.* Wiley-Interscience Series in Systems and Optimization. John Wiley & Sons, Chichester.

WILLIAMS, R. J. (2006) *Introduction to the mathematics of finance.* American Mathematical Society, Providence, RI.

WINTER, J. (2008) Optimal control of Markovian jump processes with different information structures. Ph.D. thesis, Universität Ulm.

YIN, G. and ZHOU, X. Y. (2004) Markowitz's mean-variance portfolio selection with regime switching: from discrete-time models to their continuous-time limits. *IEEE Trans. Automat. Control* **49**, 349–360.

YONG, J. and ZHOU, X. Y. (1999) *Stochastic controls.* Springer-Verlag, New York.

YUSHKEVICH, A. A. (1980) On reducing a jump controllable Markov model to a model with discrete time. *Theory Probab. Appl.* **25**, 58–69.

YUSHKEVICH, A. A. (1987) Bellman inequalities in Markov decision deterministic drift processes. *Stochastics* **23**, 25–77.

YUSHKEVICH, A. A. (1989) Verification theorems for Markov decision processes with controllable deterministic drift, gradual and impulse controls. *Teor. Veroyatnost. i Primenen.* **34**, 528–551.

ZAGST, R. (1995) The effect of information in separable Bayesian semi-Markov control models and its application to investment planning. *ZOR—Math. Methods Oper. Res.* **41**, 277–288.

ZHOU, X. Y. (2003) Markowitz's world in continuous time, and beyond. In *Stochastic modeling and optimization*, 279–309. Springer, New York.

ZHOU, X. Y. and LI, D. (2000) Continuous-time mean-variance portfolio selection: a stochastic LQ framework. *Appl. Math. Optim.* **42**, 19–33.

ZHU, S.-S., LI, D., and WANG, S.-Y. (2004) Risk control over bankruptcy in dynamic portfolio selection: a generalized mean-variance formulation. *IEEE Trans. Automat. Control* **49**, 447–457.

ZIEMBA, W. T. and VICKSON, R. G. (eds.) (2006) *Stochastic optimization models in finance.* World Scientific, New Jersey.

Index

Lightning Source UK Ltd.
Milton Keynes UK
UKOW041113130213

206233UK00003B/55/P